APPLICATION, PURIFICATION, AND RECOVERY OF IONIC LIQUIDS

APPLICATION, PURIFICATION, AND RECOVERY OF IONIC LIQUIDS

Edited by

OLGA KUZMINA
Department of Chemistry,
Imperial College London, UK

JASON P. HALLETT
Department of Chemical Engineering,
Imperial College London, UK

ELSEVIER

Amsterdam • Boston • Heidelberg • London
New York • Oxford • Paris • San Diego
San Francisco • Singapore • Sydney • Tokyo

Elsevier
Radarweg 29, PO Box 211, 1000 AE Amsterdam, Netherlands
The Boulevard, Langford Lane, Kidlington, Oxford OX5 1GB, UK
50 Hampshire Street, 5th Floor, Cambridge, MA 02139, USA

Notices

Knowledge and best practice in this field are constantly changing. As new research and experience broaden our understanding, changes in research methods, professional practices, or medical treatment may become necessary.

Practitioners and researchers must always rely on their own experience and knowledge in evaluating and using any information, methods, compounds, or experiments described herein. In using such information or methods they should be mindful of their own safety and the safety of others, including parties for whom they have a professional responsibility.

To the fullest extent of the law, neither the Publisher nor the authors, contributors, or editors, assume any liability for any injury and/or damage to persons or property as a matter of products liability, negligence or otherwise, or from any use or operation of any methods, products, instructions, or ideas contained in the material herein.

British Library Cataloguing-in-Publication Data
A catalogue record for this book is available from the British Library

Library of Congress Cataloging-in-Publication Data
A catalog record for this book is available from the Library of Congress

ISBN: 978-0-444-63713-0

CONTENTS

CONTRIBUTORS

Aleksandr M. Bochek
Institute of Applied Chemistry and Ecology, Saint-Petersburg State University of Industrial Technologies and Design, Russia

Vyacheslav Chistov
R&D Fine Chemicals Centre, Moscow, Russia

Julia Dikareva
R&D Fine Chemicals Centre, Moscow, Russia

Elena Esipova
R&D Fine Chemicals Centre, Moscow, Russia

John Gräsvik
School of Chemistry, University of Southampton, UK

Jeraime Griffith
Department of Chemistry, Imperial College London, UK

Jason P. Hallett
Department of Chemical Engineering, Imperial College London, UK

Dmitrii A. Kashirskii
Institute of Applied Chemistry and Ecology, Saint-Petersburg State University of Industrial Technologies and Design, Russia

Olga Kuzmina
Department of Chemistry, Imperial College London, UK

Alastair J.S. McIntosh
Department of Chemistry, Imperial College London, UK

Anton A. Murav'ev
Institute of Applied Chemistry and Ecology, Saint-Petersburg State University of Industrial Technologies and Design, Russia

Nikolai P. Novoselov
Institute of Applied Chemistry and Ecology, Saint-Petersburg State University of Industrial Technologies and Design, Russia

Konstantin Popov
R&D Fine Chemicals Centre, Moscow, Russia

Raquel Prado
Department of Chemistry, Imperial College London, UK

Elena S. Sashina
Institute of Applied Chemistry and Ecology, Saint-Petersburg State University of Industrial Technologies and Design, Russia

Andrei Vendilo
R&D Fine Chemicals Centre, Moscow, Russia

Cameron C. Weber
Department of Chemistry, Imperial College London, UK

INTRODUCTION

Olga Kuzmina*, Jason P. Hallett**
*Department of Chemistry, Imperial College London, UK
**Department of Chemical Engineering, Imperial College London, UK

The domain of ionic liquids (ILs) has become a broad, independent discipline with numerous published, ongoing, and potentially promising research areas. However, in our fast developing world, scientists, when looking for a comprehensive review, often face difficulties, finding out-of-date information or information which is not supported by the authors' own expertise. As of July 2015 there are more than 66,000 publications involving the concept of ILs. K.R. Seddon has mentioned in his preface to the second edition of "Ionic Liquids in Synthesis" (P. Wasserscheid and T. Welton, Wiley-VCH Verlag GmbH & Co. KGaA, 2002) the growing number of publications had reached 2000 by 2006. As is easy to calculate, the interest in ILs has resulted in more than 30 times this number of publications over the last decade.

This book summarizes the latest developments in the synthesis, application, and purification of this unique class of materials – ionic liquids. Written in the form of an overview, every chapter contains examples from the authors' own experience and therefore provides the reader with the unique and precise data accumulated in the relevant sections giving specific and comprehensive insight to the depths of this discipline:

Chapter 1, "Applications of Ionic Liquids," discusses the wide variety of IL applications, including those already known and those recently developed.

Chapter 2, "Methods of Synthesis and Purification of Ionic Liquids," gives an overview of the classical synthetic routes with the focus directed to those ionic liquids with novel moieties, and the newer synthetic routes developed since the start of the 2000s. In this chapter the reader will also find an examination of the likely impurities found in the ionic liquid at each stage of synthesis, the possible detection, and purification methods. The unique insights from the team members of a leading IL synthetic group are given to help those interested in these problems.

The perspective of using of ILs as "green solvents" is outlined in Chapter 3. Investigations into the dissolution of film-forming polysaccharides and polypeptides with imidazolium- and pyridinium-based ionic liquids (ILs) are discussed, which is one of the most promising and studied fields of IL implementation.

Chapter 4, "Formation of Metal–Ligand Complexes in Ionic Liquids and Perspective of ILs Application in Purification Technologies," presents comprehensive data on the thermodynamics of metal–ligand complex formation in hydrophobic and hydrophilic ILs, and demonstrates that ILs should be characterized not as "superpolar," but as quite common polar solvents.

Chapter 5, "Methods of IL Recovery and Destruction," and Chapter 6, "Economical Aspects of Ionic Liquid Application," focus on one of the main advantages of ionic liquids – the possibility of reusability, that is, recycling of ionic liquids, presenting a unique collection of data on the cost-estimation of ionic liquids and an overview of the economic aspects, and the problems of their industrial implementation.

We hope that with the enormous volume of collected information and unique examples of the authors' own experiences, achievements, and expertise, this book will positively contribute to the reader's knowledge (be they chemists, engineers, or material scientists interested in the domain of ionic liquids), save time on searching for the latest updates, and provide unique information helpful in their further research. As it would be impossible to contain the entire wealth of ionic liquid knowledge in one volume, we have instead chosen to familiarize readers with the most important first steps – what are they, how are they made, and what are they good for. In so doing, we hope to open up even more possibilities for this enormous class of materials in the future.

CHAPTER 1

Applications of Ionic Liquids

Raquel Prado, Cameron C. Weber
Department of Chemistry, Imperial College London, UK

Ionic liquids (ILs) are defined as low-melting salts (Hallett and Welton, 2011), with an arbitrary melting point of 100°C often used. Interest in these compounds has increased dramatically in recent years following the report of air- and water-stable ions in 1992, and as a result of growing safety and environmental concerns over the use of volatile solvents (Wilkes and Zaworotko, 1992). In this chapter we aim to highlight the diverse range of IL applications with the focus on the properties of ILs that render them suitable for each and a brief discussion of the current state of the art with regard to IL technology. As a result of the sheer number of areas where ILs can be applied, the highlighted applications do not aim to be exhaustive nor can the discussion of each area be fully comprehensive; however, we hope this summary illustrates the utility of this unique class of compounds.

ELECTROCHEMICAL APPLICATIONS

As ILs consist exclusively of ions, they are obvious candidates as electrolytes for a range of electrochemical applications. ILs offer a number of advantages over electrolytes that feature salts dissolved in molecular solvents. First, their low vapor pressures reduces their flammability, making them less of a fire hazard than electrolytes based on organic solvents (Fox et al., 2003, 2008). Their low vapor pressures also mean that they do not evaporate in open systems (Kar et al., 2014). Second, as they are composed solely of ions, ILs possess much greater concentrations of potential charge carriers relative to dilute salt solutions. Although this could be expected to lead to exceptionally high conductivities, this generally does not occur due to factors such as their substantial viscosity as well as the extent of ion aggregation and correlated ion motion (Hapiot and Lagrost, 2008; MacFarlane et al., 2007). Some ILs, such as the 1-alkyl-1-methylpyrrolidinium bis(trifluoromethanesulfonyl)imide ($[C_xC_1pyrr][NTf_2]$) class, possess very large electrochemical windows in excess of 5.5 V, which increases their compatibility with a wide variety of reagents and electrochemical processes (Hapiot and Lagrost, 2008; MacFarlane

Application, Purification, and Recovery of Ionic Liquids
http://dx.doi.org/10.1016/B978-0-444-63713-0.00001-8

et al., 1999; Plechkova and Seddon, 2008). Finally, a number of ILs possess large liquidus ranges, which enables their application over a wider range of temperatures than many conventional electrolytes (Zhang et al., 2006). These favorable properties have led to ILs being investigated as electrolytes for applications including supercapacitors, batteries, dye-sensitized solar cells (DSSCs), the electrodeposition of metals, and for sensors and sensing applications.

Electrolytes for Batteries and Supercapacitors

With increasing demand for renewable energies and portable electronics, much attention has been given to novel methods of energy storage, leading to significant advances in battery and supercapacitor technology. ILs have been intensely studied with respect to both applications in an attempt to improve the existing technology.

Supercapacitors consist of two electrodes, generally made of microporous activated carbon, separated by an ion permeable membrane coated with the electrolyte (Béguin et al., 2014). The capacitance is generated either by the adsorption of ions from the electrolyte onto the electrode surface as a result of an applied potential difference (Figure 1.1) or fast surface redox processes known as pseudocapacitance. Unlike batteries, energy storage is based primarily on physical rather than chemical processes leading to larger power densities (~10 kW kg^{-1} compared to 0.5–1 kW kg^{-1} for lithium ion batteries) as the discharging cycle is not limited by reaction kinetics (Miller and Burke, 2008). However, specific energy densities for commercial systems are

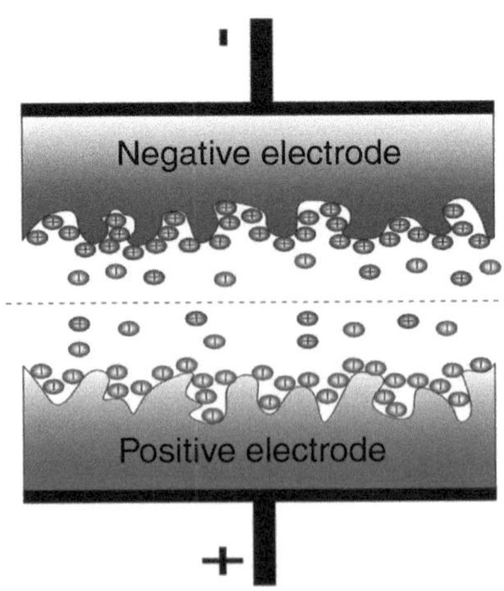

Figure 1.1 ***Representation of a Charged Symmetric Supercapacitor.*** *Reproduced with permission from Béguin et al. (2014), Copyright 2014 Wiley-VCH Verlag GmbH & Co.*

generally lower than for batteries (~5 Wh kg^{-1} compared to 70–100 Wh kg^{-1} for lithium ion batteries) as no chemical energy is stored. The energy and power densities are related to the square of the applied potential difference so the use of electrolytes with a larger electrochemical stability window will, all else being equal, result in improved supercapacitor performance.

To provide maximum power and energy densities, ideal electrolytes for supercapacitors possess low viscosities, large electrochemical windows, high conductivities, and the ability to perform over a wide range of temperatures (Nègre et al., 2015). Aqueous solutions can provide high conductivities and specific capacitances but with extremely restricted electrochemical windows, whereas organic solvents, such as acetonitrile, often offer larger electrochemical windows but at the expense of conductivity, capacitance, and safety due to their flammability (Béguin et al., 2014; Burke, 2000; Lu et al., 2009a). Owing to their low vapor pressure, which reduces the risk of fires and explosions, and their large electrochemical windows, ILs are able to satisfy many of these requirements more safely than organic solvents. The major limitation for ILs is generally their high viscosity, which reduces their conductivity and affects their low-temperature performance. Consequently, the optimal IL for use as an electrolyte will vary depending on the temperature required for the application.

Some general trends in electrochemical behavior have been identified and used to assist with IL selection for supercapacitor applications. In terms of anion selection, wide electrochemical windows and lower viscosities are generally observed for ILs with fluorinated anions such as $[NTf_2]^-$ or tris(perfluoroalkyl)trifluorophosphate ($[FAP]^-$) (Hayyan et al., 2013; Ignat'ev et al., 2005). With regard to cations, it has generally been found that electrochemical stability increases in the order imidazolium < ammonium < pyrrolidinium < phosphonium (Tian et al., 2012), and is greatest for aprotic ILs, that is, those that are not based on a protonated heteroatom. Despite their low stability imidazolium ILs have often been found to possess lower viscosities hence higher conductivities than ILs based on other cations. Imidazolium ILs are also generally liquids over a wider temperature range than those based on most other cations. Less widely used cations, such as azepanium and 3-methylpiperidinium, have been found to yield large electrochemical windows of up to 6.5 V, greater than the corresponding pyrrolidinium salts (Belhocine et al., 2011). While optimization of the electrochemical window is important for this application, recent studies have shown that this need not be the only focus (Huang et al., 2015). For supercapacitors based on graphene nanosheets, the IL $[C_4C_1pyrr][N(CN)_2]$ resulted in greater energy and power densities than its $[NTf_2]^-$ derivative despite possessing an electrochemical window that was 0.4 V more narrow. This is due to the higher conductivity and capacitance of the IL based on the smaller, less viscous $[N(CN)_2]^-$ anion, which compensates for the decrease in size of the electrochemical window. One of the leading examples of the use of IL electrolytes in supercapacitors thus far involves the use of an equimolar eutectic mixture of the ILs *N*-methyl-*N*-propylpiperidinium bis(fluorosulfonyl)imide ($[C_3C_1pip][FSI]$) and N-butyl-N-methylpyrrolidinium bis(fluorosulfonyl) imide ($[C_4C_1pyrr][FSI]$) (Lin et al., 2011). Optimization of this system with a high surface area exfoliated graphite oxide yielded capacitances greater than 100 F g^{-1} over the entire temperature range of −50 to 80°C for the carbon material indicating that the judicious choice of ILs or their mixtures can enable significant improvements in their low-temperature performance (Tsai et al., 2013). It is evident that future research in this area should focus on uncovering ILs that possess higher conductivity, lower viscosity, lower melting points, and larger electrochemical windows in tandem with the development of carbon-based materials to broaden the operational conditions available and the specific power and energy densities that can be achieved.

Many of the demands on the electrolyte for supercapacitors are comparable to the demands for batteries as high conductivities, low viscosities, and large electrochemical windows are all desirable to ensure large power and energy densities. Similarly, one of the main attractions of the use of ILs is their reduced fire and explosion risk, particularly for high-energy batteries.

The major distinction between batteries and supercapacitors is that batteries generate current as a result of redox reactions rather than purely physical ion adsorption. Correspondingly, battery systems are more complex as the electrolyte needs to also be compatible with the redox couple in terms of the potential required and the reactive species present (MacFarlane et al., 2014). The former requirement can be assisted by the formation of a solid electrolyte interphase (SEI), a surface passivating layer on the electrode composed of insoluble decomposition products arising from the solvent and electrolyte, which can enable the long-term stability of the electrolyte even under thermodynamically unfavorable conditions (Howlett et al., 2004; MacFarlane et al., 2007). This adds another variable to investigate for battery systems as SEI formation is heavily dependent on the nature of the electrolyte system used. Desirable SEIs are electrically insulating although able to conduct small ions such as Li^+, enabling these ions to migrate to or from the electrode while preventing the electrolyte from reacting at the electrode surface (Budi et al., 2012). These SEIs are of great significance for the long-term stability of batteries, particularly those based on reactive metal anodes such as lithium, which provide much greater specific energies compared to the carbon-based anodes currently used (Goodenough and Park, 2013; Howlett et al., 2004). An advantage of ILs is that they simplify SEI formation as the cation does not interact with the anode, meaning that the SEI is formed exclusively from the decomposition of the IL anion rather than the organic solvent and the electrolyte anion (Shkrob et al., 2014). ILs based on the $[FSI]^-$ and $[NTf_2]^-$ anions after cycling have been found to produce effective SEIs for lithium electrodes leading to the suppression of dendritic lithium formation, which can lead to short circuits (Bhatt et al., 2013). These electrolytes were found to enable the conduct of over 800 charge–discharge cycles of lithium metal electrodes even at high current densities of 100 mA cm^{-2} without the formation of short circuits.

An important but often underappreciated factor in the selection of ILs for battery applications is the speciation of metal cations within the IL. The speciation can influence not only the redox potential but also the transport properties of the metal species (MacFarlane et al., 2010). For example, within an $[C_2C_1im][BF_4]$ IL electrolyte for lithium ion battery applications, Li^+ was found to diffuse the slowest despite being the smallest ion and the overall conductivity of the medium was reduced (Hayamizu et al., 2004). This was rationalized on the basis of the formation of ionic complexes between Li^+ and $[BF_4]^-$, which reduced the overall mobility of Li^+. This effect is most noticeable for lithium due to its high charge density and small size, with only marginal effects being observed for $Na[BF_4]$ within a similar IL (Nikitina et al., 2012). Approaches to combat this have included the use of additives that coordinate the Li^+ such as tetrahydrofuran (THF) or ethers such as glyme (Bayley et al., 2011). Such additives have been found to reduce the extent of anion clustering around the Li^+ cation, improving its transport properties. The formation of solvate ILs based on an equimolar combination of ethers, such as glymes and crown ethers, has enabled the liquefaction of $Li[NTf_2]$ for use as an electrolyte without the need for additional cations although the conductivity of these liquids does decrease markedly in the absence of a cosolvent due to their considerable viscosity (Mandai et al., 2014; Yoshida et al., 2012; Zhang et al., 2014b). The use of these coordinating additives in small quantities allows the electrolyte to retain its nonflammability while improving the properties of the IL.

The final major advantage of ILs as electrolytes for high-energy battery applications is their lack of volatility. This prevents their evaporation and has been examined to prevent the "drying out" of electrolyte solutions for metal–air batteries including lithium–air, sodium–air, magnesium–air, zinc–air, and aluminum–air (Kar et al., 2014). Such batteries have very large theoretical capacities and use oxygen from the air as an oxidant meaning that these devices could be cost efficient and environmentally friendly if optimized. Most ILs examined for these applications are hydrophobic as this minimizes the exposure of the anodic material to moisture from the air and limits corrosion, particularly for more reactive metals such as lithium and sodium (Kuboki et al., 2005). Small amounts of water, however, can be advantageous. For example, the reversibility of the Zn/Zn^{2+} redox couple was found to improve with 2 wt% water in $[C_4C_1pyrr][NTf_2]$, with larger water volumes decreasing the electrochemical window of the solvent (Xu et al., 2013). It is worth noting that the decreased electrochemical window in the presence of water is due to the splitting of water, which increases the flammability of the system through the generation of hydrogen. Therefore, the hydrophobicity of the selected IL is another criteria that needs to be considered if they are to be employed as electrolytes for metal–air batteries.

It is clear that ILs possess attractive chemical and physical properties for use as electrolytes in supercapacitors and batteries. These properties could improve the current technology while simultaneously resulting in safer devices. Commercial application of ILs as electrolytes is dependent upon further understanding the electrochemistry of ILs and the search for ILs with higher conductivities, lower viscosities, larger electrochemical windows, and wider liquidus ranges that integrate with the restrictions imposed by each application such as the solubility of the redox species or the formation of SEIs.

Electrolytes for Dye-Sensitized Solar Cells

DSSCs have been studied as low-cost replacements for conventional photovoltaic cells (O'Regan and Grätzel, 1991). DSSCs convert solar energy to electrical energy through the use of a dye monolayer on a semiconducting nanocrystalline oxide film (known as the transparent conducting oxide (TCO) film). Photoexcitation of the dye results in the rapid injection of an electron into the conduction band of the semiconducting film (Figure 1.2). These electrons pass through a circuit before reaching the cathode where the reduction of the redox mediator, typically Co^{3+} or I_3^-, occurs. The dye is then regenerated by the redox mediator at the anode (Zakeeruddin and Grätzel, 2009). One of the concerns with these devices is that the use of

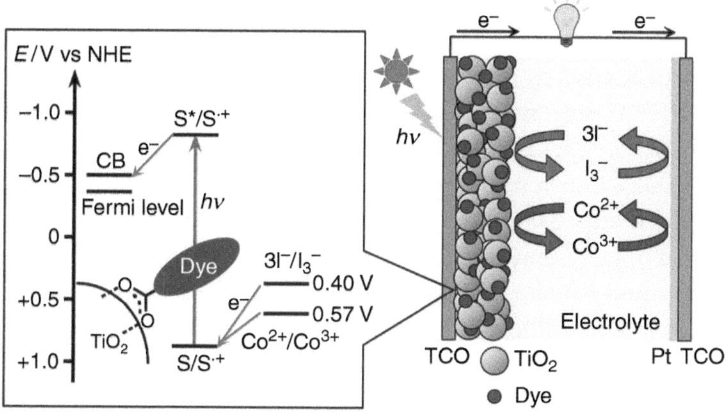

Figure 1.2 ***General Schematic of the Operation of a DSSC with I^-/I_3^- or Co^{2+}/Co^{3+} Electrolyte.*** *Reproduced with permission from Higashino and Imahiro (2015), Copyright 2014 Royal Society of Chemistry.*

organic solvents for the electrolyte requires hermetic sealing of the DSSC to prevent solvent evaporation and ensure long-term device stability. Volatile solvents can also permeate plastics and therefore impose strict requirements on the materials used to construct the cell. An alternative is the use of nonvolatile ILs as either solvents for the redox mediator or as the redox mediator itself (Gorlov and Kloo, 2008).

The electrolyte for a DSSC must be stable under the conditions of the device, ideally for decades. It should also not absorb visible light as this would reduce the efficiency of the cell (Wu et al., 2015). Finally, the electrolyte must wet the electrode material and enable the rapid diffusion of charge carriers between electrodes. The latter point is the primary limitation for the use of ILs such as $[C_3C_1im]I$ as the redox mediator and electrolyte as the high viscosity of such ILs negatively impacts their transport properties.

To address the limitations of pure I^--based ILs as electrolytes and search for alternatives, Grätzel and coworkers studied the preparation and properties of a range of 1,3-dialkylimidazolium salts including the now ubiquitous bis(trifluoromethanesulfonyl)imide class (Bonhôte et al., 1996; Papageorgiou et al., 1996). These findings in 1996 contributed greatly to the current renaissance of IL research. The original investigations used these $[NTf_2]^-$ ILs as mixtures with $[C_6C_1im]I$ to address the viscosity issues of the neat I^--based salts (Zakeeruddin and Grätzel, 2009). This approach of combining I^--based ILs with less viscous ILs has substantially improved the performance of DSSCs with the use of $[C_2C_1im][SCN]$ leading to efficiencies over 7%, the highest efficiency obtained at the time (Wang et al., 2004). An ongoing concern has been the long-term stability of the IL solvents. This was addressed by the use of a eutectic IL mixture with additives containing $[C_1C_1im]I$:$[C_2C_1im]I$:$[C_2C_1im][B(CN)_4]$:I_2:*N*-butylbenzimidazole:guanidinium thiocyanate in a 12:12:16:1.67:3.33:0.67 molar ratio (Bai et al., 2008; Shi et al., 2008). This led to good long-term stability, with 94% of its efficiency retained after 1000 h of full sun irradiation at 60°C. The maximum efficiency of this system was 8.4% also making it the most efficient IL system examined to date. This is less than the highest efficiencies of up to 13% that have been achieved recently in organic-solvent-based electrolytes (Mathew et al., 2014), although this was accomplished using novel dyes and redox mediators, which have not yet been examined using IL electrolytes.

ILs have also been used in quasi-solid-state electrolyte systems for DSSCs. These electrolytes use ILs with a gelator, reducing the risk of electrolyte leakage from the device and improving long-term stability (Wu et al., 2015). As these systems form gels, mass transport phenomena are slower than in the liquid phase resulting in reduced efficiency compared to liquid electrolytes. Nonetheless, a cell with an efficiency of over 7% was prepared using $[C_4C_1im]Cl$ as the IL, $[C_3C_1im]I$ as the redox mediator, and poly(hydroxyethyl methacrylate/glycerol) as the polymeric gelator, indicating that competitive efficiencies can be obtained from this approach (Li et al., 2014).

DSSCs featuring ILs are being actively pursued for commercialization by companies such as Dyesol and Solaronix, among others. Current progress toward this aim includes the use of DSSCs on the façade of the Swisstech Convention Centre in Lausanne by Solaronix and the partnership of Tata Steel with Dyesol to develop DSSCs for incorporation on steel roofing. To realize full commercial application of these materials, further advances in efficiency through the development of the electrolyte, electrodes, and dye technologies are required, illustrating the importance of future work within this field.

Analogous to the search for improved battery and supercapacitor electrolytes, the future of ILs as electrolytes for DSSCs involves finding more conductive, stable, and less viscous ILs to aid the diffusion of charge carriers. For this application, long-term stability of the electrolyte toward light and the dye are essential and these requirements restrict the range of ILs that can be used. The lack of volatility of ILs provides them with a significant advantage for DSSC technology given the importance of extended cell lifetimes and concerns over the evaporation of volatile organic solvents.

Unfortunately, IL systems tested so far do not possess the high efficiencies that have been observed for organic solvent electrolytes, although the use of IL electrolytes with comparable dyes and redox mediators has not yet been reported. Furthermore, their other advantageous properties may assist DSSC design such that on a cost/performance basis any reduction in solar conversion efficiency may be offset.

Electrolytes for the Electrodeposition of Metals

One of the key criteria for electrolyte selection for the electrodeposition of metals is the electrochemical window of the electrolyte system, including the relevant solvent (Simka et al., 2009). This is particularly important for reactive metals such as potassium, sodium, lithium, magnesium, and aluminum, which have very negative reduction potentials (Liu et al., 2010). The coordination environment of the metal is also important as this determines the solubility of the metal, the redox potential, and the morphology of the deposited material (Abbott et al., 2013). ILs have been studied for electrodeposition as, in addition to being natural electrolytes, they exhibit wide electrochemical windows, possess a large selection of anions, which can influence the coordination chemistry of the metal solutes and their thermal stability, and low volatility allows for higher temperature deposition than many organic solvents. ILs also offer the possibility for lower-temperature electrodeposition than is available for conventional molten salt systems (Endres, 2002). Furthermore, compared to aqueous electrolytes, ILs can eliminate the mechanical stability issues that arise from metals produced in the presence of hydrogen evolution. Metal speciation is often controlled within aqueous processing through the use of highly toxic anions, such as cyanide, which also may be averted by the use of ILs (Hartley et al., 2014).

ILs have been successfully used to electrodeposit the following metals: lithium, sodium, magnesium, zinc, iron, uranium, titanium, tantalum, manganese, chromium, molybdenum, ruthenium, rhodium, cobalt, platinum, palladium, nickel, silver, copper, gold, tin, gallium, indium, tellurium, cadmium, germanium, antimony, bismuth, lanthanum, and cerium (Armand et al., 2009; Endres, 2002; Simka et al., 2009). In addition, numerous alloys have been produced as well as semiconductors such as CdS, InSb, ZnTe, GaAs, AlSb, CdTe, and Si_xGe_{1-x} (Endres, 2002; MacFarlane et al., 2010). This diverse range of electrodeposited materials illustrates the utility of ILs as electrolytes for this application.

The specific effects that influence the successful deposition as well as the morphology and particle or grain size of the deposit are complex and vary greatly depending on the metal being deposited so only limited general comment can be made. As mentioned previously, speciation of the metal solute is of critical importance to its electrodeposition behavior and this is primarily linked to the IL anion and the metal source. For example, the reduction potential and stripping efficiency of Sn was found to depend strongly on the precursor used in [C_4C_1pyrr][NTf_2] whereas the electrochemical behavior in [C_4C_1pyrr][$N(CN)_2$] was at lower potentials and independent of the precursor, likely due to the coordinating effect of the $[N(CN)_2]^-$ anion (Martindale et al., 2010). It has also been suggested that the solubility of metal salts is enhanced when the metal shares a common anion with the IL although further study is required to verify this claim (Chiappe et al., 2010). Intricately tied to speciation are the reactions at the anode, which need to be considered if ILs are to be applied industrially. The typical anode reaction for aqueous systems is the oxidation of water to generate molecular oxygen (Abbott et al., 2013). Decomposition of the IL at the anode would not be either chemically or economically viable, requiring the use of soluble metal anodes. Understanding metal speciation in the IL is therefore essential for ensuring the rapid solubilization of the oxidized metal at the anode to prevent this process significantly limiting the rate of deposition. Obviously, the viscosity of the IL is also implicated in these transport processes and should be minimized where possible.

Interestingly, even for ILs bearing the same anion with similar metal speciation, differences have been observed in the deposited metal. For example, the deposition of aluminum from an $AlCl_3$ precursor in [C_4C_1pyrr][NTf_2] or [P_{66614}][NTf_2] was found to lead to nanocrystalline deposits whereas the use of [C_2C_1im][NTf_2] as the solvent resulted in the deposition of microcrystalline aluminum (Zein El Abedin et al., 2006). Similar grain effects were observed from the haloaluminate variants of similar IL cations (Giridhar et al., 2012). This illustrates the importance of solvation layers on the electrode surface including the IL cation and its orientation in determining the final morphology of the deposited metal irrespective of metal speciation (Endres et al., 2010).

While ILs clearly are amenable to the electrodeposition of many metals, much remains to be investigated with regard to the optimization of both conditions and IL selection for these processes. As in most electrochemical applications, low-viscosity, high-conductivity liquids with greater electrochemical windows would be ideal. However, further to these considerations, more investigations into molecular level interactions with regard to metal speciation and electrode interface dynamics are required to develop a predictive framework for the outcomes of electrodeposition processes and to enable the best IL to be chosen for a given metallic system. Nonetheless, despite the molecular complexity of electrodeposition from ILs, commercialization of these processes is being actively pursued by companies such as Scionix.

Electrolytes for Sensing Applications

Broadly speaking, chemical sensors provide qualitative and/or quantitative information regarding the presence of a compound of interest. This is generally accomplished by the use of chemical or biological recognition with the response reported through an appropriate transducer (Rehman and Zeng, 2012). Many sensors rely on the use of electrochemical methods for the recognition or response events, and consequently ILs have been investigated as electrolytes in this capacity for these systems. Again, their wide electrochemical windows, nonvolatility, and thermal stability are key features although the ability to tune solvation effects through manipulation of IL ions is also of importance as this can lead to selectivity towards dissolved substrates.

One of the initial areas of interest for ILs was in their application to gas sensors. Commercial gas sensors are typically amperometric gas sensors consisting of a gas-permeable membrane, which allows for the gas to pass through the electrolyte to the working electrode where it is detected electrochemically (Silvester, 2011). By virtue of the nature of detection, such systems are generally quite selective toward the desired gas. Commercial sensors generally use aqueous or organic solvent electrolytes, which are not able to operate in harsh conditions and have limited lifetimes due to solvent evaporation. ILs can extend the available operational conditions of these sensors and the nonvolatility of ILs eliminates concerns over the solvent drying out. Consequently, it has been proposed that the gas permeable membrane may not be required for such systems, which can improve response times by reducing the barriers toward the diffusion of gas to the working electrode (Buzzeo et al., 2004). Improving the response time is a key factor as one of the major limitations of the use of ILs has been the relatively slow response times obtained due to diffusion limitations arising from their viscosity. To further improve response times, thin films of IL in membrane-free devices with microelectrode arrays have been used, reducing the distance required for the analyte to diffuse and suggesting an approach for their future application (Huang et al., 2010). Such amperometric approaches are not confined to gas-sensing applications and IL electrolytes have been used for the analysis of a wide range of other substrates including metal oxides (Lu et al., 2010) and explosives (Guo et al., 2010). These applications require the optimal choice of IL to enable the solubility of the analyte with the electrochemical properties of the IL being the other major factor in their success.

Alternative gas sensors based on the use of a quartz crystal microbalance (QCM) have also been explored (Liang et al., 2002). These systems are more sensitive but less selective than amperometric sensors. QCM systems are nominally mass sensors that utilize a piezoelectric response from the surface of a vibrating quartz crystal that varies depending on changes to the surface coating, such as the absorption of a gas (Wei and Ivaska, 2008). It is necessary for the surface coating to provide selective absorption of the analyte to enable specificity of the measured response to the target analyte. It has been found that at lower temperatures using an IL coating, the variation of IL viscosity on absorption of the analyte is a larger determinant of the QCM response than the mass of the analyte itself. This leads to greater sensitivity for smaller analytes, such as acetonitrile, than would be expected based on their mass. The major advances with IL coatings of QCM for gas sensing is in their use for high-temperature gas applications, including the detection of flammable gas vapors, such as ethanol and heptane at 200°C with a 5% detection limit, whereas no vapor detection could be observed using a conventional solid coating at high temperatures. While ILs evidently provide significant advantages, their use as QCM coatings is somewhat limited by their physical state. Often thicker coatings are used to improve selectivity; however, the flow of the fluid affects reproducibility and the slow diffusion of analytes due to the IL viscosity reduces the sensitivity of detection (Rehman and Zeng, 2012). This has been combated by the formation of IL composites with the incorporation of ILs into polyelectrolyte films or conducting polymers to produce more rigid coatings with improved diffusion characteristics relative to inorganic solid state coatings (Rehman and Zeng, 2012).

Electrochemical biosensors have also been explored using ILs, exploiting the compatibility of many biomolecules with IL solvents and the conductive properties of ILs. In these sensors, the biomolecule is used as the recognition agent and is often immobilized onto a polyelectrolyte or carbon-based material to ensure contact with the working electrode surface (Shiddiky and Torriero, 2011). The latter utilizes the ability of various ILs to suspend carbon-based materials due to the strong interactions between the IL ions and the π systems of the carbonaceous materials (Weber et al., 2013).

Some successful electrochemical biosensing systems utilizing ILs include those for the analysis of dioxygen, hydrogen peroxide, and glucose using enzymes such as horseradish peroxidase, glucose oxidase, and myoglobin. Analysis typically involves the amperometric detection of either a reaction product, such as hydrogen peroxide in the reaction of glucose oxidase with glucose, or through direct electron transfer from the enzyme to the electrode, as has been observed for myoglobin (Ding et al., 2007; Yang et al., 2007). Such sensors are often limited by the stability of the enzyme. As ILs can provide a stabilizing environment for many enzymes, in some cases increasing their thermal stability compared to aqueous buffer systems (Patel et al., 2014), this provides another advantage for the use of ILs as electrolytes for biosensors.

ILs demonstrate real promise as electrolytes for sensing applications with their tunability and wide operational temperature range enabling their compatibility with sensing applications that cannot be achieved by conventional solvents and often with improved sensitivity compared to pure solid state sensors. Nonetheless, challenges exist, including the viscosity of these liquids, which limits their response times. Additionally, their long-term stability and compatibility with environmental contaminants, such as air and moisture, needs to be further considered.

SOLVENTS AND CATALYSIS

Interest in ILs as solvents grew from the idea that they were green alternatives to volatile organic solvents due to their negligible vapor pressures and, hence, reduced risk of exposure and atmospheric contamination arising from their use. Such a broad generalization overlooks the fact that ILs can be toxic and their lack of volatility can lead to persistence in the environment if the ions are not biodegradable (Scammells et al., 2005). Nonetheless, this motivation spurred interest in their application as alternative solvents, which has uncovered their applicability to organic, inorganic, and polymer synthesis. In some cases this has led to synthetic outcomes that would not be possible in other media, as will be outlined next.

Solvents for Organic Synthesis

While it could be envisaged that the unusual structure of ILs compared to conventional neutral organic solvents could lead to dramatic changes in reaction outcomes for all organic chemical processes, in very few cases is this observed (Hallett and Welton, 2011). A substantial number of organic reactions have now been investigated within ILs and in the majority of cases the outcomes with regard to the identity of the product formed and yields obtained coincide or can be predicted based on those obtained within polar solvents. A significant advance in the field has been the application of linear solvation energy relationships (LSERs), particularly those based on Kamlet–Taft parameters, which relate the rate constant of a reaction to solvent parameters (Eq. (1.1)) (Crowhurst et al., 2006). The solvent parameters utilized are α, the hydrogen bond acidity, β the hydrogen bond basicity, and π^*, which refers to the dipolarity and polarizability of the medium (Kamlet et al., 1977; Kamlet and Taft, 1976; Taft and Kamlet, 1976). LSERs have been extensively used to study reactions including nucleophilic substitutions (Crowhurst et al., 2006), Diels–Alder reactions (Bini et al., 2008), esterification (Wells et al., 2008), and keto-enol tautomerism (Angelini et al., 2009). Apart from the Diels–Alder reactions, where only moderate correlations were observed, the remaining studies all found good correlations between conventional organic solvents and ILs suggesting that there are no significant macroscopic differences between the use of ILs and polar organic solvents in these cases. Studies on molecular level interactions, primarily for nucleophilic substitution processes, have found that there are fundamental differences between ILs and neutral solvents when there is significant charge development (Bini et al., 2009; Yau et al., 2008). The major difference is the ability of the IL ions to independently solvate the developing charges in the transition state, which results in a more favorable enthalpic contribution at an entropic cost, overall having a negligible impact on the observed rate constant. While the macroscopic results are not indicative of a special IL effect, such findings enable the design of ILs through the rational selection of ions or the use of mixtures to optimize the rate or selectivity of a reaction. The larger number of possible ILs compared to conventional organic solvents increases the scope for tuning the solvent to the reaction to attain a desirable outcome. A typical relationship between rate constant (k) and Kamlet–Taft parameters investigated as part of an LSER is shown in Eq. (1.1) where XYZ_0, s, a, and b are fitted constants specific to a given reaction.

$$\ln(k) = XYZ_0 + s\pi^* + a\alpha + b\beta \tag{1.1}$$

Despite ILs generally leading to similar effects as polar organic solvents, there have been some unique "IL effects" that have been discovered. These are the direct result of the ionic nature of ILs, their ability to form nanosegregated structures, or the product of direct functionalization of the IL ions themselves. As ILs consist entirely of ions, their ability to interact with ionic compounds will differ from conventional organic solvents. Two notable examples where this has influenced reaction outcomes are

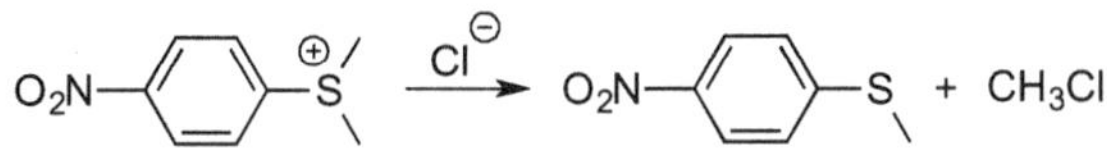

Scheme 1.1 ***Nucleophilic Substitution of a Sulfonium Cation With a Chloride Anion, as Reported by Welton and Coworkers (Hallett et al., 2009a).***

the nucleophilic substitution of a sulfonium cation with a chloride anion (Scheme 1.1) and the dediazoniation of the benzenediazonium cation (Scheme 1.2) (Bini et al., 2006; Hallett et al., 2009a). In the former case, substantially different reaction orders were observed in polar and nonpolar organic solvents compared to ILs (Hallett et al., 2009a,b). Within ILs, a bimolecular dependence on chloride and the electrophile was observed with substantially reduced reaction rates compared to organic solvents. For polar organic solvents, positive partial kinetic orders were detected whereas nonpolar solvents led to negative dependence. These relationships could be rationalized by considering the importance of ion-pairing on speciation for the substitution process. Nonpolar solvents produce an insoluble sulfonium chloride, removing chloride from the reaction system as the concentration is increased. Polar solvents lead to partially dissociated ions with the extent of ion association depending nonlinearly on concentration. ILs, however, produce a medium that completely dissociates ions (Lui et al., 2011). For the dediazoniation of the benzenediazonium cation (Scheme 1.2), the normally nonnucleophilic $[NTf_2]^-$ anion of the $[C_4C_1im][NTf_2]$ IL was found to react preferentially with the arene despite being present in an equimolar mixture with $[C_4C_1im]Br$. This was ascribed to the deactivation of the bromide anions due to hydrogen bonding and clustering around the $[C_4C_1im]^+$ cation. While there are only a few examples that indicate significantly varied reactivity due to the ionic nature of ILs, those that do exist illustrate that ILs have the potential to dramatically influence the reaction outcome when used as solvents for specific substrates and reaction types.

The effect of nanosegregated polar and nonpolar domains formed from ILs bearing long alkyl side-chains on organic reactivity has not been very widely explored (Weber et al., 2012, 2013). However, there

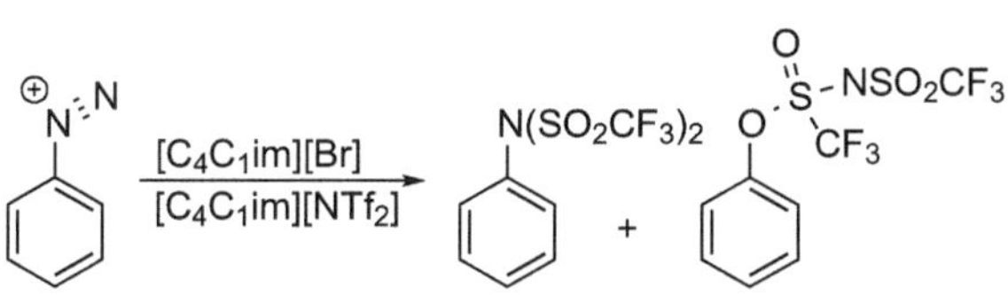

Scheme 1.2 ***Dediazoniation of Benzenediazonium to Form the $[NTf_2]^-$ Adduct, (Bini et al., 2006).***

is evidence that these domains result in reaction kinetics that bear similarity to those that would be expected from micellar systems with appropriately sized polar and apolar regions. Traditional micellar catalysis has been argued to arise from fluctuations in local concentrations with increased effective concentrations leading to faster reaction rates. Pseudo-phase models adopted from the micellar catalysis literature were successfully applied to the kinetics of the hydrolysis of a cationic trityl derivative in $[C_xC_1im][NTf_2]$ ILs to account for increased rates of reaction as the alkyl chain on the IL increased (Romsted et al., 1997; Weber et al., 2012). The primary difference between neat ILs and traditional micelles is the relative volume of the polar and nonpolar phases. In the imidazolium ILs that have been studied, polar and nonpolar volumes are of the same order of magnitude. On the other hand, the internal volume of micelles in micellar systems is generally negligibly small relative to the bulk polar (aqueous) phase. If the nanosegregated IL structures can be retained in the presence of a diluent, this could enable the relative size of each domain to be tuned through the addition of a cosolvent or the judicious choice of IL. The fact that ILs can behave as "neat surfactants," therefore, could significantly increase the scope of micellar catalysis and lead to improved organic chemical outcomes.

ILs have also been deliberately modified to imbue specific interactions for organic chemical processes. Such ILs have been denoted as "task-specific ILs" (TSILs) (Davis, 2004). Some of the modifications that have been used include the addition of acidic groups such as sulfonic acids, or basic groups such as amines, to promote acid and base-catalyzed reactions, respectively (Chiappe and Pomelli, 2014). Pendant hydrogen bonding groups such as alcohols have been used to induce stronger hydrogen bonding effects on solutes, for example, to improve the selectivity of a Diels–Alder reaction (Dzyuba and Bartsch, 2002; Tang et al., 2012). TSILs have also been employed as liquid analogues of solid support materials. This includes the use of ethers- and nitrile-functionalized ILs to coordinate metal complexes for transition metal catalyzed reactions and even the use of hydroxyl-functionalized ILs as liquid phase analogues of solid-state peptide synthesis (Miao and Chan, 2005; Tang et al., 2012). Being able to engineer the reactivity of the solvent directly through the chemical functionalization of the ions opens up new possibilities for solvent-directed control of organic reactivity. The major challenge with this approach is one often faced within ILs and that is the separation of products, catalysts, and reagents after the reaction is complete.

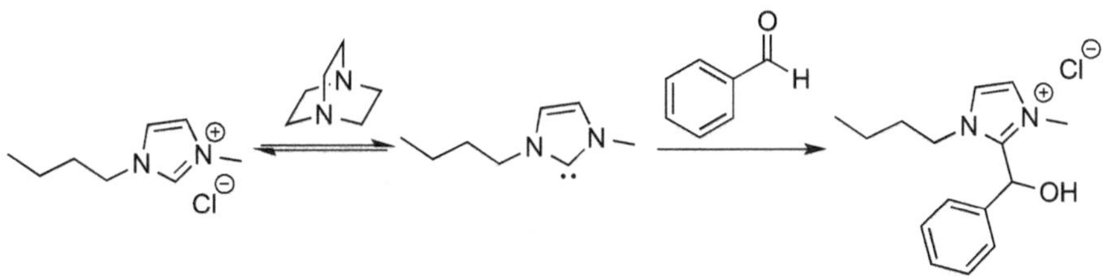

Scheme 1.3 ***The Formation of a C^2 Adduct When $[C_4C_1im]Cl$ was Used as a Solvent for the Baylis–Hillman Reaction (Aggarwal et al., 2002).***

As a corollary to the preceding discussion on TSILs, where the reactivity of ions has been employed intentionally, it is important to recognize that IL ions can be reactive even without deliberate functionalization. The most persistent problems in this regard are the hydrolysis of anions, such as $[BF_4]^-$, $[PF_6]^-$, and $[MeSO_4]^-$, to release acidic impurities or the use of bases or basic anions within ILs containing ammonium or imidazolium cations that can degrade or react through Hoffman elimination and the formation of reactive carbenes, respectively (Chowdhury et al., 2007; Scammells et al., 2005). These processes have led to side reactions such as the formation of C^2 adducts with imidazolium cations (Scheme 1.3) when such ILs were used as solvents for the Baylis–Hillman reaction (Aggarwal et al., 2002). Other notable stability issues have arisen when attempts to synthesize imidazolium ILs containing basic anions, such as hydroxide, have been made (Yuen et al., 2013). It is therefore important when using ILs for organic reactions to ensure that the ions themselves are compatible with the reaction conditions.

Industrial applications of ILs for organic synthesis have been pursued and include the Ionikylation process by PetroChina, which involves the alkylation of isobutene using chloroaluminate ILs in place of the conventional sulfuric acid catalyst (Plechkova and Seddon, 2008). The use of the IL led to increased production capacity as well as increasing the overall yield in a pilot trial of the process. Other organic reactions that have been pursued industrially include the demethylation of aryl ethers using molten [Hpy][Cl] by Eli Lilly and Co (Scheme 1.4), where the chloride anion is used as a nucleophile,

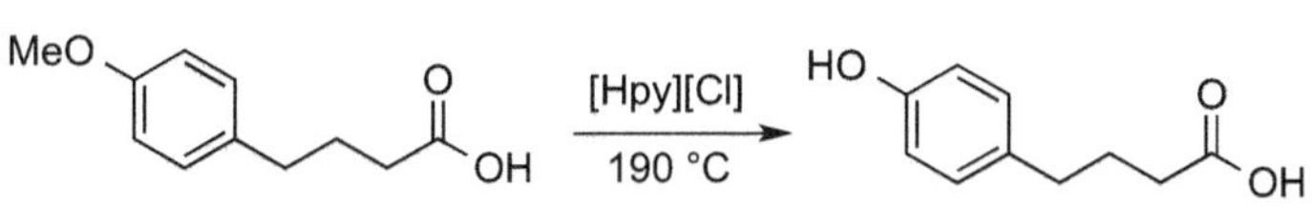

Scheme 1.4 ***The Demethylation of Aryl Ethers in [Hpy][Cl] (Schmid et al., 2004).***

and Sonogashira coupling within tetraalkylphosphonium ILs by the Central Glass Company, where the IL simply acts as a polar solvent to accelerate the reaction (Plechkova and Seddon, 2008; Schmid et al., 2004). These examples illustrate that commercialization of organic reactions within ILs is underway and encompasses processes where the IL is directly involved in the activity and those where it is simply a more effective solvent for the reaction.

While the body of literature currently indicates that unless specifically designed otherwise, ILs behave primarily as "drop-in" alternatives to polar organic solvents for the majority of organic chemical reactions, a number of specific IL effects do exist. Now that these have been identified, further directed study is required to enable these IL effects to be fully understood and their ability to influence organic reactivity optimized. Even simply as polar solvent alternatives, the sheer scope of IL choice means that improvements in reaction outcomes could be engineered through rational solvent selection. The advent of TSILs provides another dimension of creativity whereby the solvent can more actively participate in the desired reaction with the design of such solvents limited primarily by the chemist's imagination.

Solvents for Inorganic Synthesis

Further to the electrodeposition of metals discussed in the Section "Electrolytes for the Electrodeposition of Metals" where ILs were used as electrolytes to synthesize inorganic compounds, ILs have also been used as solvents in wet chemical approaches to inorganic synthesis. These synthetic approaches have been utilized for the preparation of discrete transition metals and metal clusters (Ahmed and Ruck, 2011); nanomaterials, particularly nanoparticles (Antonietti et al., 2004); framework materials such as zeolites, mesoporous silicas, and metal organic frameworks; and novel metal oxide morphologies (Ma et al., 2010). While various IL properties influence the outcome of each of these syntheses, as will be briefly discussed later, it is the ionic nature of these liquids that is at the core of these applications.

Nanoparticles are thermodynamically unstable but can be kinetically stabilized electrostatically, sterically, or by combining both methods (electrosterically). ILs act as excellent stabilizing media for nanoparticles although the precise reasons for this have not been definitively established. Nonetheless, electrostatic stabilization of the charged nanoparticle surface due to interaction with the IL ions has been identified and certainly plays a significant role (Fonseca et al., 2006). There is evidence that other effects, such as the formation of carbenes from imidazolium cations and the presence of supramolecular ion clusters providing electrosteric stabilization, also contribute depending on the IL used (Luska and Moores, 2012; Ott et al., 2005). This strong stabilizing ability combined with the relatively low interfacial energy and high viscosity of ILs, which enables rapid nucleation and restricted growth of particles, means that ILs are excellent solvents for nanoparticle synthesis. ILs even eliminate the necessity of adding capping groups to ensure the stability of the prepared nanoparticles. ILs have been used to prepare a wide range of metal nanoparticles including, but not limited to, iridium, rhodium, ruthenium, platinum, palladium, cobalt, copper, nickel, silver, chromium, molybdenum, iron, tungsten, and osmium (Ma et al., 2010). Tuning of the IL properties can influence the nanoparticle size and morphology. For example, the use of longer alkyl chains on the imidazolium IL cation can increase the size or change the morphology of iridium nanoparticles depending on whether a charged or neutral nanoparticle precursor is used (Migowski et al., 2010), although the mechanism behind this is under debate (Weber et al., 2013). The use of TSILs has also been explored in this area with the functionalization of the IL with coordinating groups such as thiols or nitriles leading to changes in nanoparticle size and morphology as well as improved stability compared to ILs simply bearing alkyl side chains (Kim et al., 2004; Ma et al., 2010).

The synthesis of porous materials, such as zeolites, has traditionally been accomplished hydrothermally with the precursors heated in water under autogenous pressure in the presence of an organic structure directing template. It was found that ILs can actually replace water and the template for the preparation of zeolites, in a method known as ionothermal synthesis (Cooper et al., 2004). This methodology makes the synthesis safer than the hydrothermal approach as the lack of volatility of the IL means that pressure vessels do not have to be used, reducing the risk of high-temperature depressurization. In addition, by using the IL as the solvent- and structure-inducing agent, any competition between these species for interaction with the growing material is eliminated, simplifying the synthetic protocol (Morris, 2009). This approach has also been studied for metal organic framework materials. In both cases, only the IL cation is included in the resultant material, although the anion has been found to inductively influence the structures formed. For example, chiral metal organic frameworks could be prepared from ILs with chiral anions even though they are not directly included in the structure (Lin et al., 2007).

As mentioned with respect to nanoparticles, ILs are capable of stabilizing surface charges through adsorption of the IL ions on the growing solid. This has been exploited for the preparation of exposed high-energy polar

Figure 1.3 ***Scanning Electron Microscope (SEM) Image of a Hexagonal ZnO Pyramid Synthesized in an IL.*** *Reproduced with permission from Zhou et al. (2005). Copyright 2005 Royal Society of Chemistry.*

surfaces of metal oxides and metal chalcogenides, among others. Generally, the polar surfaces grow the fastest due to their high surface energy and consequently they are often not observed in the final product. When an IL or molten salt is used as the synthesis media, these surfaces are stabilized, which slows their growth and leads to their presence in the final material. This produces unusual morphologies, such as the ZnO particles shown in Figure 1.3 (Zhou et al., 2005). Polar surfaces are often more active for catalysis and this synthetic approach has been used to produce more active CdS for the photocatalytic splitting of water (Lau et al., 2012). This behavior has also been observed with conventional molten salt systems, although the use of ILs enables the use of lower synthesis temperatures (Xu et al., 2009).

These examples illustrate some of the advantages that ILs can have on the synthesis of inorganic compounds. It is clear that their ionic nature, low volatility, and liquidus range can simplify and improve the safety of many synthetic procedures. Most importantly, ILs enable the preparation of a much wider variety of inorganic compounds than is accessible either with conventional solvents or high-temperature molten salt systems. Further investigation into the ability to fine-tune the morphology and size of the resultant compounds through appropriate ion selection is required to fully realize the potential of ILs to act as solvents for inorganic synthesis.

Polymer Synthesis

Due to their tunable solvation properties and highly structured nature compared to conventional solvents, ILs have been studied as alternative media for the preparation of polymers. The methodologies used traverse the entire gamut of traditional polymerization processes such as free radical, condensation, living radical, metathesis, transition metal catalyzed, and acid catalyzed polymerizations (Haddleton et al., 2008). Analogous to the discussion in the Section "Solvents for Organic Synthesis," most polymerization processes will work within an IL so long as there is no direct reactivity of the ions with the reagents and so they will not be described in detail here. The major factors influencing outcomes and some novel IL applications for polymer synthesis will be briefly discussed next.

For free-radical polymerizations, the most common industrially utilized method of polymerization, ILs have been found to increase the rate of propagation relative to termination, leading to the molecular weight of the polymers produced being up to an order of magnitude larger than within conventional organic solvents (Kowsari, 2011). This effect has been jointly attributed to the viscosity and polarity of the IL solvents, which reduces the rate of termination and increases the rate of propagation, respectively (Harrisson et al., 2003). The interplay between these factors has been studied in depth recently with trioctylphosphate, a viscous organic solvent, compared to ILs (Puttick et al., 2011, 2013). While viscosity was found to play a dominant role, other factors, such as the solvation of radical adducts by the IL anion and the formation of segregated polar and nonpolar domains, were also implicated in the behavior of the ILs (Puttick et al., 2011, 2013). This illustrates that ILs are a useful medium for such polymerizations although the molecular basis for their influence on these processes remains not thoroughly understood.

One of the often overlooked factors determining the success of a polymerization process within ILs is the solubility of the monomers and resultant polymers. The solubility of monomers is obviously required for the reaction to proceed, and the novel solvation environment provided by ILs has been exploited by Strehmel et al. (2006) to produce copolymers from nonpolar and zwitterionic methacrylate monomers, which is difficult to achieve in other solvent systems. With regard to the polymer chain produced, its solubility is one factor that governs the molecular weights attainable, with low solubilities generally resulting in lower molecular weight polymers, although this obviously aids their separation from the IL (Winterton, 2006). Given the potential to tune solvation effects using IL ions, this presents a novel method to control the degree of polymerization as well as effect the isolation and purification of the polymer from IL systems.

In addition to being employed as the medium, ILs have also been studied as monomers for the preparation of polyelectrolyte systems (Nishimura and Ohno, 2014). The resultant polymers are frequently referred to as polymerized ionic liquids or poly(ionic liquid)s (PILs) (Yuan et al., 2013). Despite ILs being known for more than a century, the concept of their polymerization has only been fully developed in the past 15 years (Nishimura and Ohno, 2014). PILs were developed as IL-based materials with a view to incorporating many of the unique properties of ILs within a macromolecular architecture. The major advantages of PILs over the monomeric IL species are their mechanical stability, ease of processing, durability, and greater control over their spatial arrangement (Yuan and Antonietti, 2011). The synthesis of PILs is generally conducted through the incorporation of a polymerizable functionality on the IL cation or anion, such as a vinyl substituent, followed by conventional polymerization procedures (Mecerreyes, 2011). An alternative preparation involves the reaction of the IL with a polymeric chain possessing acidic or basic functionality to form the PIL. The difference between these approaches has not been intensely studied to date. PILs are generally classified as cationic, anionic, or zwitterionic depending on the charge of the covalently bound ion. The variety of PIL types that can be prepared are detailed in Figure 1.4.

One of the touted advantages of PILs is the ability to produce polymers with low glass transition temperatures (T_g as low as $-60°C$), despite the presence of ionic groups in the polymer (Washiro et al., 2004). PILs can also exhibit tunable solubility that depends on ion selection, and if aprotic ILs are used then their speciation is pH-independent, which is an advantage over many conventional polyelectrolyte systems (Yuan et al., 2013). Ion conductivity is a fundamental physical property of PILs given their chemical constitution. As would be anticipated, the ionic conductivity of PILs is substantially lower than the corresponding ILs due to the lack of mobility of the polymerized ion and the reduced fluidity of the polymerized system. Strategies to improve the ionic conductivity include the addition of salts to the PILs to increase the number of mobile charge carriers and the use of aliphatic linkers to create PIL "brushes," which increase the flexibility of the polymerized ion and enable more rapid transport of the mobile ion (Nishimura and Ohno, 2014). These strategies have restricted the drop in ion mobility to an order of magnitude compared to the native ILs, compared to over two orders of

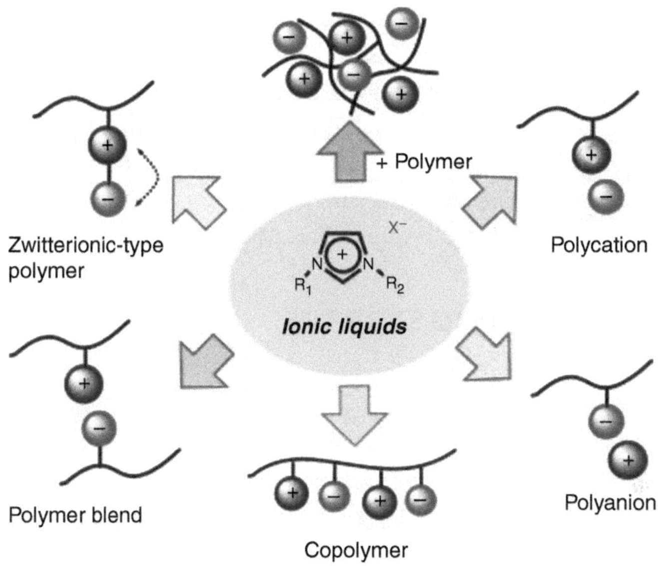

Figure 1.4 ***Schematic of the Different Types of PILs That can be Prepared Depending on the Synthetic Route.*** *Reproduced with permission from Nishimura and Ohno (2014), Copyright 2014 Elsevier Ltd.*

magnitude observed in their absence. Nonetheless, despite the drop in ion mobility, PILs possess a number of advantages over liquid electrolyte systems as they can negate some environmental and safety issues such as the risk of electrolyte leakage and flammability for applications such as batteries and fuel cells. Moreover, PILs permit the production of ion-conductive materials with defined size, shape, and geometry, such as thin films, fibers, coatings, and even complete circuits.

Analogous to some of the successful applications of ILs, PILs have been examined for catalysis, CO_2 separation, and as an absorption medium with equally satisfactory performance (Asadov et al., 2012; Kim and Chi, 2004; Tang et al., 2005). PILs have also been investigated as thermoresponsive hydrogels (Ziółkowski and Diamond, 2013) and photofunctional materials (Kadokawa, 2013). PILs have now developed into an independent area of research and its comprehensive description is beyond the scope of this book; however, readers are strongly encouraged to refer to more specific reviews (Lu et al., 2009b; Mecerreyes, 2011; Nishimura and Ohno, 2014; Yuan and Antonietti, 2011; Yuan et al., 2013).

ILs are clearly useful compounds for polymer synthesis as novel solvents and even as monomers. Their tunability and the wide scope of IL selection means that the ideal choice of ions for each application is likely to be unique and that no general "best" IL is likely to exist. For example, tuning the solubility of the polymer could enable advanced separation processes and greater molecular weight control. The potential for such specific control over polymerization processes and the resultant polymers indicates that this is a promising area for future development. It is important that the focus is also on the purification of the resultant polymers and recycling of the IL as the nonvolatility of the IL requires that novel strategies are implemented.

HEAT TRANSPORT AND STORAGE

Thermal Fluids

Heat transfer using thermal fluids is an essential component of many industrial processes, particularly for emerging energy applications such as the use of concentrating solar power and for nuclear energy applications (MacFarlane et al., 2014). Thermal fluids rely on sensible heat transfer, which requires that these compounds remain liquid within the required operational temperature range. The main properties of interest for thermal fluids are the thermal conductivity, heat capacity, thermal stability, and melting point of the fluid (Moens et al., 2003). Nonvolatile fluids are also advantageous as they simplify reactor design and a larger temperature range improves the efficiency of heat cycles (Bridges et al., 2011).

For most thermal fluid applications, it is the volumetric heat capacity rather than molar heat capacity that is of most interest with regard to design. ILs have been found to possess volumetric heat capacities of around 1.95 J K^{-1} cm^{-3} at 298 K (Gardas and Coutinho, 2008), greater than commercial heat transfer fluids such as Therminol (VP-1) (1.48 J K^{-1} cm^{-3}), which is based on more volatile organic compounds (Bridges et al., 2011). The thermal conductivities of ILs are generally between 0.1 W m^{-1} K^{-1} and 0.2 W m^{-1} K^{-1}, less than that of water but comparable or greater than neutral organic solvents and commercial organic heat transfer fluids (Ge et al., 2007). These properties have been enhanced by the use of composites involving ILs with suspended nanomaterials (Bridges et al., 2011; Ribeiro et al., 2011). For example, the use of 2.5 wt% nanoparticulate aluminum oxide led to a 30% increase in heat capacity of $[C_4C_1{}^2C_1im][NTf_2]$ with a 7% increase in thermal conductivity and no significant negative effect on thermal stability (Bridges et al., 2011).

The thermal stability of ILs governs their operational temperature and is a critical factor in their applicability as thermal fluids. The short-term stability of some ILs, particularly those based on the $[NTf_2]^-$ anion, has been measured at over 400°C (Maton et al., 2013). These measurements are typically conducted using thermogravimetric analysis at heating rates of 10 K min^{-1}. It is important to note that long-term stability can be substantially lower than the measured short-term thermal stability, although long-term thermal stabilities above 250°C have been determined for $[NTf_2]^-$ ILs (Del Sesto et al., 2009; Wooster et al., 2006). Given the low melting points of many of the simple $[NTf_2]^-$ ILs, this leads to operational temperature ranges of over 300°C where the IL is liquid and thermally stable although their viscosity becomes an issue at lower temperatures. Unfortunately, many of the other IL anions commonly used, such as Cl^-, $[OAc]^-$, and $[N(CN)_2]^-$, are more reactive and lead to reduced thermal stability and consequently a more restricted operational temperature range.

From the preceding discussion, it is evident that ILs have the potential as nonvolatile thermal fluids to be alternatives to neutral organic components for low- to medium-temperature applications. These temperatures would be too low for more thermally stable options, such as molten salts and liquid metals, due to their high melting points. As a result, thermal fluids based on ILs have the potential for use in parabolic trough concentrating solar power plants as safer replacements for the current flammable and volatile heat transfer medium (López-González et al., 2013). These power plants have operating temperatures of around 400°C, so ILs with increased thermal stability are required. The desirable specifications for thermal fluids for this application, as outlined by the US National Renewable Energy Laboratory, are high temperature stability >430°C with a freezing point <0°C and a vapor pressure below 1 atm (Van Valkenburg et al., 2005). These requirements are not met by any current organic thermal fluid. $[C_4C_1Im][BF_4]$ has been studied for this application and found to provide comparable energy density with improved thermal stability over commercial heat transfer fluids. Its viscosity and absolute thermal stability, particularly over the long term, remain a limitation with regard to the ideal properties for thermal fluids that have been postulated.

ILs clearly show promise for thermal fluid applications at temperatures below those where conventional molten salts or liquid metal systems would be accessible. Compared to neutral organic fluids, ILs provide equal or even greater thermal conductivity and heat capacity as well as reduced vapor pressures and reduced flammability. The major limitations for ILs at present are their cost and viscosity relative to other thermal fluids, although the former may be ameliorated by their increased scale of production if a large-scale application was found. Further fundamental studies into structure–activity relationships for thermal properties are still required as are applied investigations to ascertain the feasibility of ILs as thermal fluids under realistic conditions for different applications.

Sorption Cooling Media

An application that is attracting significant industrial interest for ILs is their role as part of a working fluid for absorption cooling (Preißinger et al., 2013). Absorption cooling enables the use of thermal rather than mechanical energy to drive the cooling cycle, resulting in reduced electricity consumption compared to the more widely used compression cooling (Brown and Domanski, 2014). Absorption cooling processes utilizing waste heat from solar, geothermal, or industrial processes have been proposed and offer the potential to substantially increase the utilization of energy from these renewable sources (Seiler et al., 2013). To facilitate absorption cooling, a working pair of an absorbent and a refrigerant are required. Common working pairs include lithium bromide/water and ammonia/water (Ziegler, 1999). A standard absorption cooling cycle involves four major stages (Figure 1.5). These stages include the generator where the driving heat is input to volatilize the refrigerant. The next stage is the condenser where the refrigerant is liquefied at high pressure and heat is transferred. The refrigerant is then passed through a throttling valve to reduce the pressure into

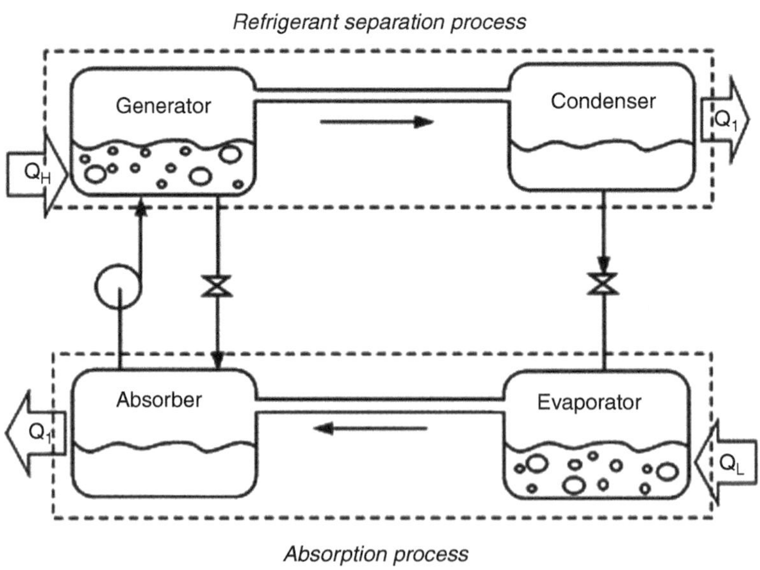

Figure 1.5 ***Schematic of the Absorption Cooling Cycle.*** *Reproduced with permission from Srikhirin et al. (2001), Copyright 2001 Elsevier Science Ltd.*

the evaporator where heat is absorbed from the surroundings and the liquid refrigerant is evaporated. Finally, the refrigerant is absorbed by the absorbent in the absorber and the resultant fluid passed back into the generator to restart the cycle (Ziegler, 1999). The figure of merit for absorption cooling is the coefficient of performance (COP), which is the ratio of the cooling capacity obtained at the evaporator divided by the heat input at the generator and work input for the pump (Srikhirin et al., 2001).

The limitations of the current working pairs are that lithium bromide is corrosive and can crystallize, while ammonia suffers from being corrosive, flammable, and toxic as well as requiring the use of high pressures (Yokozeki and Shiflett, 2010). As ILs are nonvolatile with tunable hydrophilicity, and possess the ability to be noncorrosive with large liquidus ranges, they are obvious candidates as novel sorption cooling media. Many ILs also possess long-term thermal stability at temperatures up to 423 K, which is the maximum required for most absorption cooling systems, indicating their suitability for this application (Kim et al., 2012).

For IL systems, most interest is in working pairs involving ILs as absorbents and water as the refrigerant as this is an effective working pair due to the large heat of vaporization of water and its reduced environmental impact compared to other refrigerants such as ammonia. The most successful ILs for these working pairs have been those that contain hydrophilic anions such as $[Me_2PO_4]^-$, $[MeSO_3]^-$ and glycolate as these have the highest affinity for water. Reducing the strength of cation–anion interactions through the use of bulky, weakly interacting cations has also been proposed to enhance the absorption capacity of water and the efficacy of the working fluid (Kurnia et al., 2014). In one system where the COP has been determined, $[C_2C_1Im][Me_2PO_4]$ was found to achieve a COP value as high as 0.85, greater than the lithium bromide/water working pair under the same conditions (Preißinger et al., 2013). This illustrates that not only do ILs possess properties that are attractive from an environmental perspective, but they can also perform better than existing working fluids. Further studies on the long-term chemical stability of these ILs under the chiller conditions are required to fully assess their suitability.

In addition to some published studies, the industrial application of ILs for absorption cooling processes is being actively investigated with IL-based working fluids being studied at a pilot plant scale by companies such as Iolitec (jointly with BASF SE) and Evonik Industries. The major limitation to their commercial implementation at present is their cost; however, the price will obviously decrease if required on a bulk scale for such an application.

Consequently, it is probable that ILs will see commercial application for absorption cooling in the near future as they represent a technical and environmental improvement over existing working fluids.

Phase Change Materials

Phase change materials (PCMs) store thermal energy in the form of latent heat due to a phase transition, with solid–liquid transitions being favored due to their relatively large latent heats with minimal volume change compared to liquid–gas transitions (Farid et al., 2004; Vijayraghavan et al., 2013). PCMs lead to much higher energy storage density than thermal fluids, over a smaller temperature range, and consequently are of particular interest for applications with less substantial temperature variations. As for other thermal storage applications, high thermal conductivities are advantageous, although for PCMs more important factors are the latent heat of fusion and melting temperature, which determine the thermal storage capacity and operational temperature range, respectively (MacFarlane et al., 2014). Fast crystallization kinetics are critical to the performance of PCMs as supercooling leads to an increase in the temperature swing required and reduces its reliability; unfortunately, this represents a significant limitation for many ILs. Conventional PCMs use either organic compounds, such as waxes and fatty acids, or inorganic salts and salt hydrates. Organic compounds can possess reasonable heats of fusion, up to 269 J g^{-1}, although they are often limited by relatively low thermal conductivities and by their flammability (Farid et al., 2004; Vijayraghavan et al., 2013). Inorganic salts offer higher thermal conductivities, although their melting points are often too high as anhydrous salts, requiring the use of hydrates for PCM applications. The presence of water leads to issues with phase separation and these salts are prone to supercooling. As pure ILs possess relatively low melting points, which can be tuned through appropriate ion selection without requiring the addition of cosolvents, which could lead to phase separation, they have a potential role as PCMs for low- to medium-temperature applications.

IL systems have been found to possess relatively good heats of fusion with up to 190 J g^{-1} being measured for guanidinium methanesulfonate (Vijayraghavan et al., 2013). Unfortunately, this guanidinium salt possesses a melting point close to its decomposition temperature, which limits its utility. A heat of fusion of 156 J g^{-1} was measured for [$C_{16}C_1$im][Br], which has a relatively low melting point of 63.9°C (Zhu et al., 2009). One of the major limitations of ILs is their ability to supercool and [$C_{16}C_1$im][Br] was found to consistently freeze approximately 20 K below its melting point

(Bai et al., 2011). It was suggested that supercooling could be used as an advantage for seasonal applications, where heat is stored in the warmer summer months for heating in winter, if crystallization could be induced by the addition of a nucleating agent to release the heat when required. Other work examining the functionalization of imidazolium ILs with ester, alcohol, or carboxylic acid functionalities led to improved heats of fusion relative to the aliphatic variants although the maximum value obtained was only 105 J g^{-1} (Zhang et al., 2014a). Therefore, it appears that the heat of fusion is maximized through strong hydrogen bonding interactions between ions, as in the case of guanidinium methanesulfonate, or through the development of dispersion forces between aliphatic chains, as for $[C_{16}C_1im][Br]$. These design features can be used to guide the preparation of novel ILs with larger heats of fusion.

Unfortunately, there are currently extremely few studies specifically aimed at the application of ILs as PCMs, although it is evident from those that have been conducted that some ILs possess heats of fusion that would be competitive with current PCMs. While the well-known issue of supercooling is likely to limit the use of currently known ILs for short-term thermal storage, their use as PCMs for seasonal applications is possible if nucleation can be reliably induced. It is also possible that there are superior IL candidates that have not yet been identified and further examination of the thermal properties of ILs is required to determine if this is the case.

IONIC LIQUIDS IN THE BIOREFINERY

The concept of the biorefinery is based on obtaining value-added chemicals, materials, and fuels from biomass, primarily lignocellulose. Lignocellulose is the most important source of renewable materials on Earth and is mainly composed of cellulose, hemicelluloses, lignin, and several extractives (Kumar et al., 2009). The separation of the main components of lignocellulose occurs within the primary biorefinery, and their consequent conversion into platform chemicals or value-added materials represents the secondary biorefinery (Hongbin and Lei, 2013).

Cellulose is an abundant renewable polysaccharide consisting of a linear chain of 1-4-linked β-D-glucopyranosyl units that aggregate to form a highly ordered structure divided into amorphous and crystalline domains. Hemicelluloses are mainly composed of carbohydrates based on pentose sugars, mainly xylose, as well as hexose sugars including glucose and mannose. Lignin is an amorphous, polyphenolic material arising from enzymatic dehydrogenative polymerization of three phenylpropanoid monomers, namely, coniferyl, sinapyl, and *p*-coumaryl alcohols (Kumar et al., 2009).

Due to their combination of properties and versatility of synthesis, ILs are of interest as solvents and catalysts for biorefinery processes and have been studied by several authors. ILs have been applied in different ways to processes involving lignocellulose to dissolve wood, dissolve and modify cellulose, in the selective extraction of hemicelluloses, and to dissolve, selectively extract, and depolymerize lignin (Brandt et al., 2013).

Cellulose

Cellulose is the hardest part of lignocellulose to solubilize. However, it is an important potential source of materials, biofuels, and chemicals. Solid cellulose is used in pulp and paper manufacturing and after its dissolution can be derivatized and regenerated for integration into different products (films, fibers, and composites). Glucose itself can be transformed into fructose, hydroxymethylfurfural (HMF), ethanol, and butanol, which can be used as sources of fuels and other products (Cao et al., 2009; Tan and MacFarlane, 2010; Tian et al., 2010).

After dissolution cellulose can be reprecipitated and its crystal structure typically changes from type I to type II. The cellulose type II structure is more reactive, more easily degrades and offers the possibility of crosslinking, so it is an important source of biodegradable materials (Ge and Wang, 2014).

Cellulose was chosen as the first target to try to incorporate ILs into biorefinery processes as its solubility has been studied by many researchers (Swatloski et al., 2002; Burchard, 2003; Pinkert et al., 2010). Cellulose dissolution occurs through a mechanism that involves the formation of hydrogen bonds with the IL. The solubility of cellulose depends mainly on the anion and requires that it must have a high hydrogen bonding basicity, such as that for Cl^-, $[MeCO_2]^-$, dialkylphosphates ($[R_2PO_4]^-$), alkylphosphonates ($[R_nPO_3]^{m-}$), or amino acids. However, these anions lead to ILs with low stabilities. The effect of the cation has been less well investigated but cellulose solubility appears to decrease when the alkyl chain increases and with methylation of the C^2 position in 1-alkyl-3-methylimidazolium-based ILs. In addition, water has to be avoided in the reaction media as much as possible.

$[C_4C_1im]Cl$ was the first investigated IL to dissolve cellulose (Swatloski et al., 2002); many authors have tested it in different conditions since 2002 (Stark, 2011). $[C_4C_1im]Cl$ shows very good performance for cellulose dissolution but high temperatures are needed, which can cause $[C_4C_1im]Cl$ decomposition and the release of organohalides, which are eco-toxic compounds. In addition, $[C_4C_1im]Cl$ is very hygroscopic, so the atmosphere has to be carefully controlled as water acts as an antisolvent and decreases cellulose solubility; in addition, the cellulose does not decrystallize as much in the presence of water (Brandt et al., 2013).

Other ILs have been studied including $[C_4C_1im]Br$, $[C_4C_1im][SCN]$, 1-*N*-allyl-3-methylimidazolium chloride ($[(C_1{=}C_2)C_1im]Cl$) (Cao et al., 2009), $[C_2C_1im][MePO_4]$ (Fukaya et al., 2008), and 1,3-dialkylimidazolium formate ($[C_xC_yim][HCO_2]$) (Fukaya et al., 2006). However, those with halide anions continue to have the problem of the release of toxic compounds at high temperatures. In addition, these ILs show lower cellulose solubility than $[C_4C_1im]Cl$. In order to obtain greener processes, $[C_2C_1im][MeCO_2]$ was studied and found to give satisfactory results at room temperature while being less toxic and less corrosive than those ILs based on halides (Hermanutz et al., 2008). Cellulose can be precipitated from the IL solution by the addition of water, or other solvents including ethanol and acetone (Swatloski et al., 2002).

After cellulose dissolution into the IL, cellulose fibers, films, and casts can be obtained by spinning into a bath consisting of water, and/or low molecular weight alcohols, where cellulose precipitates and the IL remains in solution. Cellulose can also be modified by adding synthetic polymers or metals to build composite materials. Its depolymerization into sugars was also tested using $[C_4C_1im][HSO_4]$, a Brønsted acidic IL, with the selectivity of the depolymerization depending strongly on the water content. Hydroxymethyl furfural, HMF, was the main product obtained with dry ILs. The depolymerization can be enhanced by the use of microwaves and ultrasound. Enzymatic reactions of cellulose in ILs have not been extensively studied and until now significant yields have not been obtained. Enzyme performance in ILs remains under investigation and has not been optimized yet (Bose et al., 2010).

Hemicellulose

Hemicelluloses are mainly composed of glucose, xylose, mannose, galactose, arabinose, and rhamnose in different proportions, depending on the plant species. These sugars can be transformed into furfural, which is an interesting platform chemical for many applications. Apart from furfural and xylitol, different acids, such as lactic acid, 3-hydroxypropionic acid, succinic acid, and galactoglucomannan, can be obtained and applied in very diverse fields, such as animal feed or emulsifiers among others (Hongbin and Lei, 2013).

Complete dissolution of the carbohydrates is achieved by using ILs with high hydrogen bond basicity in combination with the addition of acid.

Hemicelluloses can be reprecipitated by adding methanol or ethanol as an antisolvent. A problem arises, however, when removing the small oligosaccharides that comprise hemicellulose from the IL, a feat that has not yet been achieved (Stark, 2011).

Further investigation in the field toward the selective extraction of hemicellulose using switchable ILs was conducted by Anugwom et al. (2012). These switchable ILs are generally prepared by bubbling CO_2 through a mixture of 1,8-diazabicyclo-[5.4.0]-undec-7-ene (DBU) and an alcohol. The authors used an amidine in DBU with a long-chain alcohol (1-hexanol) and gaseous CO_2 under ambient pressure and at room temperature, causing an exothermic transformation of the molecular liquid mixture into an IL. These ILs have been successfully applied for the selective extraction of hemicellulose from spruce wood. The reprecipitation of the hemicellulose was achieved by adding methanol.

In general, hemicellulose reactivity in ILs has not been extensively studied; however, there has been some work done on its derivatization. Acetylation of hemicelluloses was performed by Ren et al. (2007), using $[C_4C_1im]Cl$ as solvent, iodine as catalyst, and acetic anhydride as an acetyl source; they achieved 90.8% yield of acetylated hemicelluloses with 15% of iodine as catalyst.

Lignin

Biomass delignification has been widely studied; there are several processes that achieve delignification of biomass (sulfite, Kraft, and the organosolv processes). However, all of these processes use volatile organic solvents and high-pressure equipment. With the aim of making the process greener, ILs were introduced for the delignification of biomass, through a process known as Ionosolv. High-pressure equipment is not necessary for the Ionosolv process as the reactions are done at atmospheric pressure, so the cost of the equipment is also reduced. In order to dissolve the lignin selectively, the anion of the IL should have moderate hydrogen bond basicity, such as $[MeSO_4]^-$, $[OTf]^-$, or alkylbenzenesulfonate $[RBzSO_3]^-$; these ILs are even capable of dissolving Kraft lignin up to a certain concentration. Lignin solubilization occurs via strong ion-dipole and π–π interactions between the lignin and the anion.

Kraft lignin can be solvated by $[C_1C_1im][MeSO_4]$, $[C_4C_1im][MeSO_4]$, $[C_6C_1im][OTf]$, and $[C_2C_1im][MeCO_2]$ (Fu et al., 2010; Tan and MacFarlane, 2010). Lignin was dissolved by different ILs with the same cation $[C_4C_1im]^+$; the order of lignin solubility for varying anions was

$[MeSO_4]^-$ > Cl^- > Br^- >>> $[PF_6]^-$. Noncoordinating anions, such as $[PF_6]^-$ and $[BF_4]^-$, were found to form ILs that are unsuitable as solvents for lignin (Pu et al., 2007). As with cellulose, the solubility of lignin is strongly affected by the choice of anion, although lignin dissolution requires less hydrogen bond basicity than cellulose dissolution and water content in the reaction media is not as critical, though water is still an antisolvent, so it is easier to dissolve lignin than to dissolve cellulose. Indeed, water can be added to dissolve lignin selectively against cellulose. However, the high solubility of lignin in certain ILs does not imply its direct extraction from biomass.

Other factors that can affect the solubility of lignin include temperature, the source of lignin, and the heating source. Temperature highly influences the solubility of lignin; the solubility of lignin in $[C_2C_1im][MeSO_4]$ increased from 6% to 26% when the temperature was increased to 50°C from room temperature (Pu et al., 2007). Microwaves can also enhance lignin dissolution and extraction processes, reducing reaction times from hours to minutes. Lignin was successfully extracted from hardwood by $[C_4C_1im][MeSO_4]$ in 3 min with a microwave reactor (Prado et al., 2013). The source of lignin also influences its solubility; residual softwood lignin is more soluble than Kraft lignin (Brandt et al., 2013).

A low cost IL has been tested for delignification, $[Et_3NH][HSO_4]$. This IL has good performance under mild conditions and has the advantage that it is cheap (Chen et al., 2014; George et al., 2015).

Lignin Reactivity in Ionic Liquids

Before studying lignin reactivity in ILs, it is essential to determine how its structure is affected by different cation–anion combinations. It is well known that lignin structure is affected by the extraction process, which causes chemical changes and fragmentation due to the chemical agents, temperature, and pressure used. This fact makes it especially important to study how lignin is affected by each IL. For example, Tan et al. observed that after delignification of sugarcane waste using $[C_2C_1im][RBzSO_3]$ as the IL, the obtained lignin had a lower average molecular weight than autohydrolysis lignin. Both lignins were acetylated in order to carry out a proper comparison. Acetylated lignin from $[C_2C_1im][RBzSO_3]$ had lower average molecular weight (3690 g mol^{-1}) than acetylated autohydrolysis lignin (19,300 g mol^{-1}), and the polydispersity was also lower for the 1-ethyl-3-methylimidazolium alkylbenzenesulphonate ($[C_2C_1im][ABS]$)-treated lignin compared to autohydrolysis lignin at 1.66 and 11.4, respectively (Tan et al., 2009). George et al. (2011) studied the effect of different ILs – the combination of $[C_2C_1im]^+$ with different anions: *n*-hexylsulfate $[C_6SO_4]^-$, diethylphosphate $[Et_2PO_4]^-$, *n*-butylsulfate $[BuSO_4]^-$, dimethylphosphate $[Me_2PO_4]^-$, lactate $[lac]^-$, Cl^-, $[MeCO_2]^-$, and other ILs such as $[C_4C_1im]Cl$, Cyphos 101, Cyphos 165, and 1-methyl-2-pyrrolidone chloride ($[C_1pyr]Cl$) on lignins from different sources (organosolv, alkali (NaOH), and slightly sulfonated alkali lignin). It was observed that the anion had a stronger influence on the lignin structure than the cation; the activity of the anions modifying the molecular weight followed the order: $[RSO_4]^- > [lac]^- > [MeCO_2]^- > Cl^- > [R_2PO_4]^-$. It has to be mentioned that the rate of modification of lignin by the action of IL also depends on its source, with organosolv lignin being more likely to have changes in its structure than slightly sulfonated lignin and alkali lignin.

Due to its complex structure, there are not many studies that use lignin as starting material to study its reactivity. However, there are many authors that have studied lignin model compounds in order to elucidate the reaction mechanisms and the proper conditions to extend the reactions to whole lignin. The depolymerization of lignin model compounds and lignin itself using ILs as solvents have been studied under reductive and oxidative conditions. Reductive conditions were achieved by adding Lewis and Brønsted acids as catalysts. However, these reactions were briefly studied because of the low reactivity shown by the lignin despite the good yields obtained for lignin model compounds. Oxidation has been more extensively studied; coupling a metal catalyst such as Fe, Mn, Co, V with an oxidant such as O_2 or H_2O_2 has been tested with many different ILs under different conditions. The role of ILs can be as a reaction solvent and/or a catalyst (Chatel and Rogers, 2014).

Reduction of Lignin

Binder et al. worked on reductive depolymerization of lignin model compounds (eugenol, 4-ethylguaiacol, and 2-phenylethyl ether) and organosolv lignin at moderate temperatures helped by Brønsted acid catalysts (Co/silica, Cu/alumina, $RuCl_3$, $RhCl_3$, $H_3PMo_{12}O_{40}$, $H_4SiW_{12}O_{40}$, Nafion, and several Pt catalyst) in $[C_2C_1im][OTf]$ and $[C_2C_1im]Cl$ to check its conversion to guaiacol, 4-propylguaiacol, and isoeugenol. Eugenol model compounds were treated and different products were obtained under different conditions; a 4-propylguaiacol yield of 77.5% was obtained with a 0.5% Pd/alumina (C3677) $[C_2C_1im][OTf]$ tandem catalyst, with a conversion of 99.6%, and the best guaiacol yield was achieved with Nafion/$[C_2C_1im][OTf]$ (11.6%). Applying the same optimum conditions to lignin using

Nafion as catalyst, the performance was poor and no guaiacol was obtained (Binder et al., 2009).

Oxidation of Lignin

As occurred with the reduction reactions, model compounds were used to determine the optimum conditions for reaction. Jiang et al., Zhu et al., and Lahtinen et al. worked on the reactivity of lignin model compounds under different conditions and then they applied the optimum conditions to check lignin reactivity.

Jiang et al. studied the selective oxidation of primary alcohols into aldehydes or acids. The reactivity of methoxybenzylalcohol, vanillin, and veratryl alcohol model compounds was studied and the influence of cocatalyst and IL structure was checked. When methoxybenzylalcohol was studied, the aldehyde was obtained with 99% selectivity when $[C_4C_1im][PF_6]$ was the IL and without cocatalyst. To obtain the acid, Cu(II) 2-ethylhexanoate was added as cocatalyst and $[C_6C_1im][OTf]$ was used as solvent, but the selectivity was only 87% (Jiang and Ragauskas, 2007a). When vanillin and veratryl aldehyde were studied, optimum reaction conditions were 5 mol% $VO(acac)_2$, 10 mol% DABCO, and $[C_4C_1im][PF_6]$ as solvent. For veratryl and vanillyl alcohols, the oxidation selectively stops at the aldehyde with 96% and 94% of selectivity (Jiang and Ragauskas, 2007b).

It has been demonstrated that transition metal nanoparticles have higher catalytic activity than the bulk material due to their higher surface area. Zhu et al. studied metal nanoparticles as catalysts for lignin model compounds and lignin oxidation. They studied the oxidation of benzyl alcohol derivatives in a mixture of ILs ($[C_4C_1im][MeSO_4]$:$[C_4C_1im][PF_6]$;1:2) using palladium nanoparticles as the catalyst, achieving around 85% conversion. Lignin was treated under the same conditions, achieving 72% conversion. The main products were identified as syringaldehyde, vanillin, and *p*-hydroxybenzaldehyde together with a small amount of 2,6-dimethoxy-1,4-benzoquinone (Zhu et al., 2012).

Lahtinen et al. studied the reactivity of lignin model compounds (2,6-dimethoxyphenol and conyferyl alcohol) in $[(C_1{=}C_2)C_1im]Cl$ by the enzyme *Melanocarpus albomyces* laccase. The oxidation of 2,6-dimethoxyphenol and coniferyl alcohol was achieved resulting in a mixture of different compounds. The presence of $[(C_1{=}C_2)C_1im]Cl$ affected the kind of linkages, which are formed in the polymer, enhancing the relative quantity of β–β structures and decreasing β-*O*-4 and β-5 links (Figure 1.6) (Lahtinen et al., 2013).

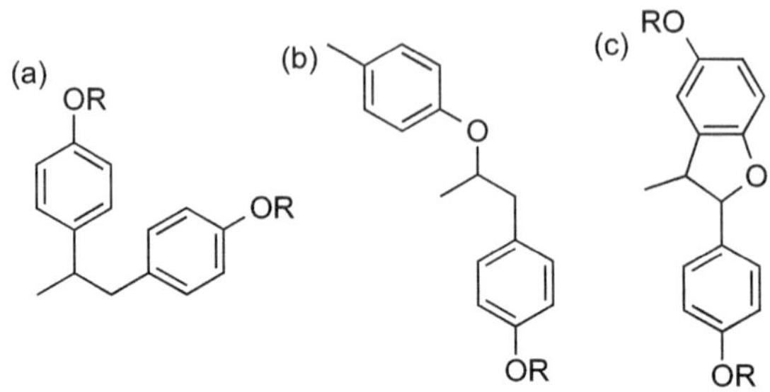

Figure 1.6 ***Lignin Structure Common Linkages.*** (a) *β*–*β*, (b) *β*-*O*-4, and (c) *β*-5.

Stärk et al., Zakzeski et al., and Liu et al. worked on the direct oxidation of lignin in ILs. Stärk et al. used Mn, Fe, and Cu salts as catalyst in four different ionic liquids: $[C_2C_1im][OTf]$, $[C_1C_1im][MeSO_4]$, $[C_2C_1im][EtSO_4]$, and $[C_2C_1im][MeSO_3]$ and different extractive solvents (toluene, dichloromethane (DCM), and ethyl acetate (EtOAc)), which extract the products from the aqueous phase after the reaction. The best performance was observed using $[C_2C_1im][OTf]$ with $Mn(NO_3)_2$, achieving 63% lignin conversion and DCM as the extracting solvent for the monomers obtaining 32% oil. Syringol, vanillin, 2,6-dimethoxy-1,4-benzoquinone (DMBQ), and syringaldehyde were determined to be the main reaction products (Stärk et al., 2010). Zakzeski et al. tried the depolymerization of organosolv and alkali lignins using a $CoCl_2{\cdot}6H_2O$ salt at mild conditions dissolved in $[C_2C_1im][Et_2PO_4]$ with EtOAc as the extraction solvent. In this case, no phenolic monomers were obtained. However, the IR-ATR spectra showed changes in the chemical structure of lignin; the lignin was oxidized, but the catalyst was not strong enough to disrupt the linkage and depolymerize it. In order to understand what was happening in the reaction media, lignin model compounds (3,3′-dimethoxy-5,5′-dimethyl-1,1′-biphenyl-2,2′-diol and 1-hydroxy-1-(4-hydroxy-3-methoxyphenyl)-2-(2,6-dimethoxyphenoxy) ethane), and veratryl alcohol) were treated under the same conditions. It was observed that 5-5′, β-*O*-4 linkages and phenolic alcohols remain intact after the reaction while benzyl and other alcohols were oxidized (Zakzeski et al., 2010). Veratryl alcohol had been previously treated with Co salen complexes in water under alkali conditions with excellent results by other authors (Kervinen et al., 2003). Sonar et al. formed salen-type complexes using a TSIL (Figure 1.7). The TSIL was used as a ligand to form the Co complexes allowing the selective oxidation of veratryl alcohol to veratraldehyde obtaining yields between 24% and 56%. No further oxidation was obtained (Sonar et al., 2012). Liu et al. also worked directly with lignin and

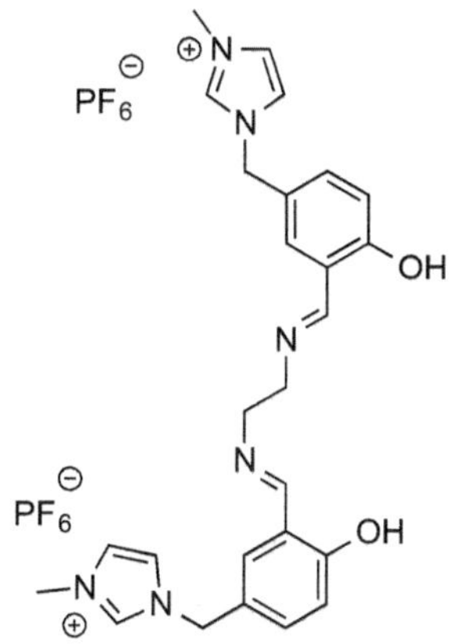

Figure 1.7 ***TSIL that Acts as Both the Salen Ligand and IL (Sonar et al., 2012).***

its oxidation conditions were optimized using different ILs. All the solvents were treated in the same conditions and good performances were obtained. The best results involved the use of $[C_1C_1im][Me_2PO_4]$ as solvent with the addition of methylisobutylketone to the reaction media, leading to lignin conversions of nearly 100%. The lignin-derived aromatic aldehydes, vanillin, syringaldehyde, and *p*-hydroxybenzaldehyde were obtained as products with a total oil yield of 29.7% (Liu et al., 2013).

ILs are highly used in electrochemistry as they are natural electrolytes. Taking advantage of this property, electrocatalytic oxidation of lignin in an IL was studied by Reichert et al. The authors prepared an electrode coated by $Ru_{0.25}V_{0.05}Ti_{0.7}O_x$ with Ag as reference electrode. $[Et_3NH][MeSO_3]$ was chosen as solvent. Electrolysis was achieved, and the products were extracted by diethyl ether. The best product yield (6%) was obtained with 1.5 V; the obtained products, benzaldehyde, 3-furaldehyde, *m*-tolualdehyde, vanillin, and acetovanillone, were identified by gas chromatography–mass spectrometry (GC-MS) (Reichert et al., 2012).

Wood Solubility

Not only have the lignocellulosic components been studied separately, but also the dissolution of whole wood itself. The first IL studied in order to dissolve lignocellulosic material was $[C_4C_1im]Cl$ as occurred with cellulose (Fort et al., 2007). Kilpeläinen et al. (2007) tried to dissolve wood with $[C_4C_1im]Cl$, $[(C_1{=}C_2)C_1im]Cl$, and 1-methyl-3-benzyl-imidazolium dicyanamide $[BzC_1im][N(CN)_2]$, with $[(C_1{=}C_2)C_1im]Cl$ being the most effective under the same conditions. The possible advantages of $[(C_1{=}C_2)C_1im]Cl$

are its lower viscosity and melting point. Regarding the drawback of the halide anions, which can release halogenated compounds due to the high temperature of the reaction, $[C_2C_1im][MeCO_2]$ as a lignocellulosic material solvent was also studied and good yields were observed. In general, to dissolve lignocellulosic material, the ILs have to possess the same properties as to dissolve cellulose, so ILs with the appropriate "net" or "effective" basicity (β–α) are required (Hauru et al., 2012).

In this case, an important drawback is also the presence of water in the solvent and wood medium. Moreover, with lignocellulose other aspects are also important in the optimization of the process such as the type of lignocellulose (e.g., grass is more easily dissolved than wood) and the particle size as ball milling time is important to prepare woodchips with adequate size; if the woodchips are too small the viscosity increases and it makes the material harder to dissolve.

In order to dissolve wood, the raw material and IL should be dried beforehand, and the dissolution should be done under an inert atmosphere (Tan and MacFarlane, 2010), and high temperatures and long dissolution times are required unless microwave radiation is used. The lignocellulosic components can be regenerated by the addition of specific antisolvents such as water, acetone, or low molecular weight alcohols.

ANALYTICAL CHEMISTRY

Analytical chemistry is a very wide field of study, so this section will be principally focused on chromatography, spectroscopy, and extraction, since these are the areas where IL utilization is most relevant at present.

Chromatography

Chromatography is a group of separation techniques, which are characterized by a separation of the different components of a homogeneous mixture between two phases, called the stationary and mobile phases, respectively. Molecules with a high tendency to stay in the stationary phase will move through the system at a lower velocity than those that are favored by the mobile phase or are less retained in the stationary phase. The chromatographic techniques can be divided depending on different factors: bed shape, physical state of mobile phase, separation mechanism, and so on.

In analytical chemistry, ILs were first applied in GC as a new class of stationary phase. Subsequently, they were used as additives to the mobile phase for liquid chromatography (LC); the small amounts used as additive for conventional solvents was little enough to not greatly affect the viscosity of the mobile phase; they were also used as stationary phase additives.

Gas Chromatography

GC is based on a partition equilibrium of an analyte between a solid or viscous liquid stationary phase (often a gel silicone-based material) and a mobile phase, which is a gas (most often helium). GC separation is always carried out in a column, which can be packed or capillary. Packed columns are cheaper and easier to use and often give adequate performance. Capillary columns generally give far superior resolution and in spite of being more expensive are widely used, especially for complex mixtures as they show better performance. Both types of columns are made from nonadsorbent and chemically inert materials. Stainless steel and glass are the usual materials for packed columns and quartz or fused silica for capillary columns.

GC is widely used in analytical chemistry; although the high temperatures used in GC make it unsuitable for high molecular weight biopolymers or proteins frequently encountered in biochemistry (heat denatures them), it is well suited for use in the petrochemical, environmental monitoring and remediation, and industrial chemical fields. It is also used extensively in chemical research.

The unique properties of ILs make them very interesting as possible stationary phases for GC. Molten salts were studied for GC by Barber et al. in 1950 for the first time. However, this first stationary phase, which was based on ethylammonium nitrate and ethylpyridinium bromide, which melt at a certain temperature, did not proved to be stable enough (Sun and Armstrong, 2010). Further investigations by Armstrong et al. (2001) found good behavior by imidazolium-based ILs as stationary phases where they coated silica capillaries with $[C_4C_1im]Cl$ and $[C_4C_1im][PF_6]$. These columns can separate polar and nonpolar analytes, and both ILs behave the same way for nonpolar compounds. However, their behavior is very different with polar analytes. Polar analytes are retained very strongly in the $[C_4C_1im][PF_6]$-based column. Anderson and Armstrong (2003) continued their work, looking for highly stable GC stationary phases and they developed $[BzC_1im][OTf]$ and 1-(4-methoxyphenyl)-3-methylimidazolium trifluoromethanesulfonate ($[(^4C_1OPh)C_1im][OTf]$), which were stable up to 220 °C and 240 °C, respectively. By increasing the cross-linking of the column filler, the range of temperatures for which it is stable increased. For example, Huang et al. developed stationary phase based on dicationic ILs where the cations were linked by a PEG or C6 chain. Different combinations of anions ($[NTf_2]^-$ and $[OTf]^-$) and imidazolium cations ($[(C_6im)_2]^{2+}$, $[(Bzim)_2]^{2+}$, and $[(C_1im)_2]^{2+}$) were tested, achieving cross-linked columns stable up to 300–400 °C. They were successfully used to separate and characterize several aromatic compounds, among them phenol, *o*-cresol, nitrobenzene, naphthalene, toluene, and pyrrole (Ho et al., 2010; Huang et al., 2007).

ILs, due to their low volatility, are good solvents for headspace GC when the analyte volatility is low and higher temperatures for the headspace are necessary. $[C_4C_1im][BF_4]$ and $[C_4C_1im][Me_2PO_4]$ had been used for pharmaceutical determination using headspace GC (Liu and Jiang, 2007).

Liquid Chromatography

LC is a separation technique in which the mobile phase is a liquid. It can be carried out either in a column or a plane. Nowadays, LC has evolved to a high-performance liquid chromatography (HPLC) that is based in columns with very small packing particles where the sample is forced by a liquid at high pressure (the mobile phase) through a column that is packed with a stationary phase composed of irregularly or spherically shaped particles, a porous monolithic layer, or a porous membrane.

HPLC is historically divided into two different subclasses based on the polarity of the mobile and stationary phases. Methods in which the stationary phase is more polar than the mobile phase (e.g., toluene as the mobile phase, silica as the stationary phase) are defined as normal phase liquid chromatography and the opposite (e.g., water–methanol mixtures as the mobile phase and C18 = octadecylsilyl as the stationary phase) is termed reversed phase liquid chromatography.

The presence of free silanol groups in the column filler is one of the main problems of LC, as these are difficult to control and can affect the retention coefficient of the analytes in the stationary phase, so reproducibility of the chromatogram can be lost. In an attempt to solve this problem, special attention was centered on imidazolium tretrafluoroborate-type ILs, which are soluble in most common solvents used for LC. $[BF_4]^-$ anions participate in Coulombic interactions and specific solute–ion interactions, especially proton donor–acceptor interactions. There are different types of LC in which ILs have been used and tested to improve selectivity, separation, and reproducibility of the technique (Skoog et al., 2007).

Thin-Layer Chromatography

Different imidazolium $[BF_4]^-$ ILs were added to the mobile phase to separate several basic compounds; these ILs showed better behavior than standard amine additives that have been used before, even with higher concentrations of amines. They observed that by adding imidazolium $[BF_4]^-$ ILs to the system the reliability of the retention coefficient as a function of eluent composition was improved, which enhanced the separation conditions. The suppression of silanol groups by the anion reduced the lack of control of the stationary phase interaction with the analyte, increasing the reproducibility (Sun and Armstrong, 2010).

High-Performance Liquid Chromatography

The silanol screening effect is always observed for $[PF_6]^-$ and $[BF_4]^-$ ILs, and the retention factor depends on the anion due to the hydrophobic or chaotropic character of the IL. Several authors checked the performance of different ILs using the Nahum and Horvath dual retention model (Eq. (1.2)) on silanol suppression where k_0 is the retention factor in absence of silanol, k is the retention factor, k_2 is the reciprocal retention factor, $[A]$ is the concentration of silanol suppresor and K_A is the silanophilic binding constant. They observed that the alkyl chain length attached to the nitrogen of the imidazolium ring was proportional to the suppression effect.

$$\frac{[A]}{k_0 - k} = \frac{1}{k_2 K_A} + \frac{[A]}{k_2} \tag{1.2}$$

In this way the retention times of several analytes decreased. However, it means that the selectivity in the separation also decreased and quantification was less precise (Marszałł et al., 2006; Wilkes and Zaworotko, 1992).

The imidazolium and pyridinium $[BF_4]^-$ ILs were successfully used as additives in the mobile phase to separate catecholamines, nucleotides, and ephedrine (Marszałł and Kaliszan, 2007). ILs have not been only used as a mobile phase additive but also as a modifier of the stationary phase; as an example, *N*-butylimidazolium was immobilized on a silica support, with $[Br]^-$ as anion. The mechanism that is involved in the separation of analytes when the ionic liquid is a dopant of the stationary phase encompasses many different interactions such as ion exchange, hydrophobic interactions, and other electrostatic interactions. There are reports that IL was immobilized on the column and tested to separate aromatics and compared with the standard phenyl-doped stationary phase (Sun and Armstrong, 2010).

C18 columns were tested to be sure that their efficiency is not affected by the addition of the ILs. It was observed that it was not harmful for the column and filler leak was not altered compared to previous mobile phases such as water and methanol.

Capillary Electrophoresis

The separation mechanism for electrophoresis is different than for LC, but once again the main problem in this type of chromatography is the free silanol group effect. The free silanol groups influence the electro-osmotic flow and transport of ions. To solve this problem, Vaher et al. (2001) tried 1-alkyl-3-imidazolium ILs in different solvents. The silanol group effect can be reversed by the covalent bonding between the silica surface and dialkyl-imidazolium cation. ILs can completely suppress the electro-osmotic flow effect depending on their concentration. After that, the negatively charged acids are not retained by the wall and their separation depends only on their electrophoretic mobility (μ_e) (Eq. (1.3)) (Marszałł and Kaliszan, 2007).

$$\mu_e = \frac{q}{6\pi\eta r} \tag{1.3}$$

Mass Spectrometry

ILs are not used in mass spectrometry (MS) directly, but they have been used in MS-related techniques to enhance the separation of ions before analysis by MS.

Electrospray Ionization

Electrospray ionization (ESI) is a technique used in mass spectrometry to produce ions using an electrospray in which a high voltage is applied to a liquid to create an aerosol. It is especially useful in producing ions from macromolecules because it overcomes the propensity of these molecules to fragment when ionized. ESI is different from other atmospheric pressure ionization processes since it may produce multiply charged ions, effectively extending the mass range of the analyzer to accommodate the kDa–MDa orders of magnitude observed in proteins and their associated polypeptide fragments. Chang et al. (2011) tried two different ILs to enhance the response, butylammonium 2,5-hydroxybenzoate and butylammonium α-cyano-4-hydroxycinnamate were added to the ESI matrix and the sensitivity was enhanced for detection of different polysaccharides. Henderson and McIndoe (2006) added a lipophilic IL to nonpolar solvents as a matrix and improved the measurement of analytes dissolved in those solvents.

Matrix-Assisted Laser Desorption/Ionization

Matrix-assisted laser desorption/ionization (MALDI) is a soft ionization technique used in MS, allowing the analysis of biomolecules and large organic molecules, which tend to be fragile and fragment when ionized by more conventional ionization methods. It is similar in character to ESI in that both techniques are relatively soft ways of obtaining large ions in the gas phase, although MALDI produces far fewer multiply charged ions.

MALDI is thought to be a three-step process. First, the sample is mixed with a suitable matrix material and applied to a metal plate. Second, a pulsed laser irradiates the sample, triggering ablation and desorption of the sample and matrix material. Finally, the analyte molecules are ionized by being protonated or deprotonated in the hot plume of ablated gases, and can then be accelerated into the mass spectrometer, which is used to analyze them. However, considerable evidence suggests that analyte ions are produced from charged particles during the ablation process. The standard matrix materials are solids or conventional liquids; for a solid matrix there are problems with heterogeneous distribution of the analyte and standard liquids are problematic when used under vacuum. These problems can be partially solved using ILs as the matrix. ILs, due to their good solvation properties and stability under vacuum conditions, seemed to be a good choice. However, ILs can vary tremendously in their ability to promote analyte ionization. Therefore, the cationic and anionic portion of the ionic matrix must be chosen with a consideration for these special requirements. The Armstrong group was the first to set up ILs as a matrix for MALDI; they successfully checked several ILs and compared them with standard matrices. They determined *N,N*-diisopropylethylammonium α-cyano-4-hydroxycinnamate and *N*-isopropyl-*N*-methyl-*t*-butylammonium α-cyano-4-hydroxycinnamate were the best matrices for proteins and peptides, while *N,N*-diisopropylethylammonium α-cyano-4-hydroxycinnamate and *N,N*-diisopropylethylammonium ferulate were the best matrices for carbohydrates (Armstrong et al., 2001; Crank and Armstrong, 2009).

Spectroscopy

Several studies have been done on IL performance in different spectroscopic analytical techniques. Only in a few techniques did ILs show a better response than the classical solvents used currently. The main reason is the poor optical properties of the ILs; as an example, imidazolium-derived ILs are not transparent in the UV region (Paul et al., 2005). Several studies on fluorescence, IR, and NMR spectroscopies were reviewed by Sun and Armstrong (2010), with no remarkable results compared to traditional solvents.

Extraction

Liquid–liquid extraction, also known as solvent extraction and partitioning, is a method for separating compounds based on their relative solubilities into two different immiscible liquids, usually water and an organic solvent. It is widely used for many applications from the purification of compounds to waste water treatment.

Liquid/Liquid Extraction

One of the main advantages of ILs is that they can be designed to create desired properties. So it is possible to adjust their polarity to prepare ILs immiscible with polar or nonpolar solvents. The anion plays the most important role in the polarity of the IL; for example, if water-immiscible ILs are required, then $[PF_6]^-$ or $[NTf_2]^-$ should be used as the anion. ILs have shown great potential for fuel desulfurization and to clean wastewater.

For fuel desulfurization, initially ILs with $[PF_6]^-$ and $[BF_4]^-$ as anions were checked by many authors, then pyridinium-derived ILs were studied by Holbrey et al. (2008) and finally, Rodríguez-Cabo et al. (2013) found that $[C_2C_1im]^+$-derived ILs had higher selectivity for extracting sulfur, specifically $[C_2C_1im][MeCO_2]$ and $[C_2C_1im][Et_2PO_4]$ showed UNIQUAC *r*-values of 8.2841 and 6.0191, respectively, whereas *n*-hexane showed 4.4998 and toluene 3.9228. The UNIQUAC equation is a derivation of the NTRL (non-random two liquid) equation used for calculating the relationship between the activity coefficients and molar fraction of the components, where the *r*-value is related to the entropic contribution of the components (Praunsnitz et al., 1999).

On the other hand, in the selectivity of extracted metals, it is the cation that plays the more important role. Visser et al. synthesized cations based on urea or thiourea, which showed good distribution ratios for Hg_2^+ and Cd_2^+. Gold complexes were removed from water by Papaiconomou et al. (2012) with $[C_8C_1im][NTf_2]$ ILs, which showed good distribution coefficients for $[AuBr_4]^-$. Lithium was successfully extracted with $[C_8C_1im][PF_6]$, $[C_8C_1im][BF_4]$, and $[C_8C_1im][NTf_2]$ containing 2,2-binaphthyldiyl-17-crown-5, with the $[NTf_2]^-$ anion being the most effective (Sun et al., 2015).

Carda-Broch et al. compared the extraction capability of $[C_4C_1im][PF_6]$ in the system $[C_4C_1im][PF_6]$/water with octanol/water for acidic, basic, and neutral compounds. The ILs had good performance extracting basic compounds whereas octanol was better for acids, and for neutral compounds there was no significant difference (Carda-Broch et al., 2003).

Solid-Phase Microextraction

Solid-phase microextraction, or SPME, is a solid-phase extraction sampling technique that involves the use of a fiber coated with an extracting phase that can be a liquid (polymer) or a solid (sorbent), which extracts different kinds of analytes (including volatile and nonvolatile) from different kinds of media that can be in the liquid or gas phase (Pawliszyn, 1997). The quantity of analyte extracted by the fiber is proportional to its concentration in the sample as long as equilibrium is reached or, in case of short-time pre-equilibrium, with the help of convection or agitation.

The high viscosity of ILs improves the quality of the fiber coating, while using the proper IL can result in high selectivity of the extraction process. Liu et al. (2005) applied ILs on SPME fibers for the first time, using $[C_8C_1im][PF_6]$, however they achieved poor results. Hsieh et al. used fibers coated with Nafion to improve IL coatings and they evaluated the performance of $[C_8C_1im][OTf]$, $[BzC_1im][OTf]$, and 1-(3-propylphenyl)-3-methylimidazolium trifluoromethanesulfonate $[C_3PhC_1im][OTf]$. $[C_8C_1im][OTf]$ showed the greatest extraction efficiency. However, principally because the durability of the coatings was very low, they had to be recoated after each analysis (Hsieh et al., 2006; Liu et al., 2005). Further investigations led to the introduction of polymeric ILs by Zhao et al. (2008) The $[(C_2 = C_3)C_{12}im]Br$ monomer was polymerized and ion exchanged to form an $[NTf_2]^-$-based polymeric IL for the separation of several esters, achieving good separation coefficients. López-Darias et al. (2010) developed poly(1-vinyl-3-hexadecylimidazolium) $[NTf_2]^-$-based coatings for separating water pollutants. Zhou et al. (2012) used a sol–gel coating of the IL ($[(C_1 = C_2)C_1im][NTf_2]$) with the hydroxyl-terminated silicone oil OH-TSO on fibers to detect and determine phthalate ester concentration in agricultural plastic films using MeOH as the extractant solvent; this IL showed good selectivity (Spietelun et al., 2013).

IONIC LIQUIDS IN TRIBOLOGY

Tribology is the science and engineering of interacting surfaces in relative motion. It includes the study and application of the principles of friction, lubrication, and wear. Tribology is a branch of mechanical engineering and materials science.

ILs have become an area of interest for novel tribology studies because of their unique properties, negligible volatility, nonflammability, high thermal stability, and good intrinsic performance. They have been investigated as lubricants and additives. Their ability, in some cases, to reduce friction and wear significantly is of most importance for this application. Their viscosity, thermal stability, and wettability were characterized in order to establish their potential as a lubricant or additive (Qu et al., 2006).

The formation of effective lubrication films depends on the alkyl chain of the cation and the anion composition for the neat ILs as lubricants and also on miscibility with the base oil and atmospheric moisture when used as an additive.

Lubricants

A lubricant is a substance introduced to reduce friction between surfaces in mutual contact, which ultimately reduces the heat generated when the surfaces move. It may also have the function of transmitting forces, transporting foreign particles, or heating or cooling the surfaces.

There are many studies about the performance of different ILs in tribology as lubricants; the ILs used in this field are composed of imidazolium, ammonium, pyridinium, and phosphonium cations, and Cl^-, Br^-, $[BF_4]^-$, and $[PF_6]^-$ anions, with different alkyl substituents on the cations. The anion seems to have more of an effect on tribology properties than the cation (Kamimura et al., 2007). The most common ILs are composed of imidazolium cations with $[BF_4]^-$ and $[PF_6]^-$ as the anion. $[Br]^-$-related ILs were discarded because of their hydrophilicity, which can increase tribocorrosion on the joints (Qu et al., 2006).

Ye et al. (2001) studied the performance of $[C_6C_1im][BF_4]$ and $[C_6C_2im][BF_4]$ as lubricants for steel (SAE-52100), aluminum (Al2024), copper, single crystal SiO_2, single crystal Si(100), and sialon (Si–Al–O–N) ceramics; these ILs showed good lubricant behavior compared with standard lubricants. Weng et al. studied several asymmetric tetraalkylphosphonium ILs' performance on steel/steel contact and compared their performance with high-temperature oils and imidazolium ILs. All the phosphonium ILs synthesized showed similar or better tribology performance but lower thermal stability in contact with the air (Weng et al., 2007). Jimenez et al. (2006) studied the performance of $[C_8C_1im][BF_4]$ and $[C_6C_1im][PF_6]$ on titanium/steel contact; they observed that $[C_6C_1im][PF_6]$ showed good tribology performance at high temperatures and when they exchange steel with ruby, tribocorrosion was avoided.

The main drawback of using ILs as lubricants is their reactivity and potential decomposition at high temperatures, which can cause tribocorrosion. In accordance with the good performance shown by the ILs at reducing friction and wear, their use as additives for conventional oils was studied in order to avoid tribocorrosion. Kamimura et al. studied the performance of additives on ILs (imidazolium, ammonium, and pyridinium). They found that the additives (tricresylphosphate and dibenzyldisulfide) helped to prevent the tribochemical decomposition of the ILs (Kamimura et al., 2007).

In general, ILs showed lower friction coefficients than conventional oils in the following order: imidazolium > ammonium > phosphonium, but similar wear coefficients as base oils.

Additives

A large number of additives are used to impart performance characteristics to lubricants. The additives can act as detergents, defoamers, antioxidants, and antiwear agents. With the use of ILs, the quantity of additives used can be reduced.

The formation of effective lubrication films depends on the alkyl chain of the cation and the anion composition and also on miscibility with the base oil and atmospheric moisture when it is used as an additive. The performance of ILs as additives is not the same as observed when ILs are used as lubricants by themselves.

Friction and wear were reduced on metallic and ceramic surfaces where a few percent of IL was used as an additive for base oils and water. While neat ILs tend to react with metallic surfaces, leading to tribocorrosion, when ILs are used as additives, the tribocorrosion decreased substantially, indicating that the ILs do not cause damage (Jiménez et al., 2006, Kamimura et al., 2007).

Cutting Fluids

Taking into account the friction-reducing ability of ILs, they have been tested as cutting fluids. Cutting fluids are used to reduce the temperature of treated surfaces and as lubricants to reduce the roughness of cut surfaces. In this way the cutting tool and material are protected. Conventional cutting fluids as ST-501, TC 1 are environmentally toxic, cause health hazards, and their recycling is very expensive. As a result, ILs have been introduced as potential novel cutting liquids. Pham et al. (2014) studied $[C_2C_1im][NTf_2]$ and $[C_4C_1im]I$ and they observed that the roughness of the surfaces obtained was reduced, especially with $[C_4C_1im]I$. On the other hand, the potential to recycle ILs makes them perfect candidates as a "greener" alternative to convenient cutting fluids.

CONCLUSIONS

ILs are clearly applicable to a wide range of different areas with substantial progress being made in each, particularly within the past decade. With the maturation or at least development of some basic structure–activity relationships in many areas, there are some clear directions for future IL research. Given the variety of ILs that can be prepared it is also important that the screening of ILs for a particular application encompasses as many different types as possible to avoid breakthroughs being missed as a result of a one-dimensional focus on a specific subclass of these substances. It is also interesting to note the rapid trajectory of ILs from compounds only of interest to a small number of academic research groups to the current progress toward their commercialization in a variety of fields. With the potential development of large-scale applications comes the benefit of bulk-scale IL supply, which will reduce the cost of these liquids and may inadvertently aid their use in completely disparate fields, as cost is one of the major hurdles the majority of ILs have yet to overcome. For some ILs it has been estimated that bulk production could enable costs as low as US$1.24 per kilogram (Chen et al., 2014), illustrating the possibility for bulk-scale production and judicious IL selection to enable their economic competitiveness with conventional solvents. Hopefully these liquids can live up to their initial promise and provide effective, environmentally friendly solutions to a number of technical challenges.

REFERENCES

Abbott, A.P., Frisch, G., Ryder, K.S., 2013. Electroplating using ionic liquids. Annu. Rev. Mater. Res. 43, 335–358.

Aggarwal, V.K., Emme, I., Mereu, A., 2002. Unexpected side reactions of imidazolium-based ionic liquids in the base-catalysed Baylis–Hillman reaction. Chem. Commun., 1612–1613.

Ahmed, E., Ruck, M., 2011. Chemistry of polynuclear transition-metal complexes in ionic liquids. Dalton Trans. 40, 9347–9357.

Anderson, J.L., Armstrong, D.W., 2003. High-stability ionic liquids. A new class of stationary phases for gas chromatography. Anal. Chem. 75 (18), 4851–4858.

Angelini, G., De Maria, P., Chiappe, C., Fontana, A., Gasbarri, C., Siani, G., 2009. The base-catalyzed keto-enol interconversion of 2-nitrocyclohexanone in ionic liquids. J. Org. Chem. 74, 6572–6576.

Antonietti, M., Kuang, D., Smarsly, B., Zhou, Y., 2004. Ionic liquids for the convenient synthesis of functional nanoparticles and other inorganic nanostructures. Angew. Chem. Int. 43, 4988–4992.

Anugwom, I., Mäki-Arvela, P., Virtanen, P., Willför, S., Sjöholm, R., Mikkola, J.-P., 2012. Selective extraction of hemicelluloses from spruce using switchable ionic liquids. Carbohyd. Polym. 87 (3), 2005–2011.

Armand, M., Endres, F., MacFarlane, D.R., Ohno, H., Scrosati, B., 2009. Ionic-liquid materials for the electrochemical challenges of the future. Nat. Mater. 8, 621–629.

Armstrong, D.W., Zhang, L.-K., He, L., Gross, M.L., 2001. Ionic liquids as matrixes for matrix-assisted laser desorption/ionization mass spectrometry. Anal. Chem. 73 (15), 3679–3686.

Asadov, Z.H., Rahimov, R.A., Poladova, T.A., Nasibova, S.M., Guliyev, A.D., Asadova, A.Z., 2012. Triethylenetetramine-based novel cationic surfactants and their complexes with anionic polyelectrolytes. J. Mol. Liq. 166, 44–48.

Bai, Y., Cao, Y., Zhang, J., Wang, M., Li, R., Wang, P., Grätzel, M., 2008. High-performance dye-sensitized solar cells based on solvent-free electrolytes produced from eutectic melts. Nat. Mater. 7, 626–630.

Bai, L., Li, X., Zhu, J., Chen, B., 2011. Effects of nucleators on the thermodynamic properties of seasonal energy storage materials based on ionic liquids. Energy Fuels 25, 1811–1816.

Bayley, P.M., Best, A.S., MacFarlane, D.R., Forsyth, M., 2011. The effect of coordinating and non-coordinating additives on the transport properties in ionic liquid electrolytes for lithium batteries. Phys. Chem. Chem. Phys. 13, 4632–4640.

Béguin, F., Presser, V., Balducci, A., Frackowiak, E., 2014. Carbons and electrolytes for advanced supercapacitors. Adv. Mater. 26, 2219–2251.

Belhocine, T., Forsyth, S.A., Gunaratne, H.Q.N., Nieuwenhuyzen, M., Puga, A.V., Seddon, K.R., Whiston, K., 2011. New ionic liquids from azepane and 3-methylpiperidine exhibiting wide electrochemical windows. Green Chem. 13, 59–63.

Bhatt, A.I., Kao, P., Best, A.S., Hollenkamp, A.F., 2013. Understanding the morphological changes of lithium surfaces during cycling in electrolyte solutions of lithium salts in an ionic liquid. J. Electrochem. Soc. 160 (8), A1171–A1180.

Binder, J.B., Gray, M.J., White, J.F., Zhang, Z.C., Holladay, J.E., 2009. Reactions of lignin model compounds in ionic liquids. Biomass Bioenerg. 33 (9), 1122–1130.

Bini, R., Chiappe, C., Marmugi, E., Pieraccini, D., 2006. The "non-nucleophilic" anion $[Tf_2N]^-$ competes with the nucleophilic Br: an unexpected trapping in the dediazoniation reaction in ionic liquids. Chem. Commun., 897–899.

Bini, R., Chiappe, C., Mestre, V.L., Pomelli, C.S., Welton, T., 2008. A rationalization of the solvent effect on the Diels–Alder reaction in ionic liquids using multiparameter linear solvation energy relationships. Org. Biomol. Chem. 6, 2522–2529.

Bini, R., Chiappe, C., Pomelli, C.S., Parisi, B., 2009. Effect of ionic liquids on the menschutkin reaction: an experimental and theoretical study. J. Org. Chem. 74, 8522–8530.

Bonhôte, P., Dias, A.-P., Papageorgiou, N., Kalyanasundaram, K., Grätzel, M., 1996. Hydrophobic, highly conductive ambient-temperature molten salts. Inorg. Chem. 35, 1168–1178.

Bose, S., Armstrong, D.W., Petrich, J.W., 2010. Enzyme-catalyzed hydrolysis of cellulose in ionic liquids: a green approach toward the production of biofuels. J. Phys. Chem. B 114 (24), 8221–8227.

Brandt, A., Gräsvik, J., Hallett, J.P., Welton, T., 2013. Deconstruction of lignocellulosic biomass with ionic liquids. Green Chem. 15 (3), 550–583.

Bridges, N.J., Visser, A.E., Fox, E.B., 2011. Potential of nanoparticle-enhanced ionic liquids (NEILs) as advanced heat-transfer fluids. Energy Fuels 25, 4862–4864.

Brown, J.S., Domanski, P.A., 2014. Review of alternative cooling technologies. Appl. Therm. Eng. 64 (1–2), 252–262.

Budi, A., Basile, A., Opletal, G., Hollenkamp, A.F., Best, A.S., Rees, R.J., Russo, S.P., 2012. Study of the initial stage of solid electrolyte interphase formation upon chemical reaction of lithium metal and *N*-methyl-*N*-propyl-pyrrolidinium-bis(fluorosulfonyl)imide. J. Phys. Chem. C 116, 19789–19797.

Burchard, W., 2003. Solubility and solution structure of cellulose derivatives. Cellulose 10 (3), 213–225.

Burke, A., 2000. Ultracapacitors: why, how and where is the technology. J. Power Sources 91 (1), 37–50.

Buzzeo, M.C., Hardacre, C., Compton, R.G., 2004. Use of room temperature ionic liquids in gas sensor design. Anal. Chem. 76, 4583–4588.

Cao, Y., Wu, J., Zhang, J., Li, H., Zhang, Y., He, J., 2009. Room temperature ionic liquids (RTILs): a new and versatile platform for cellulose processing and derivatization. Chem. Eng. J. 147 (1), 13–21.

Carda-Broch, S., Berthod, A., Armstrong, D.W., 2003. Solvent properties of the 1-butyl-3-methylimidazolium hexafluorophosphate ionic liquid. Anal. Bioanal. Chem. 375 (2), 191–199.

Chang, Y.-L., Lee, Y.-C., Yang, W.-B., Chen, C.-H., 2011. Ionic liquid-assisted electrospray ionization of polysaccharides. J. Mass Spectrom. 46 (4), 367–375.

Chatel, G., Rogers, R.D., 2014. Review: oxidation of lignin using ionic liquids – an innovative strategy to produce renewable chemicals. ACS Sustainable Chem. Eng. 2 (3), 322–339.

Chen, L., Sharifzadeh, M., Mac Dowell, N., Welton, T., Shah, N., Hallett, J.P., 2014. Inexpensive ionic liquids: $[HSO_4]^-$-based solvent production at bulk scale. Green Chem. 16 (6), 3098–3106.

Chiappe, C., Pomelli, C.S., 2014. Point-functionalization of ionic liquids: an overview of synthesis and applications. Eur. J. Org. Chem. 2014 (28), 6120–6139.
Chiappe, C., Malvaldi, M., Melai, B., Fantini, S., Bardi, U., Caporali, S., 2010. An unusual common ion effect promotes dissolution of metal salts in room-temperature ionic liquids: a strategy to obtain ionic liquids having organic–inorganic mixed cations. Green Chem. 12, 77–80.
Chowdhury, S., Mohan, R.S., Scott, J.L., 2007. Reactivity of ionic liquids. Tetrahedron 63, 2363–2389.
Cooper, E.R., Andrews, C.D., Wheatley, P.S., Webb, P.B., Wormald, P., Morris, R.E., 2004. Ionic liquids and eutectic mixtures as solvent and template in synthesis of zeolite analogues. Nature 430, 1012–1016.
Crank, J.A., Armstrong, D.W., 2009. Towards a second generation of ionic liquid matrices (ILMs) for MALDI-MS of peptides, proteins, and carbohydrates. J. Am. Soc. Mass Spectr. 20 (10), 1790–1800.
Crowhurst, L., Falcone, R., Lancaster, N.L., Llopis-Mestre, V., Welton, T., 2006. Using Kamlet–Taft solvent descriptors to explain the reactivity of anionic nucleophiles in ionic liquids. J. Org. Chem. 71, 8847–8853.
Davis, Jr., J.H., 2004. Task-specific ionic liquids. Chem. Lett. 33 (9), 1072–1077.
Del Sesto, R.E., McCleskey, T.M., Macomber, C., Ott, K.C., Koppisch, A.T., Baker, G.A., Burrell, A.K., 2009. Limited thermal stability of imidazolium and pyrrolidinium ionic liquids. Thermochim. Acta 491, 118–120.
Ding, S.-F., Xu, M.-Q., Zhao, G.-C., Wei, X.-W., 2007. Direct electrochemical response of Myoglobin using a room temperature ionic liquid, 1-(2-hydroxyethyl)-3-methyl imidazolium tetrafluoroborate, as supporting electrolyte. Electrochem. Commun. 9 (2), 216–220.
Dzyuba, S.V., Bartsch, R.A., 2002. Expanding the polarity range of ionic liquids. Tetrahedron Lett. 43 (26), 4657–4659.
Endres, F., 2002. Ionic liquids: solvents for the electrodeposition of metals and semiconductors. ChemPhysChem 3, 144–154.
Endres, F., Höfft, O., Borisenko, N., Gasparotto, L.H., Prowald, A., Al-Salman, R., Zein El Abedin, S., 2010. Do solvation layers of ionic liquids influence electrochemical reactions? Phys. Chem. Chem. Phys. 12, 1724–1732.
Farid, M.M., Khudhair, A.M., Razack, S.A.K., Al-Hallaj, S., 2004. A review on phase change energy storage: materials and applications. Energy Convers. Manage. 45 (9–10), 1597–1615.
Fonseca, G.S., Machado, G., Teixeira, S.R., Fecher, G.H., Morais, J., Alves, M.C.M., Dupont, J., 2006. Synthesis and characterization of catalytic iridium nanoparticles in imidazolium ionic liquids. J. Colloid Interf. Sci. 301, 193–204.
Fort, D.A., Remsing, R.C., Swatloski, R.P., Moyna, P., Moyna, G., Rogers, R.D., 2007. Can ionic liquids dissolve wood? Processing and analysis of lignocellulosic materials with 1-*n*-butyl-3-methylimidazolium chloride. Green Chem. 9 (1), 63–69.
Fox, D.M., Awad, W.H., Gilman, J.W., Maupin, P.H., De Long, H.C., Trulove, P.C., 2003. Flammability, thermal stability, and phase change characteristics of several trialkylimidazolium salts. Green Chem. 5, 724–727.
Fox, D.M., Gilman, J.W., Morgan, A.B., Shields, J.R., Maupin, P.H., Lyon, R.E., Trulove, P.C., 2008. Flammability and thermal analysis characterization of imidazolium-based ionic liquids. Ind. Eng. Chem. Res. 47 (16), 6327–6332.
Fu, D., Mazza, G., Tamaki, Y., 2010. Lignin extraction from straw by ionic liquids and enzymatic hydrolysis of the cellulosic residues. J. Agric. Food Chem. 58 (5), 2915–2922.
Fukaya, Y., Sugimoto, A., Ohno, H., 2006. Superior solubility of polysaccharides in low viscosity, polar, and halogen-free 1,3-dialkylimidazolium formates. Biomacromolecules 7 (12), 3295–3297.
Fukaya, Y., Hayashi, K., Wada, M., Ohno, H., 2008. Cellulose dissolution with polar ionic liquids under mild conditions: required factors for anions. Green Chem. 10 (1), 44–46.

Gardas, R.L., Coutinho, J.A.P., 2008. A group contribution method for heat capacity estimation of ionic liquids. Ind. Eng. Chem. Res. 47, 5751–5757.

Ge, W., Wang, X., 2014. Application of ionic liquids in lignocellulose biomass dissolution, fraction and chemical modification. In: Brooks, A. (Ed.), Ionic Liquids: Synthesis, Characterization and Applications. Nova Science Publishers, Inc, Hauppauge, NY, USA, pp. 153–172.

Ge, R., Hardacre, C., Nancarrow, P., Rooney, D.W., 2007. Thermal conductivities of ionic liquids over the temperature range from 293 K to 353 K. J. Chem. Eng. Data 52, 1819–1823.

George, A., Tran, K., Morgan, T.J., Benke, P.I., Berrueco, C., Lorente, E., Holmes, B.M., 2011. The effect of ionic liquid cation and anion combinations on the macromolecular structure of lignins. Green Chem. 13 (12), 3375–3385.

George, A., Brandt, A., Tran, K., Zahari, S.M.S.N.S., Klein-Marcuschamer, D., Sun, N., Hallett, J.P., 2015. Design of low-cost ionic liquids for lignocellulosic biomass pretreatment. Green Chem. 17 (3), 1728–1734.

Giridhar, P., Zein El Abedin, S., Endres, F., 2012. Electrodeposition of aluminium from 1-butyl-1-methylpyrrolidinium chloride/$AlCl_3$ and mixtures with 1-ethyl-3-methylimidazolium chloride/$AlCl_3$. Electrochim. Acta 70, 210–214.

Goodenough, J.B., Park, K.-S., 2013. The Li-ion rechargeable battery: a perspective. J. Am. Chem. Soc. 135, 1167–1176.

Gorlov, M., Kloo, L., 2008. Ionic liquid electrolytes for dye-sensitized solar cells. Dalton Trans. 2008 (20), 2655–2666.

Guo, C.X., Lu, Z.S., Lei, Y., Li, C.M., 2010. Ionic liquid–graphene composite for ultratrace explosive trinitrotoluene detection. Electrochem. Commun. 12, 1237–1240.

Haddleton, D.M., Welton, T., Carmichael, A.J., 2008. Polymer synthesis in ionic liquids. Wasserscheid, P., Welton, T. (Eds.), Ionic Liquids in Synthesis, 2, second ed. Wiley-VCH Verlags GmbH & Co. KGaA, Weinheim, pp. 619–640.

Hallett, J.P., Welton, T., 2011. Room-temperature ionic liquids. solvents for synthesis and catalysis. 2. Chem. Rev. 111, 3508–3576.

Hallett, J.P., Liotta, C.L., Ranieri, G., Welton, T., 2009a. Charge screening in the SN2 reaction of charged electrophiles and charged nucleophiles: an ionic liquid effect. J. Org. Chem. 74, 1864–1868.

Hallett, J.P., Liotta, C.L., Ranieri, G., Welton, T., 2009b. In search of an "ionic liquid effect". ECS Trans. 16 (49), 81–87.

Hapiot, P., Lagrost, C., 2008. Electrochemical reactivity in room-temperature ionic liquids. Chem. Rev. 108, 2238–2264.

Harrisson, S., Mackenzie, S.R., Haddleton, D.M., 2003. Pulsed laser polymerization in an ionic liquid: strong solvent effects on propagation and termination of methyl methacrylate. Macromolecules 36 (14), 5072–5075.

Hartley, J.M., Ip, C.-M., Forrest, G.C.H., Singh, K., Gurman, S.J., Ryder, K.S., Frisch, G., 2014. EXAFS study into the speciation of metal salts dissolved in ionic liquids and deep eutectic solvents. Inorg. Chem. 53, 6280–6288.

Hauru, L.K.J., Hummel, M., King, A.W.T., Kilpeläinen, I., Sixta, H., 2012. Role of solvent parameters in the regeneration of cellulose from ionic liquid solutions. Biomacromolecules 13 (9), 2896–2905. http://doi.org/10.1021/bm300912y

Hayamizu, K., Aihara, Y., Nakagawa, H., Nukuda, T., Price, W.S., 2004. Ionic conduction and ion diffusion in binary room-temperature ionic liquids composed of [emim][BF_4] and $LiBF_4$. J. Phys. Chem. B 108, 19527–19532.

Hayyan, M., Mjalli, F.S., Hashim, M.A., AlNashef, I.M., Mei, T.X., 2013. Investigating the electrochemical windows of ionic liquids. J. Ind. Eng. Chem. 19 (1), 106–112.

Henderson, M.A., McIndoe, J.S., 2006. Ionic liquids enable electrospray ionisation mass spectrometry in hexane. Chem. Commun. 2006, 2872–2874.

Hermanutz, F., Gähr, F., Uerdingen, E., Meister, F., Kosan, B., 2008. New developments in dissolving and processing of cellulose in ionic liquids. Macromol. Symp. 262 (1), 23–27.

Higashino, T., Imahiro, H., 2015. Porphyrins as excellent dyes for dye-sensitized solar cells: recent developments and insights. Dalton Trans. 44, 448–463.

Ho, W.-Y., Hsieh, Y.-N., Lin, W.-C., Kao, C.L., Huang, P.-C., Yeh, C.-F., Kuei, C.-H., 2010. High temperature imidazolium ionic polymer for gas chromatography. Anal. Methods 2 (5), 455–457.

Holbrey, J.D., López-Martin, I., Rothenberg, G., Seddon, K.R., Silvero, G., Zheng, X., 2008. Desulfurisation of oils using ionic liquids: selection of cationic and anionic components to enhance extraction efficiency. Green Chem. 10 (1), 87–92.

Hongbin, C., Lei, W., 2013. Lignocelluloses feedstock biorefinery as petrorefinery substitutes. In: Matovic, M.D. (Ed.), Biomass Now – Sustainable Growth and Use. InTech, Croatia, Rijeka, pp. 347–388.

Howlett, P.C., MacFarlane, D.R., Hollenkamp, A.F., 2004. High lithium metal cycling efficiency in a room-temperature ionic liquid. Electrochem. Solid State Lett. 7 (5), A97–A101.

Hsieh, Y.-N., Huang, P.-C., Sun, I.-W., Whang, T.-J., Hsu, C.-Y., Huang, H.-H., Kuei, C.-H., 2006. Nafion membrane-supported ionic liquid–solid phase microextraction for analyzing ultra trace PAHs in water samples. Anal. Chim. Acta 557 (1–2), 321–328.

Huang, K., Han, X., Zhang, X., Armstrong, D.W., 2007. PEG-linked geminal dicationic ionic liquids as selective, high-stability gas chromatographic stationary phases. Anal. Bioanal. Chem. 389 (7–8), 2265–2275.

Huang, X.-J., Aldous, L., O'Mahony, A.M., del Campo, F.J., Compton, R.G., 2010. Toward membrane-free amperometric gas sensors: a microelectrode array approach. Anal. Chem. 82 (12), 5238–5245.

Huang, P.-L., Luo, X.-F., Peng, Y.-Y., Pu, N.-W., Ger, M.-D., Yang, C.-H., Chang, J.-K., 2015. Ionic liquid electrolytes with various constituent ions for graphene-based supercapacitors. Electrochim. Acta 161, 371–377.

Ignat'ev, N.V., Welz-Biermann, U., Kucheryna, A., Bissky, G., Willner, H., 2005. New ionic liquids with tris(perfluoroalkyl)trifluorophosphate (FAP) anions. J. Fluorine Chem. 126 (8), 1150–1159.

Jiang, N., Ragauskas, A.J., 2007a. Selective aerobic oxidation of activated alcohols into acids or aldehydes in ionic liquids. J. Org. Chem. 72 (18), 7030–7033.

Jiang, N., Ragauskas, A.J., 2007b. Vanadium-catalyzed selective aerobic alcohol oxidation in ionic liquid [bmim]PF_6. Tetrahedron Lett. 48 (2), 273–276.

Jiménez, A.E., Bermúdez, M.D., Carrión, F.J., Martínez-Nicolás, G., 2006. Room Temperature ionic liquids as lubricant additives in steel-aluminium contacts: influence of sliding velocity, normal load and temperature. Wear 261, 347–359.

Kadokawa, J.-i., 2013. Polymerizable ionic liquids: development to photo functional poly(ionic liquid) materials. TCIMAIL 159, 1–6.

Kamimura, H., Kubo, T., Minami, I., Mori, S., 2007. Effect and mechanism of additives for ionic liquids as new lubricants. Tribol. Int. 40 (4), 620–625.

Kamlet, M.J., Taft, R.W., 1976. The solvatochromic comparison method. I. The β-scale of solvent hydrogen-bond acceptor (HBA) basicities. J. Am. Chem. Soc. 98 (2), 377–383.

Kamlet, M.J., Abboud, J.L., Taft, R.W., 1977. The solvatochromic comparison method. 6. The π* scale of solvent polarities. J. Am. Chem. Soc. 99 (18), 6027–6038.

Kar, M., Simons, T.J., Forsyth, M., MacFarlane, D.R., 2014. Ionic liquid electrolytes as a platform for rechargeable metal-air batteries: a perspective. Phys. Chem. Chem. Phys. 16, 18658–18674.

Kervinen, K., Korpi, H., Leskelä, M., Repo, T., 2003. Oxidation of veratryl alcohol by molecular oxygen in aqueous solution catalyzed by cobalt salen-type complexes: the effect of reaction conditions. J. Mol. Catal. A Chem. 203 (1–2), 9–19.

Kilpeläinen, I., Xie, H., King, A., Granstrom, M., Heikkinen, S., Agyropoulos, D.S., 2007. Dissolution of wood in ionic liquids. J. Agric. Food Chem. 55 (22), 9142–9148.

Kim, D.W., Chi, D.Y., 2004. Polymer-supported ionic liquids: imidazolium salts as catalysts for nucleophilic substitution reactions including fluorinations. Angew. Chem. Int. 43 (4), 483–485.

Kim, K.-S., Demberelnyamba, D., Lee, H., 2004. Size-selective synthesis of gold and platinum nanoparticles using novel thiol-functionalized ionic liquids. Langmuir 20 (3), 556–560.

Kim, Y.J., Kim, S., Joshi, Y.K., Federov, A.G., Kohl, P.A., 2012. Thermodynamic analysis of an absorption refrigeration system with ionic-liquid/refrigerant mixture as a working fluid. Energy 44, 1005–1016.

Kowsari, E., 2011. Advanced applications of ionic liquids in polymer science. In: Kokorin, A. (Ed.), Ionic Liquids: Applications and Perspectives. InTech, Rijeka, Croatia, pp. 3–28.

Kuboki, T., Okuyama, T., Ohsaki, T., Takami, N., 2005. Lithium-air batteries using hydrophobic room temperature ionic liquid electrolyte. J. Power Sources 146 (1–2), 766–769.

Kumar, P., Barrett, D.M., Delwiche, M.J., Stroeve, P., 2009. Methods for pretreatment of lignocellulosic biomass for efficient hydrolysis and biofuel production. Ind. Eng. Chem. Res. 48 (8), 3713–3729.

Kurnia, K.A., Pinho, S.P., Coutinho, J.A.P., 2014. Designing ionic liquids for absorptive cooling. Green Chem. 16, 3741–3745.

Lahtinen, M., Viikari, L., Karhunen, P., Asikkala, J., Kruus, K., Kilpeläinen, I., 2013. On the reactivity of the *Melanocarpus albomyces* laccase and formation of coniferyl alcohol dehydropolymer (DHP) in the presence of ionic liquid 1-allyl-3-methylimidazolium chloride. J. Mol. Catal. B Enzym. 85–86, 169–177.

Lau, V.W.-h., van de Water, L.G.A., Masters, A.F., Maschmeyer, T., 2012. Tuning the photocatalytic activity of CdS nanocrystals through intermolecular interactions in ionic-liquid solvent systems. Chem. Eur. J. 18, 2923–2930.

Li, Q., Tang, Q., Du, N., Qin, Y., Xiao, J., He, B., Chu, L., 2014. Employment of ionic liquid-imbibed polymer gel electrolyte for efficient quasi-solid-state dye-sensitized solar cells. J. Power Sources 248, 816–821.

Liang, C., Yuan, C.-Y., Warmack, R.J., Barnes, C.E., Dai, S., 2002. Ionic liquids: a new class of sensing materials for detection of organic vapors based on the use of a quartz crystal microbalance. Anal. Chem. 74, 2172–2176.

Lin, Z., Slawin, A.M.Z., Morris, R.E., 2007. Chiral induction in the ionothermal synthesis of a 3-D coordination polymer. J. Am. Chem. Soc. 129 (16), 4880–4881.

Lin, R., Taberna, P.-L., Fantini, S., Presser, V., Pérez, C.R., Malbosc, F., Simon, P., 2011. Capacitive energy storage from −50 to 100°C using an ionic liquid electrolyte. J. Phys. Chem. Lett. 2, 2396–2401.

Liu, F.-h., Jiang, Y., 2007. Room temperature ionic liquid as matrix medium for the determination of residual solvents in pharmaceuticals by static headspace gas chromatography. J. Chromatogr. A 1167 (1), 116–119.

Liu, J.-F., Li, N., Jiang, G.-B., Liu, J.-M., Jönsson, J.Å., Wen, M.-J., 2005. Disposable ionic liquid coating for headspace solid-phase microextraction of benzene, toluene, ethylbenzene, and xylenes in paints followed by gas chromatography-flame ionization detection. J. Chromatogr. A 1066 (1–2), 27–32.

Liu, H., Liu, Y., Li, J., 2010. Ionic liquids in surface electrochemistry. Phys. Chem. Chem. Phys. 12, 1685–1697.

Liu, S., Shi, Z., Li, L., Yu, S., Xie, C., Song, Z., 2013. Process of lignin oxidation in an ionic liquid coupled with separation. RSC Adv. 3 (17), 5789–5793.

López-Darias, J., Pino, V., Anderson, J.L., Graham, C.M., Afonso, A.M., 2010. Determination of water pollutants by direct-immersion solid-phase microextraction using polymeric ionic liquid coatings. J. Chromatogr. A 1217 (8), 1236–1243.

López-González, D., Valverde, J.L., Sánchez, P., Sanchez-Silva, L., 2013. Characterization of different heat transfer fluids and degradation study by using a pilot plant device operating at real conditions. Energy 54, 240–250.

Lu, W., Qu, L., Henry, K., Dai, L., 2009a. High performance electrochemical capacitors from aligned carbon nanotube electrodes and ionic liquid electrolytes. J. Power Sources 189 (2), 1270–1277.

Lu, J., Yan, F., Texter, J., 2009b. Advanced applications of ionic liquids in polymer science. Prog. Polym. Sci. 34 (5), 431–448.
Lu, D., Shomali, N., Shen, A., 2010. Task specific ionic liquid for direct electrochemistry of metal oxides. Electrochem. Commun. 12, 1214–1217.
Lui, M.Y., Crowhurst, L., Hallett, J.P., Hunt, P.A., Neidermeyer, H., Welton, T., 2011. Salts dissolved in salts: ionic liquid mixtures. Chem. Sci. 2, 1491–1496.
Luska, K.L., Moores, A., 2012. Ruthenium nanoparticle catalysts stabilized in phosphonium and imidazolium ionic liquids: dependence of catalyst stability and activity on the ionicity of the ionic liquid. Green Chem. 14, 1736–1742.
Ma, Z., Yu, J., Dai, S., 2010. Preparation of inorganic materials using ionic liquids. Adv. Mater. 22 (2), 261–285.
MacFarlane, D.R., Meakin, P., Sun, J., Amini, N., Forsyth, M., 1999. Pyrrolidinium imides: a new family of molten salts and conductive plastic crystal phases. J. Phys. Chem. B 103, 4164–4170.
MacFarlane, D.R., Forsyth, M., Howlett, P.C., Pringle, J.M., Sun, J., Annat, G., Izgorodina, E.I., 2007. Ionic liquids in electrochemical devices and processes: managing interfacial electrochemistry. Acc. Chem. Res. 40, 1165–1173.
MacFarlane, D.R., Pringle, J.M., Howlett, P.C., Forsyth, M., 2010. Ionic liquids and reactions at the electrochemical interface. Phys. Chem. Chem. Phys. 12, 1659–1669.
MacFarlane, D.R., Tachikawa, N., Forsyth, M., Pringle, J.M., Howlett, P.C., Elliot, G.D., Angell, C.A., 2014. Energy applications of ionic liquids. Energy Environ. Sci. 7, 232–250.
Mandai, T., Yoshida, K., Ueno, K., Dokko, K., Watanabe, M., 2014. Criteria for solvate ionic liquids. Phys. Chem. Chem. Phys. 16, 8761–8772.
Marszałł, M.P., Kaliszan, R., 2007. Application of ionic liquids in liquid chromatography. Crit. Rev. Anal. Chem. 37 (2), 127–140.
Marszałł, M.P., Baczek, T., Kaliszan, R., 2006. Evaluation of the silanol-suppressing potency of ionic liquids. J. Sep. Sci. 29 (8), 1138–1145.
Martindale, B.C.M., Ward Jones, S.E., Compton, R.G., 2010. A comparison of the cyclic voltammetry of the Sn/Sn(II) couple in the room temperature ionic liquids *N*-butyl-*N*-methylpyrrolidinium dicyanamide and *N*-butyl-*N*-methylpyrrolidinium bis(trifluoromethylsulfonyl)imide: solvent induced changes of electrode reaction mechanism. Phys. Chem. Chem. Phys. 12, 1827–1833.
Mathew, S., Yella, A., Gao, P., Humphry-Baker, R., Curchod, B.F.E., Ashari-Astani, N., Grätzel, M., 2014. Dye-sensitized solar cells with 13% efficiency achieved through the molecular engineering of porphyrin sensitizers. Nat. Chem. 6, 242–247.
Maton, C., De Vos, N., Stevens, C.V., 2013. Ionic liquid thermal stabilities: decomposition mechanisms and analysis tools. Chem. Soc. Rev. 42, 5963–5977.
Mecerreyes, D., 2011. Polymeric ionic liquids: broadening the properties and applications of polyelectrolytes. Prog. Polym. Sci. 36 (12), 1629–1648.
Miao, W., Chan, T.-H., 2005. Ionic-liquid-supported peptide synthesis demonstrated by the synthesis of Leu^5-enkephalin. J. Org. Chem. 70, 3251–3255.
Migowski, P., Zanchet, D., Machado, G., Gelesky, M.A., Teixeira, S.R., Dupont, J., 2010. Nanostructures in ionic liquids: correlation of iridium nanoparticles' size and shape with imidazolium salts' structural organization and catalytic properties. Phys. Chem. Chem. Phys. 12, 6826–6833.
Miller, J.R., Burke, A.F., 2008. Electrochemical capacitors: challenges and opportunities for real-world applications. Electrochem. Soc. Interface 17 (1), 53–57.
Moens, L., Blake, D.M., Rudnicki, D.L., Hale, M.J., 2003. Advanced thermal storage fluids for solar parabolic trough systems. J. Sol. Energy. T. ASME 125, 112–116.
Morris, R.E., 2009. Ionothermal synthesis-ionic liquids as functional solvents in the preparation of crystalline materials. Chem. Commun. 2009, 2990–2998.
Nègre, L., Daffos, B., Taberna, P.L., Simon, P., 2015. Solvent-free electrolytes for electrical double layer capacitors. J. Electrochem. Soc. 162 (5), A5037–A5040.

Nikitina, V.A., Nazet, A., Sonnleitner, T., Buchner, R., 2012. Properties of sodium tetrafluoroborate solutions in 1-butyl-3-methylimidazolium tetrafluoroborate ionic liquid. J. Chem. Eng. Data 57, 3019–3025.

Nishimura, N., Ohno, H., 2014. 15th Anniversary of polymerised ionic liquids. Polymer 55 (16), 3289–3297.

O'Regan, B., Grätzel, M., 1991. A low-cost, high-efficiency solar cell based on dye-sensitized colloidal TiO_2 films. Nature 353, 737–740.

Ott, L.S., Cline, M.L., Deetlefs, M., Seddon, K.R., Finke, R.G., 2005. Nanoclusters in ionic liquids: evidence for *n*-heterocyclic carbene formation from imidazolium-based ionic liquids detected by ^{2}H NMR. J. Am. Chem. Soc. 127, 5758–5759.

Papageorgiou, N., Athanassov, Y., Armand, M., Bonhôte, P., Pettersson, H., Azam, A., Grätzel, M., 1996. The performance and stability of ambient temperature molten salts for solar cell applications. J. Electrochem. Soc. 143 (10), 3099–3108.

Papaiconomou, N., Vite, G., Goujon, N., Lévêque, J.-M., Billard, I., 2012. Efficient removal of gold complexes from water by precipitation or liquid–liquid extraction using ionic liquids. Green Chem. 14 (7), 2050–2056.

Patel, R., Kumari, M., Khan, A.B., 2014. Recent advances in the applications of ionic liquids in protein stability and activity: a review. Appl. Biochem. Biotechnol. 172, 3701–3720.

Paul, A., Mandal, P.K., Samanta, A., 2005. On the optical properties of the imidazolium ionic liquids. J. Phys. Chem. B 109 (18), 9148–9153.

Pawliszyn, J., 1997. Solid Phase Microextraction: Theory and Practice, first ed. Wiley-VCH, Inc, New York, NY, USA.

Pham, M.-Q., Yoon, H.-S., Khare, V., Ahn, S.-H., 2014. Evaluation of ionic liquids as lubricants in micro milling – process capability and sustainability. J. Clean. Prod. 76, 167–173.

Pinkert, A., Marsh, K.N., Pang, S., 2010. Reflections on the solubility of cellulose. Ind. Eng. Chem. Res. 49 (22), 11121–11130.

Plechkova, N.V., Seddon, K.R., 2008. Applications of ionic liquids in the chemical industry. Chem. Soc. Rev. 37, 123–150.

Prado, R., Erdocia, X., Labidi, J., 2013. Lignin extraction and purification with ionic liquids. J. Chem. Technol. Biot. 88 (7), 1248–1257.

Praunsnitz, J.M., Lichtenthaler, R.N., Azevedo, E.G., 1999. Fugacities in Liquid Mxtures: Excess Functions Molecular Thermodynamics of Fluid-Phase Equilibria. Prentice Hall PTR, New Jersey, USA, pp. 213–299.

Preißinger, M., Pöllinger, S., Brüggemann, D., 2013. Ionic liquid based absorption chillers for usage of low grade waste heat in industry. Int. J. Energy Res. 37, 1382–1388.

Pu, Y., Jiang, N., Ragauskas, A.J., 2007. Ionic liquid as a green solvent for lignin. J. Wood Chem. Technol. 27 (1), 23–33.

Puttick, S., Davis, A.L., Butler, K., Lambert, L., El harfi, J., Irvine, D.J., Licence, P., 2011. NMR as a probe of nanostructured domains in ionic liquids: does domain segregation explain increased performance of free radical polymerisation? Chem. Sci. 2, 1810–1816.

Puttick, S., Davis, A.L., Butler, K., Irvine, D.J., Licence, P., Thurecht, K.J., 2013. The influence of domain segregation in ionic liquids upon controlled polymerisation mechanisms: RAFT polymerisation. Polym. Chem. 4, 1337–1344.

Qu, J., Truhan, J.J., Dai, S., Luo, H., Blau, P.J., 2006. Ionic liquids with ammonium cations as lubricants or additives. Tribol. Lett. 22 (3), 207–214.

Rehman, A., Zeng, X., 2012. Ionic liquids as green solvents and electrolytes for robust chemical sensor development. Acc. Chem. Res. 45 (10), 1667–1677.

Reichert, E., Wintringer, R., Volmer, D.A., Hempelmann, R., 2012. Electro-catalytic oxidative cleavage of lignin in a protic ionic liquid. Phys. Chem. Chem. Phys. 14 (15), 5214–5221.

Ren, J.L., Sun, R.C., Liu, C.F., Cao, Z.N., Luo, W., 2007. Acetylation of wheat straw hemicelluloses in ionic liquid using iodine as a catalyst. Carbohyd. Polym. 70 (4), 406–414.

Ribeiro, A.P.C., Vieira, S.I.C., França, J.M., Queirós, C.S., Langa, E., Lourenço, M.J.V., Nieto de Castro, C.A., 2011. Thermal properties of ionic liquids and ionanofluids. In: Kokorin, A. (Ed.), Ionic Liquids: Theory, Properties, New Approaches. InTech, Rijeka, Croatia, pp. 37–60.
Rodríguez-Cabo, B., Arce, A., Soto, A., 2013. Desulfurization of fuels by liquid–liquid extraction with 1-ethyl-3-methylimidazolium ionic liquids. Fluid Phase Equilibr. 356, 126–135.
Romsted, L.S., Bunton, C.A., Yao, J., 1997. Micellar catalysis, a useful misnomer. Curr. Opin. Colloid Interface Sci. 2 (6), 622–628.
Scammells, P.J., Scott, J.L., Singer, R.D., 2005. Ionic liquids: the neglected issues. Aust. J. Chem. 58, 155–169.
Schmid, C.R., Beck, C.A., Cronin, J.S., Staszak, M.A., 2004. Demethylation of 4-methoxyphenylbutyric acid using molten pyridinium hydrochloride on multikilogram scale. Org. Process Res. Dev. 8, 670–673.
Seiler, M., Kühn, A., Ziegler, F., Wang, X., 2013. Sustainable cooling strategies using new chemical system solutions. Ind. Eng. Chem. Res. 52, 16519–16546.
Shi, D., Pootrakulchote, N., Li, R., Guo, J., Wang, Y., Zakeeruddin, S.M., Wang, P., 2008. New efficiency records for stable dye-sensitized solar cells with low-volatility and ionic liquid electrolytes. J. Phys. Chem. C 112, 17046–17050.
Shiddiky, M.J.A., Torriero, A.A.J., 2011. Application of ionic liquids in electrochemical sensing systems. Biosens. Bioelectron. 26 (5), 1775–1787.
Shkrob, I.A., Marin, T.W., Zhu, Y., Abraham, D.P., 2014. Why bis(fluorosulfonyl)imide is a "magic anion" for electrochemistry. J. Phys. Chem. C 118, 19661–19671.
Silvester, D.S., 2011. Recent advances in the use of ionic liquids for electrochemical sensing. Analyst 136, 4871–4882.
Simka, W., Puszczyk, D., Nawrat, G., 2009. Electrodeposition of metals from non-aqueous solutions. Electrochim. Acta 54, 5307–5319.
Skoog, D.A., Holler, F.J., Crouch, S.R., 2007. Principles of Instrumental Analysis, sixth ed. Cengage Learning, México DF, Mexico.
Sonar, S., Ambrose, K., Hendsbee, A.D., Masuda, J.D., Singer, R.D., 2012. Synthesis and application of Co(salen) complexes containing proximal imidazolium ionic liquid cores. Can. J. Chem. 90 (1), 60–70.
Spietelun, A., Marcinkowski, Ł., de la Guardia, M., Namieśnik, J., 2013. Recent developments and future trends in solid phase microextraction techniques towards green analytical chemistry. J. Chromatogr. A 1321, 1–13.
Srikhirin, P., Aphornratana, S., Chungpaibulpatana, S., 2001. A review of absorption refrigeration technologies. Renew. Sust. Energ. Rev. 5 (4), 343–372.
Stark, A., 2011. Ionic liquids in the biorefinery: a critical assessment of their potential. Energy Environ. Sci. 4 (1), 19–32.
Stärk, K., Taccardi, N., Bösmann, A., Wasserscheid, P., 2010. Oxidative depolymerization of lignin in ionic liquids. ChemSusChem 3 (6), 719–723.
Strehmel, V., Laschewsky, A., Wetzel, H., 2006. Homopolymerization of a highly polar zwitterionic methacrylate in ionic liquids and its copolymerization with a non-polar methacrylate. e-Polym 6 (1), 131–140.
Sun, P., Armstrong, D.W., 2010. Ionic liquids in analytical chemistry. Anal. Chim. Acta 661 (1), 1–16.
Sun, X.-L., Zhou, W., Gu, L., Qiu, D., Ren, D.-H., Gu, Z.-G., Li, Z., 2015. Liquid-liquid extraction to lithium isotope separation based on room-temperature ionic liquids containing 2,2′-binaphthyldiyl-17-crown-5. J. Nucl. Sci. Technol. 52 (3), 332–341.
Swatloski, R.P., Spear, S.K., Holbrey, J.D., Rogers, R.D., 2002. Dissolution of cellose with ionic liquids. J. Am. Chem. Soc. 124, 4974–4975.
Taft, R.W., Kamlet, M.J., 1976. The solvatochromic comparison method. 2. The alpha-scale of solvent hydrogen-bond donor (HBD) acidities. J. Am. Chem. Soc. 98 (10), 2886–2894.
Tan, S.S.Y., MacFarlane, D.R., 2010. Ionic liquids in biomass processing. Top. Curr. Chem. 290, 311–339.

Tan, S.S.Y., MacFarlane, D.R., Upfal, J., Edye, L.A., Doherty, W.O.S., Patti, A.F., Scott, J.L., 2009. Extraction of lignin from lignocellulose at atmospheric pressure using alkylbenzenesulfonate ionic liquid. Green Chem. 11 (3), 339–345.
Tang, J., Tang, H., Sun, W., Radosz, M., Shen, Y., 2005. Low-pressure CO_2 sorption in ammonium-based poly(ionic liquid)s. Polymer 46 (26), 12460–12467.
Tang, S., Baker, G.A., Zhao, H., 2012. Ether- and alcohol-functionalized task-specific ionic liquids: attractive properties and applications. Chem. Soc. Rev. 41, 4030–4066.
Tian, J., Wang, J., Zhao, S., Jiang, C., Zhang, X., Wang, X., 2010. Hydrolysis of cellulose by the heteropoly acid $H_3PW_{12}O_{40}$. Cellulose 17 (3), 587–594.
Tian, Y.-H., Goff, G.S., Runde, W.H., Batista, E.R., 2012. Exploring electrochemical windows of room-temperature ionic liquids: a computational study. J. Phys. Chem. B 116, 11943–11952.
Tsai, W.-Y., Lin, R., Murali, S., Zhang, L.L., McDonough, J.K., Ruoff, R.S., Simon, P., 2013. Outstanding performance of activated graphene based supercapacitors in ionic liquid electrolyte from −50 to 80°C. Nano Energy 2, 403–411.
Vaher, M., Koel, M., Kaljurand, M., 2001. Non-aqueous capillary electrophoresis in acetonitrile using Ionic-liquid buffer electrolytes. Chromatographia 53 (1), S302–S306.
Van Valkenburg, M.E., Vaughn, R.L., Williams, M., Wilkes, J.S., 2005. Thermochemistry of ionic liquid heat-transfer fluids. Thermochim. Acta 425 (1–2), 181–188.
Vijayraghavan, R., Rana, U.A., Elliot, G.D., MacFarlane, D.R., 2013. Protic ionic solids and liquids based on the guanidinium cation as phase-change energy-storage materials. Energy Technol. 1, 609–612.
Wang, P., Zakeeruddin, S.M., Humphry-Baker, R., Grätzel, M., 2004. A binary ionic liquid electrolyte to achieve ≥7% power conversion efficiencies in dye-sensitized solar cells. Chem. Mater. 16, 2694–2696.
Washiro, S., Yoshizawa, M., Nakajima, H., Ohno, H., 2004. Highly ion conductive flexible films composed of network polymers based on polymerizable ionic liquids. Polymer 45 (5), 1577–1582.
Weber, C.C., Masters, A.F., Maschmeyer, T., 2012. Pseudo-encapsulation – Nano-domains for enhanced reactivity in ionic liquids. Angew. Chem. Int. 51, 11483–11486.
Weber, C.C., Masters, A.F., Maschmeyer, T., 2013. Structural features of ambient temperature ionic liquids: their consequences for material preparation and organic reactivity. Green Chem. 15 (10), 2655–2679.
Wei, D., Ivaska, A., 2008. Applications of ionic liquids in electrochemical sensors. Anal. Chim. Acta 607, 126–135.
Wells, T.P., Hallett, J.P., Williams, C.K., Welton, T., 2008. Esterification in ionic liquids: the influence of solvent basicity. J. Org. Chem. 73, 5585–5588.
Weng, L., Liu, X., Liang, Y., Xue, Q., 2007. Effect of tetraalkylphosphonium based ionic liquids as lubricants on the tribological performance of a steel-on-steel system. Tribol. Lett. 26 (1), 11–17.
Wilkes, J.S., Zaworotko, M.J., 1992. Air and water stable 1-ethyl-3-methylimidazolium based ionic liquids. J. Chem. Soc. Chem. Commun., 965–967.
Winterton, N., 2006. Solubilization of polymers by ionic liquids. J. Mater. Chem. 16, 4281–4293.
Wooster, T.J., Johanson, K.M., Fraser, K.J., MacFarlane, D.R., Scott, J.L., 2006. Thermal degradation of cyano containing ionic liquids. Green Chem. 8, 691–696.
Wu, J., Lan, Z., Lin, J., Huang, M., Huang, Y., Fan, L., Luo, G., 2015. Electrolytes in dye-sensitized solar cells. Chem. Rev. 115, 2136–2173.
Xu, T., Zhou, X., Jiang, Z.-Y., Kuang, Q., Xie, Z.-X., Zheng, L.-S., 2009. Syntheses of nano/submicrostructured metal oxides with all polar surfaces exposed via a molten salt route. Cryst. Growth Des. 9 (1), 192–196.
Xu, M., Ivey, D.G., Xie, Z., Qu, W., Dy, E., 2013. The state of water in 1-butly-1-methylpyrrolidinium bis(trifluoromethanesulfonyl)imide and its effect on Zn/Zn(II) redox behavior. Electrochim. Acta 97, 289–295.

Yang, F., Jiao, L., Shen, Y., Xu, X., Zhang, Y., Niu, L., 2007. Enhanced response induced by polyelectrolyte-functionalized ionic liquid in glucose biosensor based on sol-gel organic–inorganic hybrid material. J. Electroanal. Chem. 608 (1), 78–83.
Yau, H.M., Barnes, S.A., Hook, J.M., Youngs, T.G.A., Croft, A.K., Harper, J.B., 2008. The importance of solvent reorganisation in the effect of an ionic liquid on a unimolecular substitution process. Chem. Commun., 3576–3578.
Ye, C., Liu, W., Chen, Y., Yu, L., 2001. Room-temperature ionic liquids: a novel versatile lubricant. Chem. Commun. 2001, 2244–2245.
Yokozeki, A., Shiflett, M.B., 2010. Water solubility in ionic liquids and application to absorption cycles. Ind. Eng. Chem. Res. 49, 9496–9503.
Yoshida, K., Tsuchiya, M., Tachikawa, N., Dokko, K., Watanabe, M., 2012. Correlation between battery performance and lithium ion diffusion in glyme-lithium bis(trifluoromethanesulfonyl) amide equimolar complexes. J. Electrochem. Soc. 159 (7), A1005–A1012.
Yuan, J., Antonietti, M., 2011. Poly(ionic liquid)s: polymers expanding classical property profiles. Polymer 52 (7), 1469–1482.
Yuan, J., Mecerreyes, D., Antonietti, M., 2013. Poly(ionic liquids)s: an update. Prog. Polym. Sci. 38 (7), 1009–1036.
Yuen, A.K.L., Masters, A.F., Maschmeyer, T., 2013. 1,3-Disubstituted imidazolium hydroxides: dry salts or wet carbenes? Catal. Today 200, 9–16.
Zakeeruddin, S.M., Grätzel, M., 2009. Solvent-free ionic liquid electrolytes for mesoscopic dye-sensitized solar cells. Adv. Funct. Mater. 19, 2187–2202.
Zakzeski, J., Jongerius, A.L., Weckhuysen, B.M., 2010. Transition metal catalyzed oxidation of Alcell lignin, soda lignin, and lignin model compounds in ionic liquids. Green Chem. 12, 1225–1236.
Zein El Abedin, S., Moustafa, E.M., Hempelmann, R., Natter, H., Endres, F., 2006. Electrodeposition of nano- and microcrystalline aluminium in three different air and water stable ionic liquids. ChemPhysChem 7, 1535–1543.
Zhang, S., Sun, N., He, X., Lu, X., Zhang, X., 2006. Physical properties of ionic liquids: database and evaluation. J. Phys. Chem. Ref. Data 35 (4), 1475–1517.
Zhang, Z., Salih, A.A.M., Li, M., Yang, B., 2014a. Synthesis and characterization of functionalized ionic liquids for thermal storage. Energy Fuels 28, 2802–2810.
Zhang, C., Ueno, K., Yamazaki, A., Yoshida, K., Moon, H., Mandai, T., Watanabe, M., 2014b. Chelate effects in glyme/lithium bis(trifluoromethanesulfonyl)amide solvate ionic liquids. I. Stability of solvate cations and correlation with electrolyte properties. J. Phys. Chem. B 118, 5144–5153.
Zhao, F., Meng, Y., Anderson, J.L., 2008. Polymeric ionic liquids as selective coatings for the extraction of esters using solid-phase microextraction. J. Chromatogr. A 1208 (1–2), 1–9.
Zhou, X., Xie, Z.-X., Jiang, Z.-Y., Kuang, Q., Zhang, S.-H., Xu, T., Zheng, L.-S., 2005. Formation of ZnO hexagonal micro-pyramids: a successful control of the exposed polar surfaces with the assistance of an ionic liquid. Chem. Commun., 5572–5574.
Zhou, X., Shao, X., Shu, J.-J., Liu, M.-M., Liu, H.-L., Feng, X.-H., Liu, F., 2012. Thermally stable ionic liquid-based sol–gel coating for ultrasonic extraction-solid-phase microextraction-gas chromatography determination of phthalate esters in agricultural plastic films. Talanta 89, 129–135.
Zhu, J., Bai, L., Chen, B., Fei, W., 2009. Thermodynamical properties of phase change materials based on ionic liquids. Chem. Eng. J. 147, 58–62.
Zhu, Y., Chuanzhao, L., Sudarmadji, M., Min, N.H., Biying, A.O., Maguire, J.A., Hosmane, N.S., 2012. An efficient and recyclable catalytic system comprising nanopalladium(0) and a pyridinium salt of iron bis(dicarbollide) for oxidation of substituted benzyl alcohol and lignin. ChemistryOpen 1 (2), 67–70.
Ziegler, F., 1999. Recent developments and future prospects of sorption heat pump systems. Int. J. Therm. Sci. 38, 191–208.
Ziółkowski, B., Diamond, D., 2013. Thermoresponsive poly(ionic liquid) hydrogels. Chem. Commun. 49, 10308–10310.

CHAPTER 2

Methods of Synthesis and Purification of Ionic Liquids

Alastair J.S. McIntosh*, Jeraime Griffith*, John Gräsvik**
*Department of Chemistry, Imperial College London, UK
**School of Chemistry, University of Southampton, UK

SYNTHESIS OF IONIC LIQUIDS

Introduction

Most ionic liquid (IL) syntheses lack a fully detailed description of their preparation and purification. Though the chemistry involved is generally very simple the propensity for colored impurities and runaway reactions is evident from the nature of the reactants involved. A detailed step-by-step account should be the *de rigueur* for any synthetic method. Additionally, it may be advantageous to include a photo of the pure ionic liquid alongside supplementary information as a visual cue of purity. Outlined here is an overview of the techniques employed for ionic liquid synthesis along with their merits and demerits.

The synthesis of ethylammonium nitrate in 1914 is generally regarded as the first ionic liquid to be made (Walden, 1914). The acid–base neutralization reaction of concentrated nitric acid and ethylamine provided, after removal of water by distillation, a room temperature ionic liquid with a melting point of 12°C. Indeed, the simplest ionic liquids can be made in an analogous manner via the protonation of amines and phosphines. Ionic liquids produced in this manner, however, generally suffer degradation through deprotonation and have found limited application.

Of more utility are the ionic liquids of cations depicted in Figure 2.1. These cations encompass the vast majority of ionic liquids encountered to date.

This chapter will outline the synthesis of ionic liquids containing these cations; however, major focus will be given to the 1-alkyl-3-methylimidazolium cations, as they are the most commonly encountered. The properties and applications of ionic liquids will be dealt with elsewhere.

Ionic liquids are generally made via one of two routes: through simple alkylation or through protonation.

Application, Purification, and Recovery of Ionic Liquids
http://dx.doi.org/10.1016/B978-0-444-63713-0.00002-X

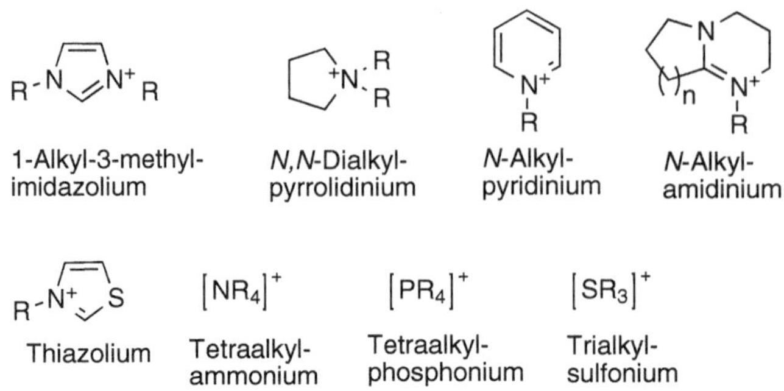

Figure 2.1 ***Cations Exemplified in Ionic Liquids.***

Alkylation Reactions

The most commonly used method for ionic liquid synthesis involves the alkylation of 1-methylimidazole with a haloalkane. The main advantages of this route are the ready availability of cheap haloalkanes and the relatively mild temperatures employed. Reactivity of the haloalkanes increases from chloride < bromide < iodide, with the iodides being very reactive. Use of iodoalkanes requires low temperatures and high dilution conditions to prevent rapid formation of the product, which can cause localized hot-spots and runaway reactions. Additionally, they have to be shielded from light as the iodide anion is photoactive, making for a very troublesome purification of the ionic liquid. For these reasons they are generally avoided. The very strong C—F bond of fluoroalkanes precludes their use in ionic liquid synthesis.

The ease and generality of the alkylation method, addition of the haloalkane to a stirred solution of the 1-alkylimidazole with heating, lends itself to the smooth quaternization of a range of amines such as, amidines (Nowicki et al., 2014), trialkylamines (Sun et al., 1997), pyrrolidines (MacFarlane and Meakin, 1999), pyridines (Lethesh et al., 2011), and imidazoles (Ab Rani et al., 2011). Phosphines can also be made in this way. To obtain the best results and cleanest product, all starting reagents and solvents should be thoroughly dried and purified before use – see Section "Impurities and Purification of Ionic Liquids" for further details. The alkylating reagent should be added slowly to a solution of 1-methylimidazole (the most commonly encountered amine) cooled in an ice bath under an inert atmosphere. Following warming to room temperature, the mixture is heated to a temperature generally below 80°C to avoid decolorization due to thermal decomposition impurities. The amount of time required for completion of the reaction depends on the haloalkane chain length, with the equivalent bromide requiring a shorter time compared to the chloride. Due to the reactivity of the bromides and iodides, heating of the reaction mixture is seldom required. The choice of ethyl acetate or toluene as solvent makes for a simple workup of washing and decanting under Schlenk conditions, as the ionic liquid and solvent are mutually immiscible. Despite these precautions, if a colored product is obtained, treatment with activated decolorizing charcoal and/or recrystallization generally helps. Heating under high vacuum with vigorous stirring removes the last vestiges of solvent to yield a pure product. The ionic liquids thus obtained are usually white solids, with decreasing melting points as the chain length increases, to clear viscous oils. This trend generally holds true; however, one author's experience with 1-methyl-3-propylimidazolium chloride ($[C_3C_1im]Cl$) finds this to be a clear viscous oil/semisolid at room temperature while 1-butyl-3-methylimidazolium chloride $[C_4C_1im]Cl$ is generally a crystalline solid. A glovebox is advisable for storing the synthesized ionic liquids for any extended period. They remain of high purity and can be used without further purification even 10 years after they were made.*

One of the most commonly encountered cations is 1-ethyl-3-methylimidazolium, $[C_2C_1im]^+$. In an academic laboratory setting it is made via the reaction between 1-methylimidazole and 1-bromoethane. 1-Bromoethane being a liquid is easier to handle as opposed to the gaseous 1-chloroethane. In this setting simple round-bottomed flasks and reflux condensers are all that are required. Whereas, reactions involving 1-chloroethane require complex gas handling apparatus and an autoclave as undertaken industrially (Seddon et al., 2000). 1-Chloroethane is cooled to below its boiling point of 12°C before addition to 1-methylimidazole in a pressure vessel (Pitula and Mudring, 2010). At the end of the reaction, increasing the pressure allows the halide to boil off and the recovered material can be recycled into the next reaction.

Microwave Irradiation

A solvent-free microwave irradiation route can also be used in the synthesis of ionic liquids with higher boiling haloalkanes. Under microwave irradiation the newly formed ionic liquid increases the polarity of the mixture thus increasing the rate of microwave absorption. This often leads to charred products and in some cases runaway reactions. However, if the reaction is throttled, by alternating heating and stirring, high yields can be obtained

* Personal experience of one of the authors using decade old $[C_2C_1im]Br$.

in a very short time (Varma and Namboodiri, 2001). Using this method Varma et al. were able to produce a variety of alkylimidazolium-based ionic liquids with slightly higher yields than those achieved via conventional methods in a much shorter timescale. The reaction can be monitored by observing the change in appearance of the reaction mixture from clear to opaque then back to clear. Though issues with reproducibility were encountered, these were easily overcome by using a water-moderated method (Law et al., 2002). The water acted as a heat sink and moderated the amount of microwave radiation reaching the reaction vessel. With microwave ovens specifically made for laboratory synthesis, a finer-tuned throttling of the reaction conditions may be achieved through a programmed method. All aspects of the reaction can be monitored in real-time aiding in the reproducibility of the results. The microwave process represents a viable option for the synthesis of ionic liquids where a slightly discolored product can be tolerated in light of the ease of performing the reactions and the solvent-free conditions. A purer product can easily be obtained by treating with decolorizing charcoal as mentioned earlier.

Sonication

Namboodiri and Varma, (2002) have also explored the sonochemical synthesis of ionic liquids. Their synthesis of mono and dicationic 1-alkyl-3-methylimidazolium halides was performed, solvent-free, in a standard laboratory ultrasonic cleaning bath. The progress was followed visually. As in their microwave-assisted synthesis, the mixture went from clear to opaque before the formation of a clear, viscous, oil or precipitation as a solid. With the final bath temperature rising to 40°C from ambient, and the internal temperature of the mixtures reaching as high as 125°C, the reactions with bromoalkanes were complete in 2 h. In general, yields of greater than 90% were standard except for those involving chloroalkanes. The reactions with chloroalkanes were much more sluggish giving a 24% yield of $[C_8C_1im]Cl$ after 6 h in contrast to 94% of $[C_8C_1im]Br$ after 2 h. A significant improvement of yield from 24% to 86% of $[C_8C_1im]Cl$ was achieved when an ultrasonic probe was placed directly into the reaction mixture. In a clever adaptation, after the synthesis of the ionic liquid, reagents for the palladium acetate–catalyzed Suzuki reaction of iodobenzene with 4-fluoro-phenylboronic acid were added. A one-pot, two-operation, strategy such as this should be equally applicable to other coupling procedures, such as the Heck reaction, and could be sustainable as the ionic liquid can be recycled (Chatel and MacFarlane, 2014).

Macrocyclic Ionic Liquids

Though linear 1-alkyl-3-methylimidazolium ionic liquids have been known for more than a quarter of a century, research into macrocyclic 1,3-dialkyl-imidazolium ionic liquids is a fledgling field. The ionic liquid depicted in Figure 2.2a, possesses planar chirality due to the inhibition of rotation of the imidazolium ring because of the methyl groups at the C-2 and C-4 positions (Ishida et al., 2002). Chiral ionic liquids are attractive because of their potential application for chiral discrimination, optical resolution of racemates and asymmetric synthesis. Ishida et al. carried out the synthesis of the macrocycle under high dilution conditions over 20 days. These conditions were necessary to minimize oligo formation through intermolecular reactions. The racemic cyclophane-type imidazolium ionic liquid forms diastereomic salts with a chiral anion, silver(I) (1S)-(+)-10-camphorsulfonate, proving their ability to act as chiral resolving agents in their enantiopure form. The same group later developed pseudocrown ether variants of the planar chiral ionic liquids where the imidazolium nitrogen atoms are connected with a tris(oxoethylene) chain (Figure 2.2b) (Ishida et al., 2004).

The tris(oxoethylene) group showed favorable selectivity in bridging the nitrogen atoms such that no oligomers resulting from the intermolecular reaction were observed. This arises from the preference of the vicinal oxygen atoms in adopting a *gauche* conformation. Consequently, high dilution conditions were not needed saving 99% of the solvent required for their synthesis compared with the alkyl-bridged variant. The Lewis-basic oxygen atoms play a crucial role in the molecular recognition of Lewis-acidic substrates. Evidence for this was observed from the europium tris(β-diketonate)

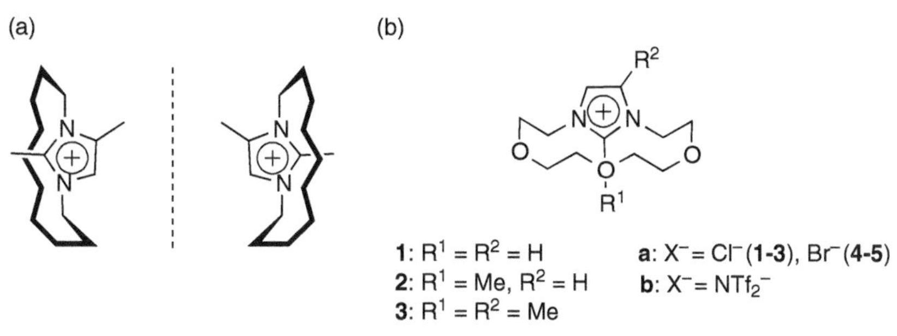

Figure 2.2 Imidazolium cyclophane-type ionic liquid (a) and imidazolium-based ionic liquid with a pseudocrown ether moiety (b). *Part (a) reprinted with permission from Ishida et al. (2002). The Royal Society of Chemistry; Part (b) reprinted with permission from Ishida et al. (2004). Copyright (2004), Elsevier.*

complexes of **3** (Figure 2.2b), where the signal attributed to one of the OCH_2 protons showed a notable downfield shift.

The enantiopure synthesis of planar chiral imidazolium-based ionic liquids was achieved in 2006, 4 years after the last work of Ishida et al. (2006). Here, the imidazolium ring was built up from first principles using acetaldehyde, NH_4MeCO_2 and glyoxal with (*S*)-valinol contributing the second nitrogen as well as the stereogenic center (Figure 2.3). Alkylation of the alcohol and the imidazole followed giving the planar chiral imidazolium salt with high diastereoselectivity (d.r. = 95:5). The major diastereomer (S_p,*S*)-**8**-Cl could be further resolved to d.r. ≥ 99:1 by simple recrystallization. The high diastereoselectivity observed arises from having the methyl group at the C-2 position of the imidazolium ring, and the isopropyl group at the C-1 position, oriented in opposite directions. This alleviates steric hindrance and is more thermodynamically and kinetically favorable.

Crown Ethers

Crown ethers are useful in phase transfer catalysis due to their ability to complex inorganic salts in order to solubilize them in a wide range of nonpolar solvents. Lüttringhaus and Ziegler (1937) first described them and their name is attributed to their ability to "crown" metal cations (Pedersen, 1967).

The combination of a large cyclic ether cation and a large anion results in complexes with low melting points and high decomposition temperatures. This idea has been applied to the synthesis of a range of crown ether complex ionic liquids (CECILs) (Song et al., 2011). These ionic liquids were made by mixing equimolar quantities of 15-crown-5, or 18-crown-6, with organic or inorganic salts. Of the 17 CECILs described about half of them have melting points below 100°C, with [15-crown-5Na][OH] having a melting point at −54°C. The same group later developed aCECILs (aza-crown ether analogues), for example, Figure 2.4. The aCECILs all display melting points below 0°C with decomposition temperatures as high as 245°C (Song et al., 2014). The design of these ILs is such that they are protic and possess dicationic and dianionic centers. By judicious choice of the acid and metal salt the anions can be altered, effectively making these aCECILs highly tunable.

Ring-Opening Reactions

Yoshizawa et al. (2001) have pursued zwitterionic imidazolium sulfonates as electrolytes for developing conductive polymeric materials. The molten

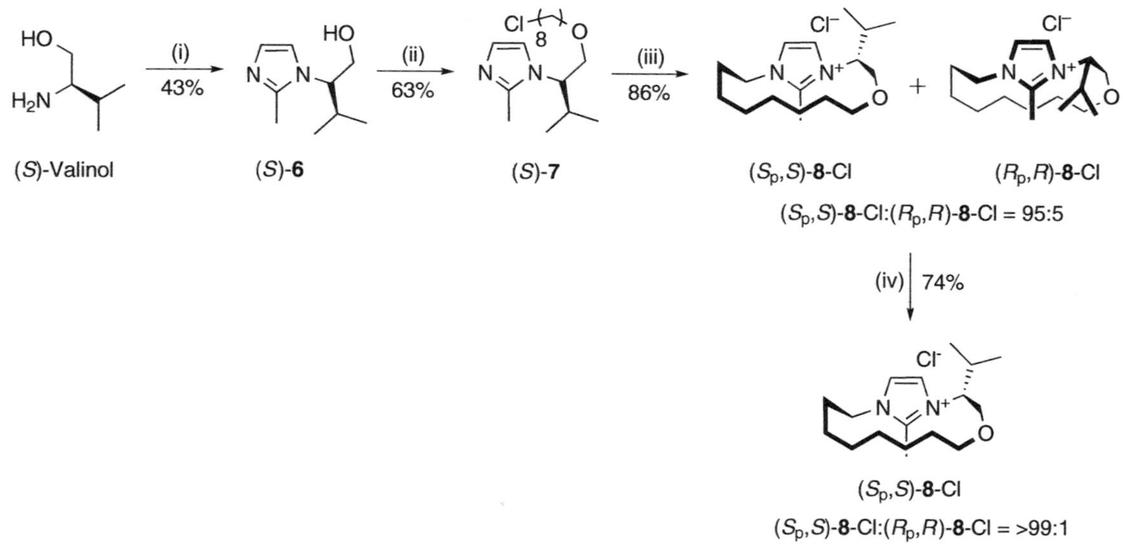

Figure 2.3 ***Synthesis of the Planar–Chiral Imidazolium Salt (S_p,S)-8-Cl.*** Reagents and conditions: (i) CH_3CHO (5.0 equiv), glyoxal (1.0 equiv), NH_4MeCO_2 (1.0 equiv), MeOH, rt; (ii) NaH (1.2 equiv), 1,8-dichlorooctane (3.0 equiv), DMF, 0°C then 50°C; (iii) DMAc, 150°C; (iv) recrystallization from CH_2Cl_2/AcOEt. *Adapted with permission from Ishida et al. (2006). Copyright (2006), Elsevier.*

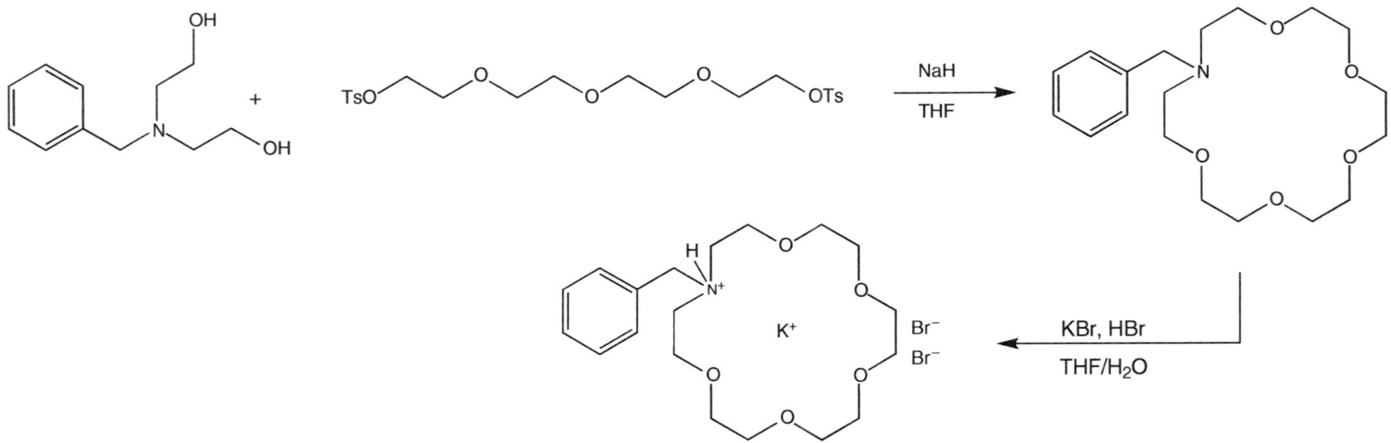

Figure 2.4 ***Illustrative Synthesis of an Aza-Crown Ether Based Ionic Liquid.*** *Reprinted with permission from Song et al. (2011). Copyright (2014), Wiley-VCH Verlag GmbH & Co. KGaA, Weinheim.*

salts were obtained via the ring-opening reaction of the amine base on the sulfone to give a covalently tethered sulfonate. By the addition of an equimolar amount of an acid, with sufficiently low pK_a, the sulfonate was converted to the corresponding Brønsted acid (Cole et al., 2002). The two-step synthesis was quantitative making the production of these ionic liquids 100% atom efficient. The use of these sulfonic acid functionalized ionic liquids have become pervasive in synthesis and material chemistry due to their ease of manufacture, metal-free nature, and recyclability (Amarasekara and Owereh, 2010; Cheng and Jing, 2014; Han et al., 2012; Peng et al., 2014; Shi et al., 2013; Zhuo et al., 2015).

Acid–Base Neutralization

The neutralization reaction of a Brønsted base and a Brønsted acid affords protic ionic liquids. The key difference between protic ionic liquids (PILs) and Brønsted acidic ionic liquids is that the proton is transferred to the heteroatom of the Brønsted base. In doing so the ionic liquid consists of proton acceptor and proton donor sites (Greaves and Drummond, 2008). The degree of protonation is related to the relative strengths (i.e., pK_a's) of the acids and bases involved, with a stronger acid affording a greater degree of proton-transfer. In an ideal scenario proton transfer will be 100% efficient but this is seldom the case, resulting in the presence of aggregates of acidic and neutral species, and other ions. A Walden plot relates the mobility of the ionic liquid (molar conductivity, Λ) to its fluidity (viscosity, η). On this plot an "ideal" line, determined using aqueous potassium chloride solutions, represents an ideal electrolytic solution where ions are noninteracting, that is, no ion–ion interactions (Figure 2.5) (Yoshizawa et al., 2003). It is often used in combination with ^{1}H NMR and IR spectroscopy to probe the ideality of ionic liquids giving an insight into the composition of the ionic liquid and the degree of protonation (Austen Angell et al., 2012; MacFarlane et al., 2009; Schreiner et al., 2010). In general, the larger the difference in pK_a values between the Brønsted acid and the Brønsted base the closer to ideal Walden behavior and the greater likelihood of efficient protonation.

1,1,3,3-Tetramethylguanidine (TMG), 1,5-diazabicyclo[4.3.0]non-5-ene (DBN), 1,8-diazabicyclo[5.4.0]undec-7-ene (DBU), 7-methyl-1,5,7-triazabicyclo[4.4.0]dec-5-ene (MTBD), and 1,5,7-triazabicyclo[4.4.0]dec-5-ene (TBD) typify organic superbases in having high pK_a's in the range of 19–26 (Ishikawa, 2009). Protic ionic liquids made from these amines are stable toward deprotonation by strong alkaline bases. Indeed, [MTBDH][NTf_2] is stable to a 1N solution of KOH, while in contrast [Et_3NH][NTf_2] degrades

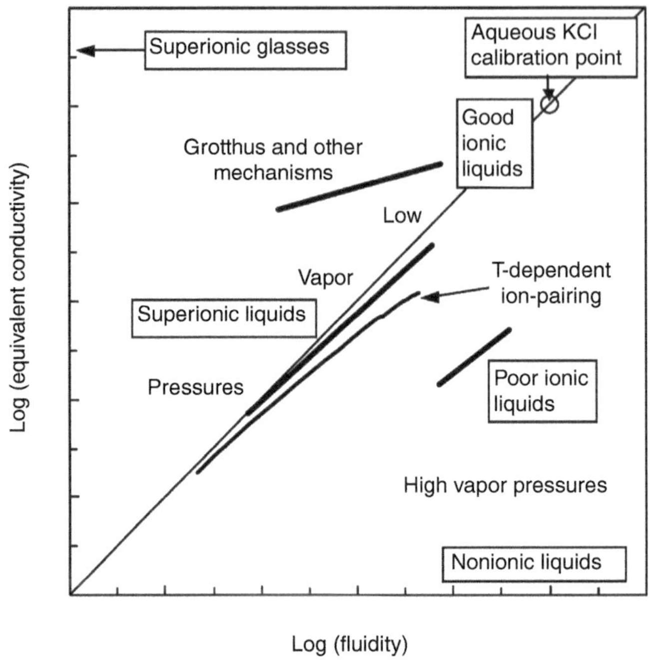

Figure 2.5 ***Classification Walden Plot for Ionic Liquids, and Exemplar Deviations From the "Ideal" Line.*** "Good ionic liquids" are those where the ions are fully dissociated with no (or minimal) ion-ion interaction i.e., the ions act independently from each other. In comparison, "Poor ionic liquids" are those that have a higher degree of ion-ion interaction e.g. ion-pairing or aggregation. *Reprinted with permission from Yoshizawa et al. (2003). Copyright (2003), American Chemical Society.*

readily in this environment (Luo et al., 2009). Some PILs also exhibit enhanced thermal stability and are candidates for high-temperature fuel cells.

In a further application, the bases listed were employed in the preparation of PILs for cellulose dissolution. The PIL [DBNH][$EtCO_2$] was found to possess a viscosity close to, if not lower than, 1-ethyl-3-methylimidazolium acetate, [C_2C_1im][$MeCO_2$]. Its cellulose dissolution activity is also similar – 18 wt% versus 16 wt% of prehydrolysis kraft pulp. Additionally, [DBNH][$EtCO_2$] was found to be distillable at 170°C and 1 mbar, though there was a slight excess of the cation over the propionate anion, indicating fractional distillation of the unpaired conjugated acid and base (Parviainen et al., 2013). The [C_2C_1im]$^+$ cation is not a simple bystander under typical biomass processing conditions at 120°C. Under these conditions the basic acetate anion deprotonates the C-2 position of the imidazolium cation producing a carbene, which reacts with the reducing end of the sugar removing a molecule of formaldehyde (Figure 2.6) (Clough et al., 2015).

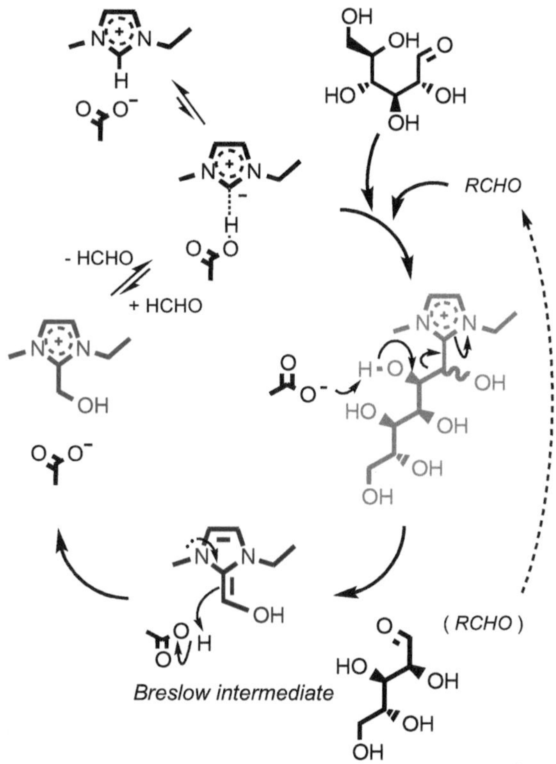

Figure 2.6 ***Proposed Mechanism for the Condensation of [C$_2$C$_1$im][MeCO$_2$] With the Reducing End of D-(+)-Glucose.*** *Reprinted with permission from Clough et al. (2015). Published by The Royal Society of Chemistry.*

The physicochemical attributes of PILs from superbases bodes well for their candidacy as replacements for 1-ethyl-3-methylimidazolium acetate, [C_2C_1im][$MeCO_2$], in biomass processing as the change in the cation inhibits carbene formation and prevents destruction of the cellulose.

The substitution of the C-2 position of the [C_2C_1im]$^+$ cation with a methyl group, while removing the possibility of carbene formation and subsequent degradation of cellulose, increases the melting point, viscosity, and thermal instability of the ionic liquid (Clough et al., 2013).

A cost-based analysis of the simple PILs triethylammonium hydrogensulfate ([Et_3NH][HSO_4]) and 1-methylimidazolium hydrogensulfate ([HC_1im][HSO_4]) was performed by Chen et al. (2014) as a means of determining their practicability of replacing typical organic solvents. Similar to the superbase PILs given earlier, these ionic liquids are made in a single step by the mixing of the base and acid in equimolar ratios. The cost of production of [Et_3NH][HSO_4] and [HC_1im][HSO_4] were estimated at \$1.24 kg^{-1}

and \$2.96 kg^{-1}, respectively. This analysis proves that these ionic liquids can indeed be price-competitive with existing molecular solvents such as ethyl acetate, acetone, and may even be cheaper than toluene. The cost efficiency of ionic liquids is described further in Chapter 6 of this book.

Anion Reactions

The properties of ionic liquids can be fine-tuned by reactions involving the anion counterpart of the salt through the addition of metal salts to give Lewis acidic ionic liquids and the metathesis of the anion to afford a new ionic liquid.

Lewis Acidic Ionic Liquids

In 1951, Frank H. Hurley and Thomas P. Wier, Jr. reported one of the earliest mentions of a Lewis acidic ionic liquid formed by the mixing of ethylpyridinium bromide and $AlCl_3$. They were able to electrodeposit aluminum from a 1:2 mixture of ethylpyridinium bromide-$AlCl_3$ observing it as gray shiny plates (Hurley and Wier, 1951). A series of species in equilibrium can be observed as the ionic liquid mixture takes on a more acidic character by varying the molar ratio of the dissolved $AlCl_3$ (Robinson and Osteryoung, 1979):

$$[C_4py]Cl + AlCl_3 \rightleftharpoons [C_4py]AlCl_4^-$$
$$[C_4py]AlCl_4^- + AlCl_3 \rightleftharpoons [C_4py]Al_2Cl_7^-$$

C_4 = denotes the alkyl chain length; py = pyridinium

Acidic ionic liquids such as these, where the parent ionic liquid and metal salt can be varied to suit the application, produce a considerable exotherm when the components are mixed. Additionally, they tend to be moisture sensitive and would be best prepared in a glove box. The metal salt can be added in portions allowing the heat to dissipate before adding further quantities. Dissipating the heat adequately is extremely important in ensuring there is no thermal degradation/decomposition of the ionic liquid. Alternatively, the reaction can be carried out under Schlenk conditions in a fume hood using a suitable solvent as a heat sink, coupled with a cooling bath. At the end of the reaction the solvent can be removed under inert conditions and the remaining ionic liquid dried under vacuum and stored in a glove box.

In replacing the imidazolium cation and reactive aluminum chloride with simple quaternary ammonium salts and zinc chloride respectively, Abbott et al. (2001) have prepared low-cost moisture-stable ionic liquids. It was found that the functionalized quaternary ammonium salt, choline chloride, gives ionic liquids with freezing temperatures of 23–25°C. These ionic liquids were prepared by mixing the components in air and heating the mixture to 150°C, in a manner contradictory to that of the chloroaluminate ionic liquids. These reactions appear to be endothermic and the resulting ionic liquids are as stable as the chloroaluminates.

From these early days, a large number of transition metal salts of iron (Clark et al., 2015; De Pedro et al., 2011; Funasako et al., 2013; Gao et al., 2014; Lee et al., 2007; Li et al., 2009; Nacham et al., 2015), manganese (Del Sesto et al., 2008; Pitula and Mudring, 2010), cobalt/chromium (Branco et al., 2011; Chen et al., 2012), terbium (Getsis and Mudring, 2011), and copper/nickel (Funasako et al., 2012) have been used to make Lewis acidic ionic liquids. They encompass a wide range of properties and applications from switchable luminescence, extraction of DNA from aqueous solutions, and desulfurization of fuels, but their magnetic properties are of extreme interest (Santos et al., 2014).

Anion Metathesis

The inherent insolubility of silver halides was recognized early on to be an efficient way of expanding the number of accessible ionic liquids through anion metathesis. The ionic liquids $[C_2C_1im][NO_3]$, $[C_2C_1im][NO_2]$, $[C_2C_1im][BF_4]$, $[C_2C_1im][MeCO_2]$, $[C_2C_1im]_2[SO_4]$-H_2O were prepared by mixing $[C_2C_1im]I$ and the corresponding silver salt in a methanol or methanol–water mixture at room temperature. The products obtained were of high purity and represented some of the earliest air and water stable ionic liquids (Wilkes and Zaworotko, 1992).

Since then anion metathesis has become one of the most utilized methods for the synthesis of ionic liquids. This is one of the simplest methods for making ionic liquids on a laboratory scale and it demands very little when it comes to laboratory equipment. As such there are several examples and procedures in the literature for how to make a particular ionic liquid but in general the procedure does not vary significantly. The common theme for all the procedures is to separate the desired ionic liquid from the byproduct salt. This is usually done via phase transfer and the exact method used varies depending on if the desired ionic liquid exhibits hydrophobic or hydrophilic properties.

If the desired IL is hydrophobic then water can be used as a solvent (Bonhôte et al., 1996). In this case a stoichiometric amount of the two salts are mixed together in a water solution and are allowed to stir for up to 48 h. As the desired IL is formed it will separate out of the aqueous phase into a separate ionic liquid (or organic) layer, or alternately precipitate out depending on the ionic liquid's melting point. This layer, or precipitate, can then be easily separated and the aqueous phase washed with an appropriate organic solvent like dichloromethane to extract any remaining ionic liquid.

For hydrophilic ILs the procedure is a bit more challenging and it varies depending on the properties of the by-product salt. In cases like these the solvent used should be water-immiscible but it should also be able to dissolve the desired hydrophilic ionic liquid. Furthermore, the by-product salt that is formed should preferably be an inorganic alkali metal halide salt due to the fact that they have, in general, a very low solubility in organic solvents and a high affinity for water which makes their removal easier. Again, a general procedure for this involves suspending/dissolving a stoichiometric amount of the two salts in a water-immiscible solvent such as dichloromethane and stirring for up to 48 h (Cammarata et al., 2001). As the salt is formed it precipitates out of the liquid phase and can easily be filtered off. The filtrate is then washed several times with a minimal volume of water to remove the last traces of the salt. The challenge with the aqueous washing is to only extract the by-product salt and not the ionic liquid. This is a tedious process and can, in many cases, preferably be done using a continuous extractor.

1-Ethyl-3-methylimidazolium hexafluorophosphate, $[C_2C_1im][PF_6]$, was one of the first water-immiscible ionic liquid to be described employing anion metathesis. It was made by combining equimolar amounts of $[C_2C_1im]Cl$ and 60% aqueous HPF_6 in water over 2 h in an ice bath (Fuller et al., 1994). It is not always necessary to use a strong acid; the main requirement is that the pK_a difference between the acid used and the acid formed should be sufficiently large enough so that complete deprotonation occurs.

The nature of the cation plays a significant role in the water solubility of the ionic liquid. While the low partition coefficient of $[C_2C_1im][PF_6]$ in a water–DCM mixture prevents its extraction, Lancaster et al. (2001) were able to partition $[C_4C_1im][BF_4]$ from water into DCM in good yields (69%). The degree of symmetry in the cation decreases as the alkyl chain length increases. This reduces the association of the ion pairs and so influences water solubility. Holbrey and Seddon (1999a) synthesized 1-alkyl-3-methylimidazolium tetrafluoroborate ionic liquids $[C_nC_1im][BF_4]$ ($n = 0$–18) to analyze their physical properties. All the ionic liquids in the series were synthesized from aqueous solutions of the respective chloride and bromide 1-alkyl-3-methylimidazolium salts with either HBF_4, $NaBF_4$, or $AgBF_4$. Those with chain lengths less than $n = 3$, due to their low water:DCM partition coefficient, were not recoverable from the aqueous phase. Their synthesis relied on metathesis of the halide salts with $AgBF_4$ and the insolubility of the formed silver(I) halide salts. $[C_5C_1im][BF_4]$ can also be partitioned into DCM while those of $n = 6–10$ separates from the aqueous phase into a viscous dense oil aiding purification and recovery giving yields above 90%. Those of chain lengths greater than $n = 10$ were recovered as solids and readily recrystallized from methanol, as low melting solids, in good yields.

Though silver(I) salts (Crawford et al., 2009; Larsen et al., 2000; MacFarlane et al., 2001; Wilkes and Zaworotko, 1992) have proven to be very efficient surrogates for the preparation of a wide range of ionic liquids. However, their prohibitive cost and the significant heavy metal waste produced means that they are not viable for use in large-scale ionic liquid production. The large-scale synthesis of water-immiscible ionic liquids can be readily achieved using alkali metal (Na, K, Li) salts with the desired anion (e.g., $[C_2B_{10}H_{11}]^-$,$[C(CN_3)]^-$, $[B(CN_4)]^-$, $[NTf_2]^-$, and $[Azolate]^-$) (Bonhôte et al., 1996; Marszalek et al., 2011; Matsumi et al., 2009; Nieuwenhuyzen et al., 2009; Smiglak et al., 2010). There are also cases where cheaper organic salts can be used. Ionic liquids, such as $[C_2C_1im][BF_4]$, which previously have been made using $[C_2C_1im]I$ and $Ag[BF_4]$, can be synthesized using $[NH_4][BF_4]$ in an acetone solution (Fuller, 1997).

Employing an appropriate cation, as mentioned earlier, the ionic liquid readily partitions into a suitable organic solvent and the alkali metal salt is washed into the aqueous layer giving a pure, and halide-free, product. The negative charge distribution over the atoms of the anion reduces its coordination capability. This effectively reduces the Coulombic interaction with the cation and increases the hydrophobicity of the ionic liquid.

Halides are much harder to eliminate from water-soluble ionic liquids. Some researchers have relied on the relative insolubility of the alkali metal halides in dry media such as DCM, but the salts are measurably more soluble in DCM/ionic liquid mixtures. The ionic liquid 1-butyl-3-methylimidazolium thiocyanate, $[C_4C_1im][SCN]$, was prepared by the anion exchange of 1-butyl-3-methylimidazolium chloride $[C_4C_1im]Cl$ with KSCN in acetone for 48 h at room temperature (Kamal and Chouhan, 2005). The mixture was filtered to remove the halide salt, potassium chloride (KCl), then concentrated and taken up in DCM before cooling to precipitate residual KCl. Though their analysis of the ionic liquid by mass spectrometry showed no presence of chloride ions, it is generally accepted that this method is ineffective in removing the last vestiges of this ion. Ion chromatography and ion selective electrodes are normally used when a quantitative measure of the impurity is required.

The basic ionic liquids (BILs), 1-butyl-3-methylimidazolium hydroxide ($[C_4C_1im][OH]$), 1-butyl-3-methylimidazolium phosphate ($[C_4C_1im][PO_4]$), 1-butyl-3-methylimidazolium carbonate ($[C_4C_1im]_2[CO_3]$), and 1-butyl-3-methylimidazolium bicarbonate ($[C_4C_1im][HCO_3]$) were made in a similar fashion (Choi et al., 2011; Deng et al., 2012). Here, no mention was made of any qualitative halide test being performed. In their synthesis of $[C_6C_1im][HSO_4]$ and $[C_4C_1im][H_2PO_4]$ Fraga-Dubreuil et al. (2002) removed the chloride ion *in situ* upon treatment of the 1-alkyl-3-methylimidazolium chloride with concentrated sulfuric and phosphoric acids. The chloride ion was distilled as hydrogen chloride gas and monitored by titrating against a sodium hydroxide solution of known concentration. In an analogous method, Priede et al. (2014) obtained chloride-free dimethylphosphate ionic liquids via the metastatic release of chloromethane from a reaction of imidazolium, cholinium, and pyrrolidinium chlorides and trimethylphosphate. Researchers in Belgium described a "phenolate platform" for the synthesis of water-soluble ionic liquids. This principle relies first on preparing the 4-*tert*-butylphenolate salt of the quaternary amine followed by reaction with a Brønsted acid to afford the desired ionic liquid. This method was applied to the synthesis of ionic liquids of 1-butyl-3-methylimidazolium, tetrabutylammonium, tetrabutylphosphonium, and 1-butyl-1-methylpyrrolidinium cations with a range of anions (acetate, methylsulfate, nitrate, dihydrogen phosphate, nitrate, tosylate, formate, hydrogensulfate, hydrogen dipicolinate, isonicotinate, and picolinate). Their optimized method gave a negative silver(I) nitrate test and a chloride content of between 0.03 and 0.05 mol kg^{-1} by the Volhard titration method (Lethesh et al., 2012).

Ranu and Banerjee (2005) first described the synthesis of 1-butyl-3-methylimidazolium hydroxide, $[C_4C_1im][OH]$. The ionic liquid was concentrated and dried at 90°C for 10 h to give a "pure" product. Yuen et al. (2013) did not agree with this "pure" claim since in the concentrated form and at elevated temperature the imidazolium carbene is readily formed. $[C_4C_1im][OH]$ is however stable at low aqueous concentrations. Given its varied applications, its role in synthesis, and as a useful synthetic intermediate for making a diverse array of ionic liquids, it was vital that it be obtained as a halide-free aqueous solution.

These hydroxide ionic liquids, [Cation][OH], can be synthesized by mixing hydrogensulfate anion ionic liquids, $[Cation][HSO_4]$, with an equal amount of barium hydroxide in water resulting in the formation of barium sulfate, which precipitates out of solution (Clough et al., 2013). However, since the hydroxide anion ionic liquids are very basic and cannot exist in a concentrated state without reacting with themselves, they cannot be further purified after synthesis. This means that the ionic liquid needs to be neutralized immediately without knowing the exact conversion ratio/yield. A situation that creates yet another challenge when it comes to calculating the amount of acid needed to neutralize the hydroxide anion. Many of the weak acids used are not easily separated from their corresponding ionic liquids. It is therefore crucial that the stoichiometry between the hydroxide anion ionic liquid, [Cation][OH], and the acid is as accurate as possible.

It is possible to also use sacrificial anions such as the methyl carbonate anion ($[MeCO_3]^-$) to avoid using ionic liquid halide salts. Further details of which are given later in this chapter.

Ion-Exchange Resin

One of the synthetic difficulties of any metathesis, or neutralization reaction, is the removal of the by-product salt or acid. Even when meticulous precision is taken to ensure the correct stoichiometry between the two salts, or salt and acid, some form of purification will still be required. An ion-exchange reaction never gives a 100% conversion nor is any given salt totally insoluble in a chosen solvent. It is therefore often common to have either acid or salt contamination in ionic liquids made via metathesis reactions – see Section "Impurities and Purification of Ionic Liquids" for further details.

A solution to this problem is to use an ion-exchange resin. These resins do not dissolve in solvents and can therefore be flushed with large volumes of solvent as required. Furthermore, this type of solid phase reaction ensures that only the ionic liquid will remain in solution, which removes the need for additional purification steps. The main factor that needs to be taken into account is the conversion ratio between the precursor ionic liquid and the desired ionic liquid product. If the conversion is not complete the resulting product may be a mixture of both. If this is the case however, it is easily rectified simply by repeatedly running the ionic liquid through the ion-exchange column until the desired conversion ratio is reached.

The resin can be added directly to the precursor ionic liquid in a chosen solvent, and left to react before decanting or filtered off once conversion is complete. This tends to lead to poor conversion and the resin beads can be crushed during stirring causing contamination of the ionic liquid. The benefits of using the resin as a stationary phase is that no crushing of the beads will occur, the column provides a concentration gradient against which

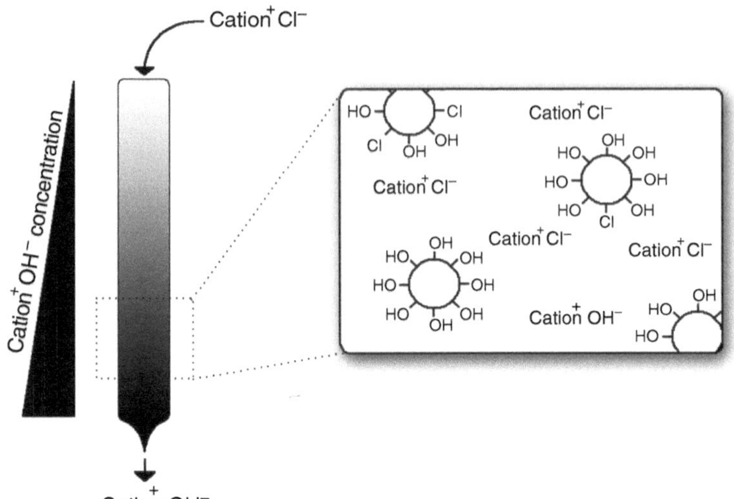

Figure 2.7 ***Schematic Representation of an Ion-Exchange Column.*** As the precursor ionic liquid, [Cation]Cl, passes down the column, it will constantly run in to a higher concentration of $[OH]^-$ resin resulting in a concentration gradient that theoretically will result in a 100% conversion to the corresponding hydroxide ionic liquid. *Reprinted with permission from Gräsvik (2013).*

the ionic liquid is flowing, which results in a more efficient conversion (Figure 2.7), and once the column has been used it can be rejuvenated with the desired anion by simply flushing with a solution of the corresponding metal salt, for example, NaOH or $Na[MeCO_2]$.

This ion-exchange chromatography technique has been used to make $[C_4C_1im][MeCO_2]$ and $[C{=}C_2C_1im][MeCO_2]$ using $[C_4C_1im]Cl$ and $[C{=}C_2C_1im]Cl$, respectively, as precursor ionic liquids (Gräsvik et al., 2014a). A column of Amberlite IRA400OH resin was loaded with $[MeCO_2]^-$ by flushing the column with a 10 equivalents $Na[MeCO_2]$ solution. A 0.75 charge equivalent of the precursor ionic liquid ($[C_4C_1im]Cl$ and $[C{=}C_2C_1im]Cl$) was then run down the column resulting in greater than 99% conversion to the corresponding acetate ionic liquid.

Peng et al. (2009) developed a two-step hybrid method for the production of pure aqueous solutions of $[C_4C_1im][OH]$ (Figure 2.8) involving the use of an ion-exchange column. The first step involves the anion metathesis of $[C_4C_1im]Cl$ with NaOH in dry tetrahydrofuran. This solvent was chosen as others gave colored mixtures. The precipitated sodium chloride was then filtered off and the filtrate concentrated. An aqueous solution of the crude

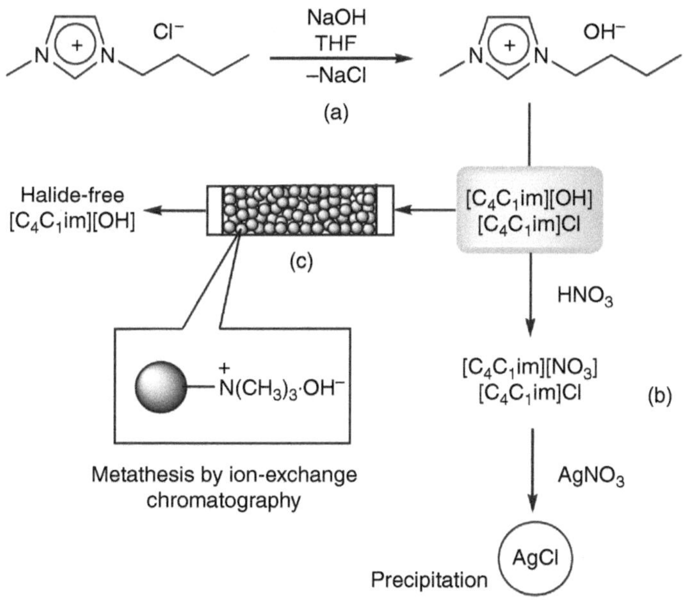

Figure 2.8 ***Synthesis of Halide-Free 1-Butyl-3-Methylimidazolium Hydroxide, [C_4C_1im][OH].*** *Reprinted with permission from Peng et al. (2009). Copyright (2009), Elsevier.*

[C_4C_1im][OH] was then passed through an ion-exchange column charged with the hydroxide anion to fully convert any residual [C_4C_1im]Cl. The product obtained was free from chloride and stored as an aqueous 30–40 wt% solution. As documented, while anion exchange to the hydroxide can be done directly using an ion-exchange resin, the volume needed to produce large-scale quantities is significant and comes with a corresponding financial and time cost. The benefits of this two-step method lies in the quick access to crude [C_4C_1im][OH] and the much smaller volume of ion-exchange resin required as only a small amount of [C_4C_1im]Cl remains to be converted.

With halide-free [C_4C_1im][OH] in hand the authors were able to produce a variety of water-miscible ionic liquids by treatment of the basic ionic liquid solution with ammonium salts. The ammonium hydroxide byproduct can be removed, along with water, by rotary evaporation under vacuum. The use of ammonium salts alleviates the need for silver(I) metal salts, though the ammonia byproduct must be treated before disposal. Furthermore, the unintentional production of HF through the use of unstable and dangerous acids such as HBF_4 and HPF_6 is avoided.

IMPURITIES AND PURIFICATION OF IONIC LIQUIDS

Introduction

With the increased commercial availability of ionic liquids, the number of research groups investing in ionic liquids is ever increasing. For some it is the synthesis of "task-specific" ionic liquids that is their focus, whereas for others it is the determination of their properties.

Does the purity of the ionic liquid matter?

Ultimately the purity level required for the ionic liquid depends upon the application it is going to be used for. For many applications a simple cost (which in this case is time and effort) benefit (level of purity) analysis can be undertaken. However, this question must be asked, and answered, before undertaking the synthesis or attempting to purify ionic liquids that have already been produced.

Academically, most research groups will always aim to have "ultra-pure" ionic liquids so that they are sure the property/effect being investigated is due to the ionic liquid itself and not an impurity present within it.

For example, if the liquid is to be used with an optical spectroscopy based method then the requirement for the purity level is for it to be optically pure. One of the authors synthesized 1-butylimidazolium hydrogen sulfate, $[HC_4im][HSO_4]$, which was found to have an almost unnoticeable green tint to it. For the majority of applications this would have not been an issue. In this case however, the ionic liquid was to be used for an acidity study using the UV–Vis spectra of Hammett acidity dyes. This green tint blocked out the lower wavelengths of the spectra therefore inhibiting the measurement being performed.

Other examples include: the use of the ionic liquid in electrochemical systems where the presence of halide ions, or indeed any free and unassociated ions, within the liquid can be detected and inhibit the process being investigated; silver ions left over (depending on your chosen synthetic route) could have a detrimental effect on biological-based experiments due to their antibacterial properties (Dorjnamjin et al., 2008); even "simple" water contamination, when undertaking biomass pretreatment, shows that in the presence of 1–3 wt% water the ability of 1-ethyl-3-methylimidazolium acetate, $[C_2C_1im][MeCO_2]$, to dissolve cellulose is significantly inhibited (Padmanabhan et al., 2011).

The impurities that are most troublesome, however, are those that have a significant impact on the physical properties of an ionic liquid but are not immediately noticeable. This can result in physical properties of the same ionic liquid being reported that contradict one another. For example, the ionic liquid $[C_2C_1im][BF_4]$ has a melting point value, in two different publications, which differs by almost 10°C (Holbrey and Seddon, 1999b; Wilkes and Zaworotko, 1992), and $[C_4C_1im][PF_6]$ has been published with a viscosity value deviation of 140 cP (mPa·s) (Branco et al., 2002; Suarez et al., 1998).

This lack of consistent data is of major concern from an industrial point of view and impedes the implementation of some ionic liquids in industrial processes (Buzzeo et al., 2004). Therefore, it is of particular importance that as pure ionic liquids as possible are used when mapping their physical properties, and that not only their purity is reported but also what kind of impurities are present.

Ionic Liquid Impurities

The impurities present in ionic liquids can be broadly classed into three categories:

1. Contamination from the starting materials:
   Impurities that were in the starting materials before the reaction, for example, side products from their production, hydrochloric acid in the case of most chloride-based alkanes, or their corresponding alkyl-alcohol.
2. Contamination from the undertaken reactions:
   a. Side products from alkylation steps, for example, elimination reactions;
   b. Decomposition upon heating;
   c. Starting materials, for example, unreacted methylimidazole, alkylating agents, acids/bases or salts;
   d. Left over by-products, for example, alkali metal halide salts.
3. Unwanted solvents:
   These are present from the reactions undertaken. Solvents, or antisolvents, are used during the synthesis and processing steps of the ionic liquid without directly reacting with the liquid.

Solvents are often used in excess and therefore the removal of solvents is a different challenge when compared to the removal of small quantities of other impurities. This is key when the ionic liquid is being used in a process where recycling of the ionic liquid (up to possibly 50 times) is occurring (Plechkova and Seddon, 2008). Purification of recycled ionic liquids is not being considered within this chapter however, it is described within Chapter 5 of this book.

Purification can be undertaken in two ways to minimize the listed impurities present in the final ionic liquid:

1. Purification of the final ionic liquid product, or
2. Purification of the ionic liquid during each stage of the synthesis.

Various techniques can be used to do this and what follows are the standard purification methods that can be used to treat the ionic liquid and its precursors, the impurities common at each stage, and the detection methods for them that are currently used.

Purification of Starting Materials

Before the synthesis of the ionic liquid even begins the impurities in the starting materials have to be considered. As shown by the ever popular and expanding "Purification of Laboratory Chemicals" book by Armarego and Chai (2013), the purification of even simple chemicals can be complex and time-consuming.

All starting materials should be purified before use to remove any impurities present before starting any reactions with them. If an alkylation reaction is being performed with a haloalkane the main impurity removed from the starting materials is normally water. Furthermore, starting materials can degrade over time and therefore they should be used as soon as they are purified. It should be assumed that any older than a week or so will need to be purified again.

For acid–base neutralization reactions, where an acid is added to neutralize the base therefore forming the ionic liquid directly, the acid concentration must first be determined. This is performed by calculating a titration curve against a base of a known concentration.

Purification of Ionic Liquid Halide Salts

For most ionic liquids the cation is formed via an alkylation reaction with a haloalkane, for example, Figure 2.9.

The impurities present in the ionic liquid halide salts at this stage are normally unreacted starting materials and side-reaction by-products as there are very few reactions that give a 100% yield.

Short-chain haloalkanes have low boiling points, for example, 1-chlorobutane has a boiling point of 78°C, and can be removed by simple heating under vacuum or by liquid–liquid extraction using a nonpolar solvent. In

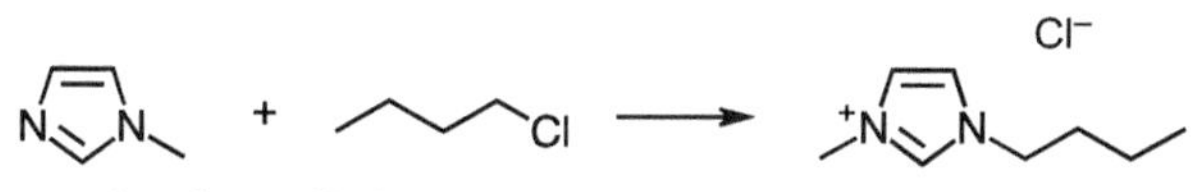

Figure 2.9 ***Example of an Alkylation Reaction.*** Alkylation of 1-methylimidazole with 1-chlorobutane to form 1-butyl-3-methylimidazolium chloride, $[C_4C_1im]Cl$.

comparison, amines possess a high affinity toward the ionic liquid, and in many cases have high boiling points, for example, 1-methylimidazole's boiling point is 198°C, which makes them difficult to quantitatively remove using liquid extractive methods or vacuum treatment. Therefore, when undertaking the alkylation step it should be performed using a small molar excess of the alkylating agent to minimize the unreacted amine/phosphine/sulfone present in the intermediate.

When the halide salt is a viscous gel-like solid, multiple washing steps should first be performed. In some cases it is the presence of unreacted starting materials, which means that the halide salt does not crystallize. The multiple washing steps remove the unreacted starting materials and this can cause the crystallization of the salt to occur. If the halide salt does not crystallize the best purification method is to simply continue washing until the ^1H NMR spectrum is shown to be free of unreacted starting materials.

The standard purification approach for a solid is recrystallization from a suitable solvent. The basis of this purification method is that the solubility of the desired product and the impurities are different in the chosen solvent or the mixture of different solvents being used (Furniss et al., 1989). Dry acetonitrile is used in the majority of cases, with prolonged cooling to −21°C, for the recrystallization of ionic liquid halide salts. If this does not result in crystallization dry ethyl acetate can slowly be added as an anti-solvent until crystallization commences. Multiple recrystallizations may be required to ensure the highest purity of intermediate possible. Figure 2.10 shows the increase in optical purity within the 400–600 nm wavelength region when 1-ethyl-3-methylimidazolium bromide, $[C_2C_1im]Br$, was recrystallized.

Once the ionic liquid halide salt has been washed and recrystallized (if possible or appropriate) it should be dried under high vacuum with minimal heating. This ensures that any remaining, and unwanted, solvents are removed.

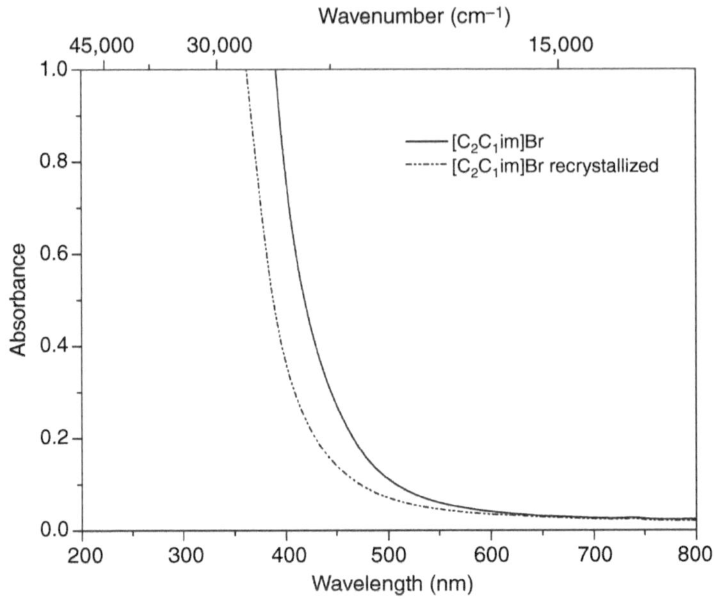

Figure 2.10 Absorption spectrum of $[C_2C_1im]Br$ synthesized at low temperature (thick black line) and after recrystallization (dotted line) in a quartz cuvette, path length of 1 cm. *Reprinted with permission from Nockemann et al. (2005). Copyright (2005), Elsevier.*

The presence of unreacted starting materials is easily detectable by NMR spectroscopy with a limit of detection of around 1 mol%. Trace amount of 1-methylimidazole can be detected through the addition of a copper(II) solution and colorimetric detection (Holbrey et al., 2001). It has also been shown that zone electrophoresis can be used to detect, quantify, and purify ionic liquids with trace amounts of imidazole contaminants (Markuszewski et al., 2004). Moreover, it is possible to use the standard addition method coupled with gas chromatography (GC) to determine the 1-methylimidazole contamination concentration (Stark et al., 2008).

It has been reported that activated charcoal could be used at this stage to aid in the decolorization and removal of impurities from ionic liquid halide salts (Nockemann et al., 2005). The activated charcoal is added to the hot solvent once the crystals have been dissolved; it is then filtered while hot before allowing the halide salt to recrystallize from the cooling filtrate. However, the effectiveness of the activated charcoal when used with ionic liquid halide salts has been questioned by Seddon and Welton (Earle et al., 2007). They found that ionic liquids with strongly hydrogen bonding ionic liquid anions (or if the ionic liquid cation were to interact strongly with the charcoal, i.e., carbon) were not effectively decolorized. Hence, they recommend that activated charcoal only be used with the final ionic liquid product.

If the starting materials have been purified, and the reaction temperature kept low, the degree of coloration will have been minimized and the recrystallization step (or steps) will have ensured that coloration of the ionic liquid halide salt is limited. This avoids the requirement for any activated charcoal treatment.

Where halide impurities could be detrimental to a process or experiment, for example, use in electrochemistry, and the conjugate acid of the desired ionic liquid anion exists it may be possible to use the methyl carbonate route discovered by Rogers and coworkers (Holbrey et al., 2003).

Figure 2.11 illustrates the methyl carbonate route where dimethylcarbonate is used to methylate 1-butylimidazole forming 1-butyl-3-methylimidazolium methylcarbonate, $[C_4C_1im][MeCO_3]$. The methyl carbonate anion is then decomposed via the addition of the acid, H[Y], to form carbon dioxide (which drives the equilibrium to the right), methanol, and the desired anion, Y^-. Any remaining methanol can be removed through slight heating and the application of high vacuum. This also has the added benefit of driving the reaction to completion. In the absence of an excess of methanol the zwitterionic carboxylate is formed. Here, the dimethylcarbonate upon alkylation of the imidazole forms the methyl carbonate anion. The anion abstracts the C-2 hydrogen giving a carbene and the unstable MeO(O)COH, which decomposes to methanol and CO_2. Nucleophilic attack by the carbene then forms the zwitterion. Reaction of the zwitterion with an acid, H[Y], causes decarboxylation forming the conjugate base of the acid, Y^-, i.e., the desired anion (Smiglak et al., 2007).

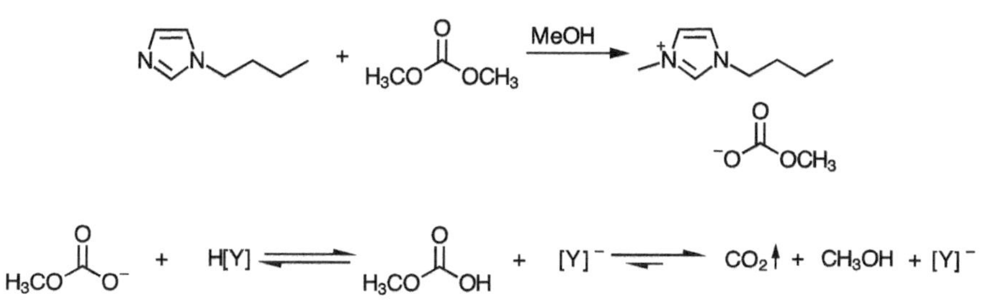

Figure 2.11 Illustrative reaction forming 1-butyl-3-methylimidazolium methylcarbonate $[C_4C_1im][MeCO_3]$ and the decomposition of the methyl carbonate anion via addition of the acid H[Y].

Purification of the Final Ionic Liquid Product

Depending on the ionic liquid being synthesized, the final step may indeed be the only synthetic step being undertaken to form the desired ionic liquid, for example, an acid–base neutralization reaction (Figure 2.12).

In the case of the ionic liquid halide salts the final step is normally to undertake an ion-exchange metathesis reaction (Figure 2.13). The impurities present after the metathesis are usually:

- Unreacted [Cation]X (where X^- is a halide ion: Cl^-, Br^-, or I^-) starting material;
- The inorganic salt (M[Y]) where M is typically an alkali metal, for example, lithium or sodium, and $[Y]^-$ is the desired anion of the ionic liquid, for example, $[NTf_2]^-$, bis(trifluoromethylsulfonyl)imide;
- The produced metal halide salts (M[X]) or ammonium salts ($[NH_4][X]$);
- Acidic impurities (H[X]) when a conjugate acid of the desired anion has been used (H[Y]);
- Organic solvents;
- Water.

With judicial choice of the solvent for the reaction, or different salt analogues, the inorganic metal halide salt or ammonium halide salts will precipitate out of solution. The bulk of this salt can then be removed via simple filtration.

After filtration, aqueous washings using distilled water to remove any remaining salt, in the case of the alkali metal and ammonium salts, from the ionic liquid is normally performed. The aqueous washing has two functions here: (1) to remove the alkali metal or ammonium halide salts and (2) to remove any unreacted ionic liquid halide salt ([Cation]X), which is water soluble, while the IL product remains in the organic solvent (the reaction medium) (Ab Rani et al., 2011). Studies have shown that halide contamination

$$\mathrm{HA} + \mathrm{B} \rightleftharpoons [\mathrm{BH}][\mathrm{A}]$$

Figure 2.12 ***Synthesis of Ionic Liquid [BH][A] From the Brønsted Acid, HA, and Brønsted Base, B.***

Figure 2.13 ***Ionic Liquid Metathesis Route With an Inorganic Metal Salt (M[Y]), a Conjugate Acid of the Desired Anion (H[Y]) or With an Ammonium Salt ($[NH_4][Y]$).*** *Reprinted with permission from Hallett and Welton (2011). Copyright (2007), American Chemical Society.*

Table 2.1 Water solubilities for silver halide salts at 25°C (Holleman and Wiberg, 1985)

Silver halide salt	Solubility in water (mol L^{-1})
AgCl	1.3×10^{-5}
AgBr	7.1×10^{-7}
AgI	1.3×10^{-8}

of the final ionic liquid usually comes from unreacted halide salt rather than the inorganic alkali metal salt (Branco et al., 2002; Seddon et al., 2000).

For hydrophilic ionic liquids a minimum volume of water should be used to ensure that any loss of the ionic liquid is minimal (Hallett and Welton, 2011). The initial filtration aids this, and the washing is continued until the salt is undetectable.

The halide concentration is normally monitored by testing the washings with 0.1 M (aq) silver nitrate ($AgNO_3$), which is simply the first half of the Volhard back-titration method for halide detection (Jeffery et al., 1989). Once treated with silver nitrate it is normally worth waiting for 15 min to ensure no positive result (if a precipitate is not immediately obvious).

The use of silver nitrate is a qualitative indication of the presence of halide ions, and for most applications it is enough to state that the concentration of the halide in the ionic liquid is below that given in Table 2.1. The silver nitrate test is sensitive; however, it is not specific and can give a large number of false positives. For example, hydroxide, sulfate, and carbonate ions also all precipitate out in water when treated with silver nitrate. It should be noted that the silver nitrate test can also be used to detect ammonium halides ($[NH_4][X]$) when an ammonium salt, $[NH_4][Y]$, is used.

For quantitative measurements an ion-selective electrode (Seddon et al., 2000) or ion chromatography can be used, which gives a limit of detection of <10 ppm (Gräsvik et al., 2014a; Stark et al., 2008).

The silver nitrate test highlights one of the other synthetic methods that is possible for ion-exchange metathesis especially for water-miscible ionic liquids – that is the use of silver salts. This metathesis makes use of the low water solubility of the silver halide salts, for example, Figure 2.14. Here, additional [Cation]X is added until no further precipitation of the Ag[X] salt

$$[C_2C_1im]Br_{(aq)} + Ag[MeCO_2]_{(aq)} \longrightarrow [C_2C_1im][MeCO_2]_{(aq)} + AgBr_{(s)}$$

Figure 2.14 Synthesis of 1-ethyl-3-methylimidazolium acetate, $[C_2C_1im][MeCO_2]$, from aqueous solutions of 1-ethyl-3-methylimidazolium bromide, $[C_2C_1im]Br$, and silver acetate, $Ag[MeCO_2]$.

is observed. Standard filtration is used to remove the precipitated salt, and the presence of any remaining silver ions can be tested for by using hydrochloric acid, which forms a silver chloride precipitate. If a silver salt of the desired anion, $[Y]^-$, is not available it can be synthesized from silver nitrate as long as the Ag[Y] salt is insoluble in water. This synthetic route tends to be expensive due to the use of silver. Additionally, due to the submicrometer size of the silver ions there is a possibility of contamination of the ionic liquid even after filtration (Kirchner, 2010).

In the cases where the conjugate acid of the desire anion (H[Y]) has been used, such as the synthesis of 1-butyl-3-methylimidazolium hexafluorophosphate, $[C_4C_1im][PF_6]$ (Fuller et al., 1994), the unreacted acid can be removed through aqueous washing (which also removes any remaining [Cation]X) when the ionic liquid product is hydrophobic. The washings are then periodically tested until their pH is found to be neutral. When the product is hydrophilic unreacted acidic starting materials can be removed via steam distillation (Seddon et al., 2000). Due to the nature of hydrophilic ionic liquids they are not always pH neutral in a water solution and as such pH cannot be solely relied upon as a detection method. Instead the use of a proton-sensitive test is recommended (Wasserscheid and Welton, 2008). At this stage the product can be run over a short pad of neutral, or basic, alumina to aid drying and also to ensure complete removal of acidic starting materials (Williams and Lawton, 2010). To avoid acidic impurities when synthesizing ionic liquids with a $[PF_6]^-$ or $[BF_4]^-$ anion it is recommended to use the sodium or lithium salts of the anion rather than the conjugate acid.

Ionic liquids can be synthesized via the direct equimolar combination of a Brønsted acid, HA, with a Brønsted base, B (Figure 2.12). The neutralization is an exothermic reaction and to minimize the formation of any decomposition product (and with it coloration) the acid should be added in a drop-wise manner to a diluted solution of the base. Due to the equilibrium nature of the reaction, the ionic liquid tends to contain a mixture of unreacted acid or base along with the desired product, [BH][A]. Excess acid can be removed through the use of a basic alumina column, and excess base through the use of an acidic alumina column (Gräsvik et al., 2014b).

At this stage of the synthesis, even with multiple washings, recrystallizations, washings of the final product (depending upon the route taken), and evaporation of solvents, the ionic liquid can be colored. The colored impurities present in the ionic liquid are usually undetectable by most analytical techniques (apart from UV–Vis spectroscopy) and are therefore only in trace quantities. The impurities causing the coloration are likely to be a combination of the following (Wasserscheid and Welton, 2008):

- Thermal decomposition products found when the liquid undergoes unwanted side reactions either by reacting with itself during heating (Clough et al., 2013) or by reacting with other compounds present during the synthesis (Nockemann et al., 2005) or process (Clough et al., 2015) steps.
- Oxidation products.
- Trace impurities from the starting materials.

In their 2007 paper, Seddon and Welton proposed that rapid decolorization of ionic liquids could be undertaken using an activated charcoal column over which the ionic liquid is run (Figure 2.15) (Earle et al., 2007).

Here the ionic liquid, dissolved up in dichloromethane, is run over the charcoal column – aided by a flow of nitrogen over the top of the column. The pad of silica gel acts as a drying agent (Williams and Lawton, 2010) and helps to minimize the charcoal carry-over along with the 1 cm pad of Celite as the bottom phase of the column. The column itself is washed with fresh dichloromethane once the ionic liquid has been eluted to ensure minimal losses. Once the ionic liquid is reisolated the solvent is removed via the use of a rotary evaporator.

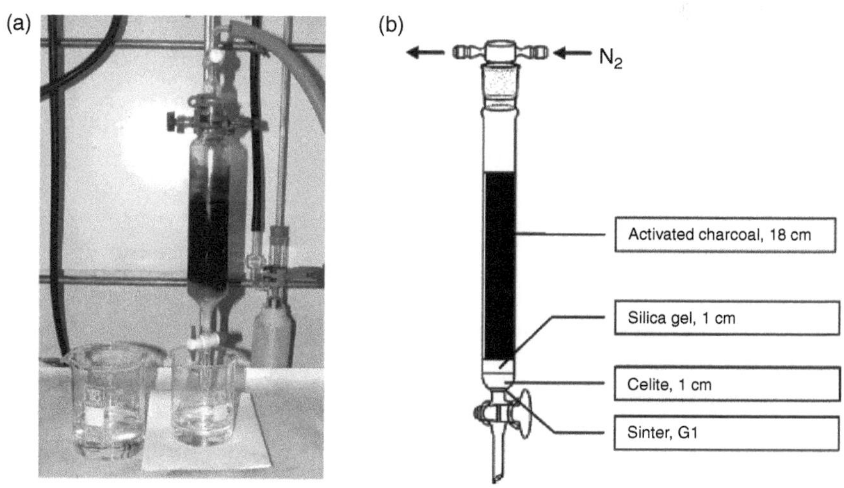

Figure 2.15 Activated charcoal column used to decolorize ionic liquids (a) and the schematic diagram of the column indicating the layers of charcoal, silica, and celite (b). *Reprinted with permission from Earle et al. (2007). Copyright (2007), American Chemical Society.*

This approach works well to decolorize the ionic liquid and can indeed be a relatively quick procedure to undertake. It should be noted that depending on the level of coloration multiple runs of the ionic liquid over fresh charcoal might be required. This is due to the relatively fast movement of the ionic liquid solution over the charcoal column and is also dependent upon the width/length of the column itself. In a number of research groups the ionic liquid is first stirred overnight with the charcoal before being poured into the column, thereby maximizing the charcoal contact time and adsorption of impurities from the ionic liquid.

Decolorization of the ionic liquid can be also attempted with adsorption of the colored impurities through the use of C18 columns (Figure 2.16) or acidic/basic/neutral alumina columns (i.e., a rough chromatographic column).

MacFarlane and coworkers, however, have reported the contamination of ionic liquids with the typical sorbent materials (alumina, silica, and activated charcoal) used for the purification of ionic liquids even after 200 nm filtration (Clare et al., 2008). Figure 2.17 shows the multimodal size distribution of particles in silica-treated 1-hexyl-3-methylimidazolium bromide, $[C_6C_1im]Br$, measured via dynamic light scattering, after filtration through Celite and after 200 nm filtration. This shows that while the number of particles with effective diameters greater than 300 nm is reduced with a 200 nm filtration there is a substantial bimodal size distribution of particles below 300 nm. The average count rate was reported to be 20 before exposure to the silica, and 184 after treatment and 200 nm filtration. Inductively coupled plasma optical emission spectroscopy (ICP-OES) analysis has showed that 200 nm filtration can lower the contamination level from alumina, or silica, to less than 10 ppm.

(a) (b) (c)

Figure 2.16 The ionic liquid 1-butylimidazolium hydrogensulfate, $[HC_4im][HSO_4]$, turned red upon drying (a). By dissolving it in water and running it over a prepacked C18 column (×3) the impurities causing the red coloration were trapped on the column (b) resulting in a clear and colorless liquid (c). The C18 column was then rejuvenated by flushing with IMS or methanol.

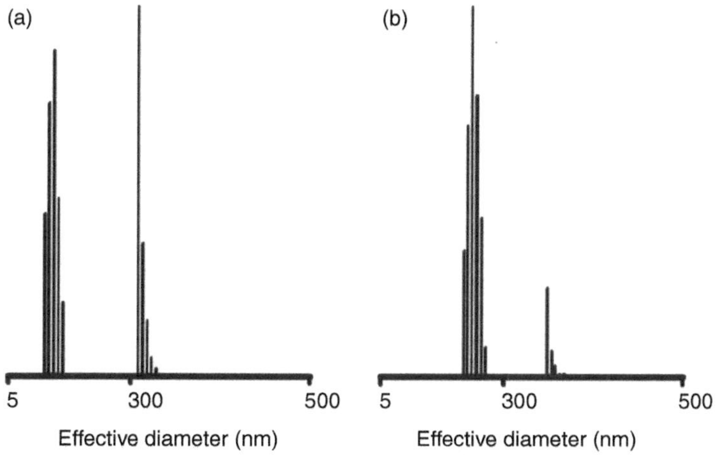

Figure 2.17 Dynamic light scattering determined multimodal size distribution for silica treated [C_6C_1im]Br after filtration through Celite (a) and after 200 nm filtration (b). *Reprinted with permission from Clare et al. (2008), The Royal Society of Chemistry.*

The removal of particles in this size range is notoriously difficult and would require the use of distillation (Earle et al., 2006; Widegren et al., 2007) or zone melting (Choudhury et al., 2005). To avoid this it is recommended that these sorbents be simply avoided. This means, however, that the desired ionic liquid would have to be pure and colorless without their use. Figure 2.18 shows the absorption spectrum for 1-hexyl-3-methylimidazolium bromide, [C_6C_1im]Br, synthesized via standard "high" temperature conditions (65°C), which led to a yellow-colored ionic liquid halide salt, and the spectra after treatment with the typical sorbent materials of alumina, charcoal, and silica in comparison to the low-temperature method (continual cooling with an ice bath). This highlights that it is possible to achieve ionic liquids with, at least, optical purity via simple synthetic and purification methods without the use of sorbent materials.

Distillation

Distillation is a standard purification method used for the purification of chemicals and is likely to have been used to purify the starting materials of the ionic liquid. Due to the low volatile nature of ionic liquids it has been assumed that an ionic liquid cannot be distilled. Earle et al. (2006) demonstrated the distillation of a number of ionic liquids through the use of a Kugelrohr oven and via a sublimation method. They found that even with the use of high temperatures (300°C) decomposition of the ionic liquid, and distillate, was limited. This process, however, was found to be extremely

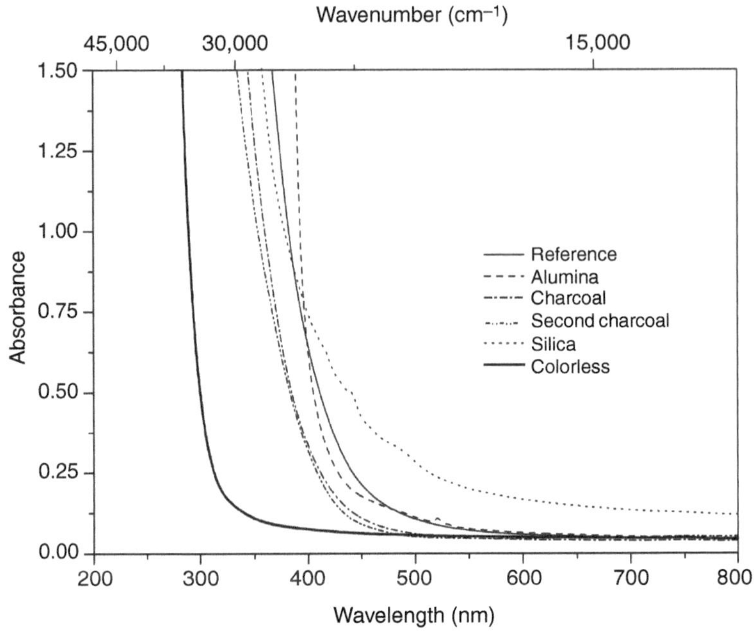

Figure 2.18 ***Absorption Spectra of [C_6C_1im]Br.*** Prepared via standard conditions (reference), after treatment with alumina (dashed line), charcoal (dash dot line), a second treatment with charcoal (dash, dot, dot line), and with silica (dotted line). The thick solid line represents the absorption spectrum of [C_6C_1im]Br synthesized under low temperature conditions without any sorbent treatments. *Reprinted with permission from Nockemann et al. (2005). Copyright (2005), Elsevier.*

slow (0.12 g h^{-1} for [C_2C_1im][NTf_2]) and therefore unfeasible for large volumes. In the case of desiring only a small quantity of ultrapure ionic liquid distillation of the final product should be considered.

Drying of the Ionic Liquid

At the final stage of the synthesis there tends to be one persistent impurity remaining – water. Some ionic liquids will absorb 1–3.5 molar equivalents of water when allowed to equilibrate with the atmosphere (Brandt, 2011), and high affinity to water is shown by hydrophilic carbonate ionic liquids when they release up to 11 kJ mol^{-1} of energy when mixed with water (Brandt et al., 2013).

The drying of the ionic liquid is performed by stirring the ionic liquid while heating under the application of high vacuum. It should be noted that in addition to a correlation between the basicity of an ionic liquid and its thermal stability, there is also a correlation between the basicity of the ionic liquid and its affinity for water absorption (Figure 2.19). This makes the drying of some ionic liquids particularly challenging since those liquids that need the most energy in the form of applied vacuum and heating are also those that are more prone to decomposition at elevated temperatures, for example, those with an acetate anion, $[MeCO_2]^-$.

Ionic liquids have long been looked at as inert and thermally stable but lately research has shown that ionic liquids are not as stable as previously thought. This is especially the case when it comes to long-term degradation or decomposition on the application of heat. Continuing with the example of the acetate anion, 1-ethyl-3-methylimidazolium acetate, $[C_2C_1im][MeCO_2]$ is known to have a decomposition temperature just above 200°C, but recent studies have shown that it still undergoes slow decomposition of approximately 1% every 10 h when heated at 102°C (Clough et al., 2013). This decomposition may occur via either heterocyclic carbene formation, S_N2 nucleophilic substitution, or an E2 Hoffman elimination reaction (Figure 2.20).

In all these cases the decomposition is facilitated by the anion, with the carbene formation and Hoffman elimination controlled by its basicity, and the speed of both controlled by its nucleophilicity. Therefore, the anion will play a major role in the thermal stability of the ionic liquid and will influence the temperatures that an ionic liquid can be heated to when removing water.

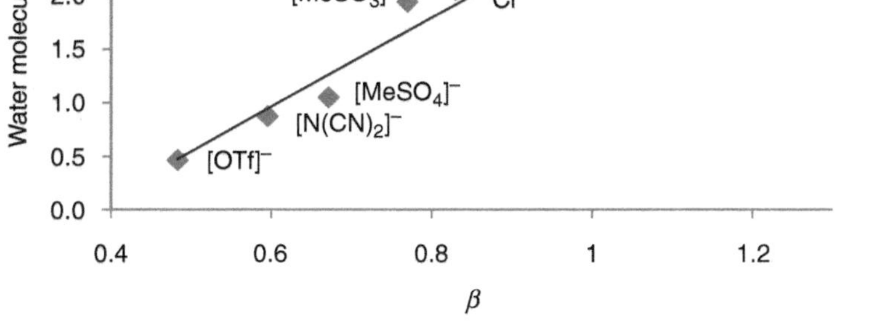

Figure 2.19 Plot illustrating the correlation of the Kamlet–Taft β value (or basicity) of dry 1-butyl-3-methylimidazolium ionic liquids with the number of water molecules associated with each ion pair after exposure to the atmosphere. *Reprinted with permission from Brandt (2011).*

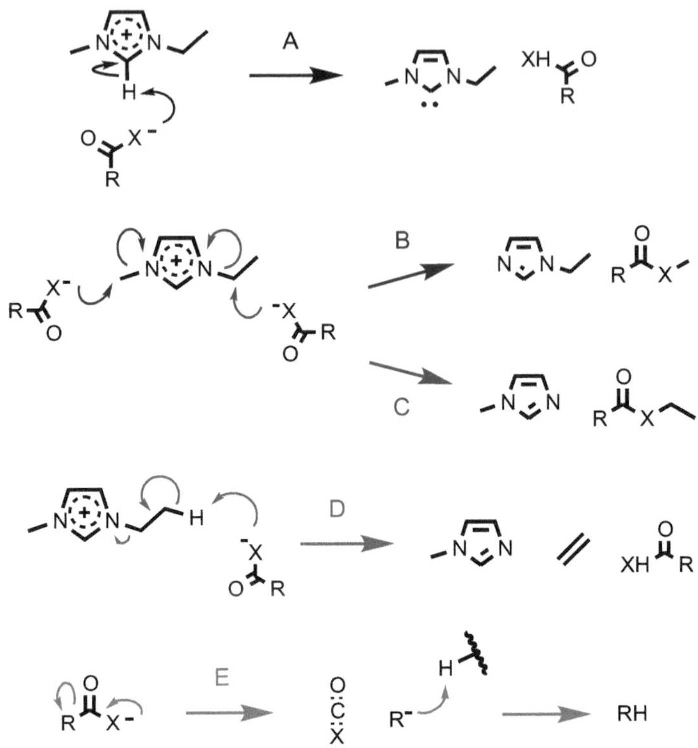

Figure 2.20 ***Postulated Decomposition Mechanisms for [C_2C_1im][$MeCO_2$].*** (A) N-heterocyclic carbene formation; (B and C) S_N2 nucleophilic substitution; (D) E2 Hofmann elimination; (E) decarboxylation of the anion, where R = Me, and X = O. *Reprinted with permission from Clough et al. (2013). Published by the PCCP Owner Societies.*

Due to the uptake of water that is possible, by even hydrophobic ionic liquids, it is recommended that all ionic liquids are dried immediately before use. Where heating is likely to decompose the ionic liquid, azeotropic distillation with toluene or propan-2-ol is possible (Gordon and Ford, 1972). It is also possible to expose the ionic liquid to a di-phosphorous pentoxide environment to aid drying, this has been found to work extremely well for acetate based ionic liquids. The use of molecular sieves is not recommended due to the possibility of contamination from particulates.

The ionic liquid's water content can be easily determined via Karl–Fischer titration, which can detect parts-per-million (ppm) levels of water as long as a large enough sample volume is used (Stark et al., 2008). Water content can also be determined using GC analysis; however, the equipment must be free of moisture from the air to ensure an accurate reading (Stark et al., 2008).

RECOMMENDATIONS

The chosen purification methods or approaches used depend upon the synthetic route undertaken and the impurities that are present, or have been formed, during the synthesis. The recommended approach is to always purify the ionic liquid/ionic liquid precursor after each stage of its production, as this will minimize the build-up of different types of impurities that can complicate the purification process. As a minimum the following is recommended to obtain clear, colorless ionic liquids with low impurities contamination:

1. Freshly purify all starting materials and solvents being used in the synthesis of the ionic liquid.
2. Undertake quaternization in a dry environment using either nitrogen or argon, and heating should be kept to a minimum.
3. Wash and recrystallize any solid products until a colorless mother liquor is obtained.
4. Perform appropriate washing until acidic/ion/salt impurities are removed.
5. Keep the drying temperature for the final product low.

When planning your ionic liquid synthesis a cost–benefit analysis should be undertaken; this must look not only at the purification methods that can be applied but also the synthetic route undertaken.

For example, your ionic liquid is going to be used for an electrochemical process. Avoiding possible ion contamination is therefore a key requirement. This means that a halide-free synthetic route would be preferred. In comparison, if coloration of the ionic liquid will not impact your process, high temperatures can be used to complete reactions faster, and the use of a charcoal purification step can be avoided.

CONCLUSION

What has been discussed here aims to give the reader a brief overview of the synthetic methodologies that are available for the synthesis of a range of the classical ionic liquids while highlighting the new and novel synthetic approaches that are being developed as the field of ionic liquids continues to expand. The "class" of ionic liquids now being synthesized is markedly different to those originally envisaged in the 1990s, and as the possible application areas continue to grow we can but expect an onward push to more varied and exciting ionic liquids – and the synthetic routes to form them.

For some researchers the ability to buy ionic liquids from a number of commercial companies is the best approach, as they may not have the facilities, capabilities, desire, or need to synthesize the required ionic liquid *in-house*. If this is the case we believe that the synthetic route, along with the purification methods used, must be determined before use of the liquid. This will ensure that the researchers have a reasonable chance at identifying the possible impurities (if any) remaining in the ionic liquids they are using and will highlight the possible additional purification methods that can be applied before their use.

We aimed for this chapter to give an overview of the synthesis, purification, impurities, and detection methodologies that can be found in the field of ionic liquids and will end with one consideration, the old adage:

"If you want something done right, do it yourself."

REFERENCES

Ab Rani, M.A., Brant, A., Crowhurst, L., Dolan, A., Lui, M., Hassan, N.H., Wilding, R., 2011. Understanding the polarity of ionic liquids. Phys. Chem. Chem. Phys. 13 (37), 16831–16840.

Abbott, A.P., Capper, G., Davies, D.L., Munro, H.L., Rasheed, R.K., Tambyrajah, V., 2001. Preparation of novel, moisture-stable, Lewis-acidic ionic liquids containing quaternary ammonium salts with functional side chains. Chem. Commun. (19), 2010–2011.

Amarasekara, A.S., Owereh, O.S., 2010. Synthesis of a sulfonic acid functionalized acidic ionic liquid modified silica catalyst and applications in the hydrolysis of cellulose. Catal. Commun. 11 (13), 1072–1075.

Armarego, W.L.F., Chai, C.L.L., 2013. Purification of Laboratory Chemicals, seventh ed. Butterworth-Heinemann, Oxford, UK.

Austen Angell, C., Ansari, Y., Zhao, Z., 2012. Ionic liquids: past, present and future. Faraday Discuss. 154 (November), 9.

Bonhôte, P., Dias, A.-P., Armand, M., Papageorgiou, N., Kalyanasundaram, K., Grätzel, M., 1996. Hydrophobic, highly conductive ambient-temperature molten salts. Inorg. Chem. 35 (5), 1168–1178.

Branco, L.C., Rosa, J.N., Moura Ramos, J.J., Afonso, C.A.M., 2002. Preparation and characterization of new room temperature ionic liquids. Chemistry 8 (16), 3671–3677.

Branco, A., Branco, L.C., Pina, F., 2011. Electrochromic and magnetic ionic liquids. Chem. Commun. 47 (8), 2300–2302.

Brandt, A., 2011. Ionic liquid pretreatment of lignocellulosic biomass. PhD Thesis. Imperial College London.

Brandt, A., Gräsvik, J., Hallett, J.P., Welton, T., 2013. Deconstruction of lignocellulosic biomass with ionic liquids. Green Chem. 15 (3), 550–583.

Buzzeo, M.C., Evans, R.G., Compton, R.G., 2004. Non-haloaluminate room-temperature ionic liquids in electrochemistry – A review. ChemPhysChem 5 (8), 1106–1120.

Cammarata, L., Kazarian, S.G., Salter, P.A., Welton, T., 2001. Molecular states of water in room temperature ionic liquids. Phys. Chem. Chem. Phys. 3 (23), 5192–5200.

Chatel, G., MacFarlane, D.R., 2014. Ionic liquids and ultrasound in combination: synergies and challenges. Chem. Soc. Rev. 43 (23), 8132–8149.

Chen, X., Song, D., Asumana, C., Yu, G., 2012. Deep oxidative desulfurization of diesel fuels by Lewis acidic ionic liquids based on 1-*n*-butyl-3-methylimidazolium metal chloride. J. Mol. Catal. A Chem. 359, 8–13.

Chen, L., Sharifzadeh, M., Mac Dowell, N., Welton, T., Shah, N., Hallett, J.P., 2014. Inexpensive ionic liquids: $[HSO4]^-$-based solvent production at bulk scale. Green Chem. 16 (6), 3098.

Cheng, C., Jing, H., 2014. Brønsted acidic ionic liquids of aza-crown ether complex cations: preparation and applications in organic reactions. RSC Adv. 4 (65), 34325–34331.

Choi, Y.S., Shim, Y.N., Lee, J., Yoon, J.H., Hong, C.S., Cheong, M., Lee, J.S., et al., 2011. Ionic liquids as benign catalysts for the carbonylation of amines to formamides. Appl. Catal. A General 404, 87–92.

Choudhury, A.R., Winterton, N., Steiner, A., Cooper, A.I., Johnson, K.A., 2005. *In situ* crystallization of low-melting ionic liquids. J. Am. Chem. Soc. 127 (48), 16792–16793.

Clare, B.R., Bayley, P.M., Best, A.S., Forsyth, M., MacFarlane, D.R., 2008. Purification or contamination? The effect of sorbents on ionic liquids. Chem. Commun. 23, 2689–2691.

Clark, K.D., Nacham, O., Yu, H., Li, T., Yamsek, M.M., Ronning, D.R., Anderson, J.L., 2015. Extraction of DNA by magnetic ionic liquids: tunable solvents for rapid and selective DNA analysis. Anal. Chem. 87 (3), 1552–1559.

Clough, M.T., Geyer, K., Hunt, P.A., Mertes, J., Welton, T., 2013. Thermal decomposition of carboxylate ionic liquids: trends and mechanisms. Phys. Chem. Chem. Phys. 15 (47), 20480–20495.

Clough, M.T., Geyer, K., Hunt, P.A., Son, S., Vagt, U., Welton, T., 2015. Ionic liquids: not always innocent solvents for cellulose. Green Chem. 17 (1), 231–243.

Cole, A.C., Jensen, J.L., Ntai, I., Weaver, K.J., Forbes, D.C., Tran, K.L., et al., 2002. Novel Brønsted Acidic Ionic Liquids and Their Use as Dual Solvent – Catalysts. Am. Chem. Soc. 124, 5962–5963.

Crawford, M.J., Karaghiosoff, K., Klapötke, T.M., Martin, F.A., 2009. Synthesis and characterization of 4,5-dicyano-2*H*-1,2,3-triazole and its sodium, ammonium, and guanidinium salts. Inorg. Chem. 48 (4), 1731–1743.

De Pedro, I., Rojas, D.P., Blanco, J.A., Fernández, J.R., 2011. Antiferromagnetic ordering in magnetic ionic liquid Emim[FeCl4]. J. Magn. Magn. Mater. 323 (10), 1254–1257.

Del Sesto, R.E., McCleskey, T.M., Burrell, A.K., Baker, G.A., Thompson, J.D., Scott, B.L., Williams, P., et al., 2008. Structure and magnetic behavior of transition metal based ionic liquids. Chem. Commun. 4, 447–449.

Deng, Z., Guo, J., Qiu, L., Zhou, Y., Xia, L., Yan, F., 2012. Basic ionic liquids: a new type of ligand and catalyst for the AGET ATRP of methyl methacrylate. Polym. Chem. 3 (9), 2436.

Dorjnamjin, D., Ariunaa, M., Shim, Y.K., 2008. Synthesis of silver nanoparticles using hydroxyl functionalized ionic liquids and their antimicrobial activity. Int. J. Mol. Sci. 9 (5), 807–819.

Earle, M.J., Esperança, J.M.S.S., Gilea, M.A., Lopes, J.N.C., Rebelo, L.P.N., Magee, J.W., Widegren, J.A., et al., 2006. The distillation and volatility of ionic liquids. Nature 439 (7078), 831–834.

Earle, M.J., Gordon, C.M., Plechkova, N.V., Seddon, K.R., Welton, T., 2007. Decolorization of ionic liquids for spectroscopy. Anal. Chem. 79 (2), 758–764.

Fraga-Dubreuil, J., Bourahla, K., Rahmouni, M., Hamelin, J., et al., 2002. Catalysed esteri cations in room temperature ionic liquids with acidic counteranion as recyclable reaction media. Catal. Commun. 3, 185–190.

Fuller, J., 1997. The room temperature ionic liquid 1-ethyl-3-methylimidazolium tetrafluoroborate: electrochemical couples and physical properties. J. Electrochem. Soc. 144 (11), 3881.

Fuller, J., Carlin, R.T., De Long, H.C., Haworth, D., 1994. Structure of 1-ethyl-3-methylimidazolium hexafluorophosphate: model for room temperature molten salts. J. Chem. Soc., Chem. Commun. 3, 299.

Funasako, Y., Mochida, T., Takahashi, K., Sakurai, T., Ohta, H., 2012. Vapochromic ionic liquids from metal–chelate complexes exhibiting reversible changes in color, thermal, and magnetic properties. Chemistry 18 (38), 11929–11936.

Funasako, Y., Inagaki, T., Mochida, T., Sakurai, T., Ohta, H., Furukawa, K., Nakamura, T., 2013. Organometallic ionic liquids from alkyloctamethylferrocenium cations: thermal properties, crystal structures, and magnetic properties. Dalton Trans. 42 (23), 8317–8327.

Furniss, B.S., Hannaford, A.J., Smith, P.W.G., Tatchell, A.R. (Eds.), 1989. Vogel's Textbook of Practical Organic Chemistry. fifth ed. Longman Group UK Ltd., Harlow, UK.

Gao, H., Zeng, S., He, H., Dong, H., Nie, Y., Zhang, X., Zhang, S., 2014. Deep desulfurization of gasoline fuel using $FeCl_3$-containing Lewis-acidic ionic liquids. Sep. Sci. Technol. 49 (8), 1208–1214.

Getsis, A., Mudring, A.V., 2011. Switchable green and white luminescence in terbium-based ionic liquid crystals. Eur. J. Inorg. Chem. 21, 3207–3213.

Gordon, A.J., Ford, R.A., 1972. The Chemist's Companion - A Handbook of Practical Data, Techniques and References. John Wiley & Sons Inc., New York.

Gräsvik, J., 2013. Ionic liquids in bio-refining: synthesis and applications. PhD thesis. Umeå University.

Gräsvik, J., Winestrand, S., Normark, M., Jönsson, L.J., Mikkola, J.-P., 2014a. Evaluation of four ionic liquids for pretreatment of lignocellulosic biomass. BMC Biotechnol. 14, 34.

Gräsvik, J., Hallett, J.P., To, T.Q., Welton, T., 2014b. A quick, simple, robust method to measure the acidity of ionic liquids. Chem. Commun. 50 (55), 7258–7261.

Greaves, T.L., Drummond, C.J., 2008. Protic ionic liquids: properties and applications. Chem. Rev. 108, 206–237.

Hallett, J.P., Welton, T., May 11, 2011. Room-temperature ionic liquids: solvents for synthesis and catalysis. 2. Chem. Rev. 111, 3508–3576.

Han, F., Yang, L., Li, Z., Xia, C., 2012. Sulfonic acid-functionalized ionic liquids as metal-free, efficient and reusable catalysts for direct amination of alcohols. Adv. Synth. Catal. 354 (6), 1052–1060.

Holbrey, J.D., Seddon, K.R., 1999a. The phase behaviour of 1-alkyl-3-methylimidazolium tetrafluoroborates; ionic liquids and ionic liquid crystals. J. Chem. Soc. Dalton Trans. 13, 2133–2140.

Holbrey, J.D., Seddon, K.R., 1999b. Ionic liquids. Clean Technol. Environ. Policy 1, 223–236.

Holbrey, J.D., Seddon, K.R., Wareing, R., 2001. A simple colorimetric method for the quality control of 1-alkyl-3-methylimidazolium ionic liquid precursors. Green Chem. 3 (1), 33–36.

Holbrey, J.D., Reichert, W.M., Tkatchenko, I., Bouajila, E., Walter, O., Tommasi, I., Rogers, R.D., 2003. 1,3-dimethylimidazolium-2-carboxylate: the unexpected synthesis of an ionic liquid precursor and carbene-CO_2 adduct. Chem. Commun. 347 (1), 28–29.

Holleman, A.F., Wiberg, N., 1985. Lehrbuch der Anorganischen Chemie, thirty-third ed. Walter de Guyter & Co, Berlin.

Hurley, F.H., Wier, T.P., 1951. Electrodeposition of metals from fused quaternary ammonium salts. J. Electrochem. Soc. 98 (5), 203–206.

Ishida, Y., Miyauchi, H., Saigo, K., 2002. Design and synthesis of a novel imidazolium-based ionic liquid with planar chirality. Chem. Commun. 19, 2240–2241.

Ishida, Y., Sasaki, D., Miyauchi, H., Saigo, K., 2004. Design and synthesis of novel imidazolium-based ionic liquids with a pseudo crown-ether moiety: diastereomeric interaction of a racemic ionic liquid with enantiopure europium complexes. Tetrahedron Lett. 45 (51), 9455–9459.

Ishida, Y., Sasaki, D., Miyauchi, H., Saigo, K., 2006. Synthesis and properties of a diastereopure ionic liquid with planar chirality. Tetrahedron Lett. 47 (45), 7973–7976.

Ishikawa, T. (Ed.), 2009. Superbases for Organic Synthesis: Guanidines, Amidines, Phosphazenes and Related Organocatalysts. John Wiley & Sons, Ltd., Chichester, UK.

Jeffery, G.H., Bassett, J., Mendham, J., Denney, R.C., 1989. Vogel's - Textbook of Quantitative Chemical Analysis, fifth ed. Longman Group UK Ltd, Harlow, UK.

Kamal, A., Chouhan, G., 2005. A task-specific ionic liquid [bmim]SCN for the conversion of alkyl halides to alkyl thiocyanates at room temperature. Tetrahedron Lett. 46 (9), 1489–1491.
Kirchner, B. (Ed.), 2010. Ionic liquids. Topics in Current Chemistry, vol. 290. Springer, Berlin Heidelberg.
Lancaster, N.L., Welton, T., Young, G.B., 2001. A study of halide nucleophilicity in ionic liquids. J. Chem. Soc., Perkin Trans. 2 (12), 2267–2270.
Larsen, A.S., Holbrey, J.D., Tham, F.S., Reed, C.A., 2000. Designing ionic liquids: imidazolium melts with inert carborane anions. J. Am. Chem. Soc. 122 (30), 7264–7272.
Law, M.C., Wong, K.-Y., Chan, T.H., 2002. Solvent-free route to ionic liquid precursors using a water-moderated microwave process. Green Chem. 4 (4), 328–330.
Lee, S.H., Ha, S.H., Ha, S.S., Jin, H.B., You, C.Y., Koo, Y.M., 2007. Magnetic behavior of mixture of magnetic ionic liquid [bmim] Fe Cl4 and water. J. Appl. Phys. 101 (9), 127–130.
Lethesh, K.C., Van Hecke, K., Van Meervelt, L., Nockemann, P., Kirchner, B., Zahn, S., Binnemans, K., et al., 2011. Nitrile-functionalized pyridinium, pyrrolidinium, and piperidinium ionic liquids. J. Phys. Chem. B 115 (26), 8424–8438.
Lethesh, K.C., Parmentier, D., Dehaen, W., Binnemans, K., 2012. Phenolate platform for anion exchange in ionic liquids. RSC Adv., 2, 11936–11943.
Li, M., De Rooy, S.L., Bwambok, D.K., El-Zahab, B., Ditusa, J.F., Warner, I.M., 2009. Magnetic chiral ionic liquids derived from amino acids. Chem. Commun. 45, 6922–6924.
Luo, H., Baker, G.A., Lee, J.S., Pagni, R.M., Dai, S., 2009. Ultrastable superbase-derived protic ionic liquids. J. Phys. Chem. B 113 (13), 4181–4183.
Lüttringhaus, A., Ziegler, K., 1937. Über vielgliedrige Ringsysteme: VIII. Über eine neue Anwendung des Verdünnungsprinzips. Justus Liebig's Annalen Der Chemie 528 (1), 155–161.
MacFarlane, D., Meakin, P., 1999. Pyrrolidinium imides: a new family of molten salts and conductive plastic crystal phases. J. Phys. Chem. B 103 (20), 4164–4170.
MacFarlane, D.R., Golding, J., Forsyth, S., Forsyth, M., Deacon, G.B., 2001. Low viscosity ionic liquids based on organic salts of the dicyanamide anion. Chem. Commun. 16, 1430–1431.
MacFarlane, D.R., Forsyth, M., Izgorodina, E.I., Abbott, A.P., Annat, G., Fraser, K., 2009. On the concept of ionicity in ionic liquids. Phys. Chem. Chem. Phys. 11 (25), 4962–4967.
Markuszewski, M.J., Stepnowski, P., Marszałł, M.P., 2004. Capillary electrophoretic separation of cationic constituents of imidazolium ionic liquids. Electrophoresis 25 (20), 3450–3454.
Marszalek, M., Fei, Z., Zhu, D.R., Scopelliti, R., Dyson, P.J., Zakeeruddin, S.M., Grätzel, M., 2011. Application of ionic liquids containing tricyanomethanide [C (CN) 3]$^-$ or tetracyanoborate [B (CN) 4]$^-$ anions in dye-sensitized solar cells. Inorg. Chem. 50 (22), 11561–11567.
Matsumi, N., Miyamoto, M., Aoi, K., 2009. Preparation of ionic liquids bearing *o*-carborane anion via *N,N′*-dialkylimidazol-2-ylidene carbene. J. Organomet. Chem. 694 (11), 1612–1616.
Nacham, O., Clark, K.D., Yu, H., Anderson, J.L., 2015. Synthetic strategies for tailoring the physicochemical and magnetic properties of hydrophobic magnetic ionic liquids. Chem. Mater. 27 (3), 923–931.
Namboodiri, V.V., Varma, R.S., 2002. Solvent-free sonochemical preparation of ionic liquids. Org. Lett. 4 (18), 3161–3163.
Nieuwenhuyzen, M., Seddon, K.R., Teixidor, F., Puga, A.V., Viñas, C., 2009. Ionic liquids containing boron cluster anions. Inorg. Chem. 48 (3), 889–901.
Nockemann, P., Binnemans, K., Driesen, K., 2005. Purification of imidazolium ionic liquids for spectroscopic applications. Chem. Phys. Lett. 415 (1–3), 131–136.
Nowicki, J., Muszyński, M., Gryglewicz, S., 2014. Novel basic ionic liquids from cyclic guanidines and amidines – new catalysts for transesterification of oleochemicals. J. Chem. Technol. Biotechnol. 89 (1), 48–55.

Padmanabhan, S., Kim, M., Blanch, H.W., Prausnitz, J.M., 2011. Solubility and rate of dissolution for *Miscanthus* in hydrophilic ionic liquids. Fluid Phase Equilib. 309 (1), 89–96.
Parviainen, A., King, A.W.T., Mutikainen, I., Hummel, M., Selg, C., Hauru, L.K.J., Kilpeläinen, I., et al., 2013. Predicting cellulose solvating capabilities of acid-base conjugate ionic liquids. ChemSusChem 6 (11), 2161–2169.
Pedersen, C.J., 1967. Cyclic polyethers and their complexes with metal slats. J. Am. Chem. Soc. 89 (26), 7017–7036.
Peng, Y., Li, G., Li, J., Yu, S., 2009. Convenient synthesis of various ionic liquids from onium hydroxides and ammonium salts. Tetrahedron Lett. 50 (29), 4286–4288.
Peng, Q., Mahmood, K., Wu, Y., Wang, L., Liang, Y., Shen, J., Liu, Z., 2014. A facile route to realize the copolymerization of L-lactic acid and ε-caprolactone: sulfonic acid-functionalized Brønsted acidic ionic liquids as both solvents and catalysts. Green Chem. 16 (4), 2234.
Pitula, S., Mudring, A.-V., 2010. Synthesis, structure, and physico-optical properties of manganate(II)-based ionic liquids. Chemistry 16 (11), 3355–3365.
Plechkova, N.V., Seddon, K.R., 2008. Applications of ionic liquids in the chemical industry. Chem. Soc. Rev. 37 (1), 123–150.
Priede, E., Bakis, E., Zicmanis, A., 2014. When chlorides are the most reactive: a simple route towards diverse mono- and dicationic dimethyl phosphate ionic liquids. Synlett 25 (17), 2447–2450.
Ranu, B.C., Banerjee, S., 2005. Ionic liquid as catalyst and reaction medium. The dramatic influence of a task-specific ionic liquid, [bmim]OH, in Michael addition of active methylene compounds to conjugated ketones, carboxylic esters, and nitriles. Org. Lett. 7 (14), 3049–3052.
Robinson, J., Osteryoung, R.A., 1979. An electrochemical and spectroscopic study of some aromatic hydrocarbons in the room temperature molten salt system aluminum chloride-*n*-butylpyridinium chloride. J. Am. Chem. Soc. 101 (2), 323–327.
Santos, E., Albo, J., Irabien, A., 2014. Magnetic ionic liquids: synthesis, properties and applications. RSC Adv. 4 (75), 40008.
Schreiner, C., Zugmann, S., Hartl, R., Gores, H.J., 2010. Fractional Walden rule for ionic liquids: examples from recent measurements and a critique of the so-called ideal KCl line for the Walden plot. J. Chem. Eng. Data 55 (5), 1784–1788.
Seddon, K.R., Stark, A., Torres, M.-J., 2000. Influence of chloride, water, and organic solvents on the physical properties of ionic liquids. Pure Appl. Chem. 72 (12), 2275–2287.
Shi, J., Gao, H., Xia, Y., Li, W., Wang, H., Zheng, C., 2013. Efficient process for the direct transformation of cellulose and carbohydrates to 5-(hydroxymenthyl)furfural with dual-core sulfonic acid ionic liquids and co-catalysts. RSC Adv. 3 (21), 7782.
Smiglak, M., Holbrey, J.D., Griffin, S.T., Reichert, W.M., Swatloski, R.P., Katritzky, A.R., Rogers, R.D., et al., 2007. Ionic liquids via reaction of the zwitterionic 1,3-dimethylimidazolium-2-carboxylate with protic acids. Overcoming synthetic limitations and establishing new halide free protocols for the formation of ILs. Green Chem. 9 (1), 90.
Smiglak, M., Corey Hines, C., Singh, S., Wilsion, T.B., Vincek, A.S., Kirichenko, K., Rogers, R.D., et al., 2010. Ionic liquids based on azolate anions. Chemistry 16 (5), 1572–1584.
Song, Y., Jing, H., Li, B., Bai, D., 2011. Crown ether complex cation ionic liquids: preparation and applications in organic reactions. Chemistry 17 (31), 8731–8738.
Song, Y., Cheng, C., Jing, H., 2014. Aza-crown ether complex cation ionic liquids: preparation and applications in organic reactions. Chemistry 20 (40), 12894–12900.
Stark, A., Behrend, P., Braun, O., Müller, A., Ranke, J., Ondruschka, B., Jastorff, B., 2008. Purity specification methods for ionic liquids. Green Chem. 10 (11), 1152.
Suarez, P.A.Z., Einloft, S., Dullius, J.E.L., de Souza, R.F., Dupont, J., 1998. Synthesis and physical–chemical properties of ionic liquids based on 1- *n* -butyl-3-methylimidazolium cation. J. Chim. Phys. Phys. Chim. Biol. 95, 1626–1639.

Sun, J., MacFarlane, D.R., Forsyth, M., 1997. Synthesis and properties of ambient temperature molten salts based on the quaternary ammonium ion. Ionics. 3 (5), 356–362.
Varma, R.S., Namboodiri, V.V., 2001. An expeditious solvent-free route to ionic liquids using microwaves. Chem. Commun. 7, 643–644.
Walden, P., 1914. Ueber die Molekulargrösse und elektrische Leitfähigkeit einiger geschmolzenen Salze. Bulletin de l'Académie Impériale Des Sciences de St. -Pétersbourg 8, 405–422.
Wasserscheid, P.P., Welton, T. (Eds.), 2008. Ionic Liquids in Synthesis. second ed. Wiley-VCH Verlag GmbH & Co KGaA, Weinheim.
Widegren, J.A., Wang, Y.M., Henderson, W.A., Magee, J.W., 2007. Relative volatilities of ionic liquids by vacuum distillation of mixtures. J. Phys. Chem. B 111 (30), 8959–8964.
Wilkes, J.S., Zaworotko, M.J., 1992. Air and water stable 1-ethyl-3-methylimidazolium based ionic liquids. J. Chem. Soc. Chem. Commun., 965–967.
Williams, D.B.G., Lawton, M., 2010. Drying of organic solvents: quantitative evaluation of the efficiency of several desiccants. J. Org. Chem. 75 (24), 8351–8354.
Yoshizawa, M., Hirao, M., Ito-Akita, K., Ohno, H., 2001. Ion conduction in zwitterionic-type molten salts and their polymers. J. Mater. Chem. 11 (4), 1057–1062.
Yoshizawa, M., Xu, W., Angell, C.A., 2003. Ionic liquids by proton transfer: vapor pressure, conductivity, and the relevance of ΔpK_a from aqueous solutions. J. Am. Chem. Soc. 125 (50), 15411–15419.
Yuen, A.K.L., Masters, A.F., Maschmeyer, T., 2013. 1,3-Disubstituted imidazolium hydroxides: dry salts or wet carbenes? Catal. Today 200 (1), 9–16.
Zhuo, K., Du, Q., Bai, G., Wang, C., Chen, Y., Wang, J., 2015. Hydrolysis of cellulose catalyzed by novel acidic ionic liquids. Carbohydr. Polym. 115, 49–53.

CHAPTER 3

Perspective of Using ILs as "Green Solvents"

Elena S. Sashina, Aleksandr M. Bochek, Nikolai P. Novoselov, Dmitrii A. Kashirskii, Anton A. Murav'ev
Institute of Applied Chemistry and Ecology, Saint-Petersburg State University of Industrial Technologies and Design, Russia

One of the most important applications of ionic liquids (ILs) is their use as a dissolution medium for natural polymer processing, in particular, for the dissolution of polysaccharides (cellulose, chitin, and chitosan) and polypeptides (fibroin, collagen, and keratin) for the subsequent production of new materials, films, and fibers. Extensive research on this topic started after the publication of Swatloski et al. (2002), where the possibility of cellulose dissolution with a series of 1-alkyl-3-methylimidazolium-based ionic liquids was first demonstrated. Nevertheless, among the wide variety of possible combinations of cation–anion ionic pairs, constituting ILs, only a few of them are able to dissolve even a small amount of cellulose.

Numerous papers have been devoted to the search for new solvents for natural polymers. The main focus of these is to understand the relationship between the IL structure (the nature of the anion and cation species, number, and type of alkyl-chain substituents) and physico-chemical properties. Depending on the intended application of the IL as a solvent, it is necessary to achieve a compromise between important parameters (such as melting point and toxicity) and the cost of IL synthesis.

This chapter discusses the possibilities of the practical application of ILs for the dissolution of selected natural polymers. The rational choice of ILs for use in particular processes, the features of dissolution methods for natural polymers (including cellulose, chitin, and fibroin), and the preparation of blends from the solutions of polymers with ILs are discussed.

Application, Purification, and Recovery of Ionic Liquids
http://dx.doi.org/10.1016/B978-0-444-63713-0.00003-1

APPLICATION OF IONIC LIQUIDS FOR NATURAL POLYMER PROCESSING

Effect of the Structure of ILs on Their Dissolving Power With Respect to Natural Polymers

It is well established that good solvents for cellulose are ILs with nitrogen-containing heterocyclic cations such as imidazolium and pyridinium. Various ammonium-based ILs can dissolve cellulose at relatively low concentrations (Basa, 2010; Sashina et al., 2012b; Ohira et al., 2012). The nature of the anion significantly affects the properties and the dissolving power of ILs. In general, the dissolving power of ILs depends on the type of anion and decreases as follows: $[OAc]^- > Cl^- \gg Br^- \approx [SCN]^-$ (Sashina and Novoselov, 2009). The structure and number of substituents on the cation also strongly affect the IL's dissolving power. Thus, the best solvents for cellulose among imidazolium- and pyridinium-based ILs are those having substituents in positions 1 and 3 (see Figure 3.1). The cations having substituents in other positions usually make the ILs unsuitable solvents for natural polymers (Basa, 2010; Sashina and Novoselov, 2009).

Increasing the length of the alkyl-substituent chains on the cation usually decreases the IL's dissolving power with respect to polysaccharides and polypeptides (Sashina and Novoselov, 2009; Erdmenger et al., 2007; Sashina et al., 2012b), and this is believed to happen due to a decrease in the polar character of the ILs with the nonpolar alkyl chain lengthening (Zherenkova and Khalatur, 2010). Our quantum chemical calculations at the DFT:B3LYP/6-311+G(d,p) level for 1-alkyl-3-methylpyridinium chloride- and bromide-based models of ILs confirmed that the electric dipole moment μ values, which are characteristics of the electron density asymmetry, are gradually reduced in chloride-containing ILs, when the

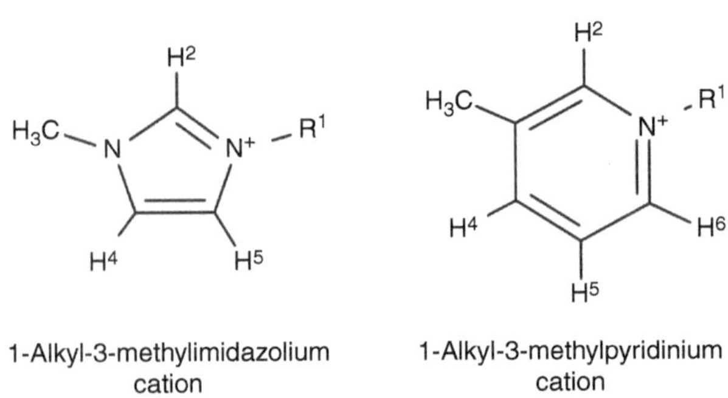

Figure 3.1 ***The Structures of the 1-Alkyl-3-Methylimidazolium ($[C_nC_1im]^+$) and the 1-Alkyl-3-Methylpyridinim IL's ($[C_nC_1py]^+$) Cations.***

Table 3.1 Selected physical and chemical parameters calculated by DFT:B3LYP/6-311 + G(d,p) level for the series of pyridinium-based ILs (Sashina et al., in press)

IL's models	$d_{H^2...X}$ (Å)	μ (D)	E_P (kJ mol^{-1})
$[C_2C_1py]Cl$	2.056	13.19	−400.7
$[C_3C_1py]Cl$	2.110	13.11	−366.1
$[C_4C_1py]Cl$	2.113	12.98	−364.9
$[C_5C_1py]Cl$	2.112	12.91	−364.0
$[C_6C_1py]Cl$	2.111	12.86	−363.6
$[C_7C_1py]Cl$	2.111	12.87	−363.2
$[C_8C_1py]Cl$	2.115	12.79	−362.8
$[C_9C_1py]Cl$	2.113	12.67	−362.9
$[C_{10}C_1py]Cl$	2.112	12.80	−362.5
$[AC_1py]Cl$	2.096	13.32	−413.8
$[C_2C_1py]Br$	2.238	13.80	−378.9
$[C_4C_1py]Br$	2.259	13.44	−369.9

chain length of the substituents is increased (see Table 3.1). The displacement of the vector of the dipole moment along the alkyl chain takes place gradually as shown in Figure 3.2.

The decrease in the calculated values of the dipole moments directly corresponds to the dissolving power with respect to cellulose among 1-alkyl-3-methylpyridiniun chloride-based ILs. Figure 3.3 shows the maximum concentration of cellulose dissolved in ILs at 110°C and 120°C after 12 h versus the number of carbon atoms in the alkyl-substituent chains. For the chain lengths up to C_6, it is possible to dissolve up to 20 wt% of cellulose. The further elongation of the alkyl chain substituent gradually decreases the dissolving power of ILs. Similarly, the replacement of the Cl^- anion with the Br^- anion significantly decreases the dissolving power of ILs with respect to the natural polymer due to the different charge density (Sashina et al., 2012b).

For the right choice of ILs we should find a relationship between their dissolving power and the individual physico-chemical parameters. There are a number of indicators for the reasonable choice of solvents for natural polymers. For example, one can use the Kamlet–Taft solvatochromic parameters (Taft and Kamlet, 1976; Kamlet and Taft, 1976; Kamlet et al., 1977). They characterize selected contributions to the polarity of solvents (in this case for ILs). These include such parameters as α, the hydrogen-bond donor (HBD) parameter; β, the hydrogen-bond acceptor (HBA) parameter; and π^*, a measure of the polarity/polarizability of the solvent. A comparison of the characteristics of the donor and acceptor properties of ILs shows that the

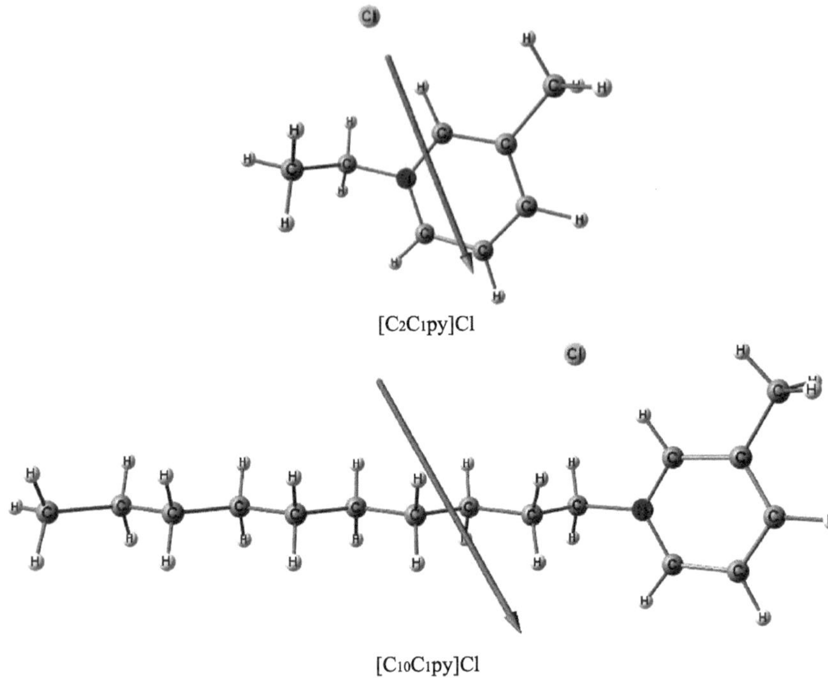

Figure 3.2 ***The Displacement of the Vector of the Dipole Moments for [C_2C_1py]Cl and [$C_{10}C_1$py]Cl (Sashina et al., in press).***

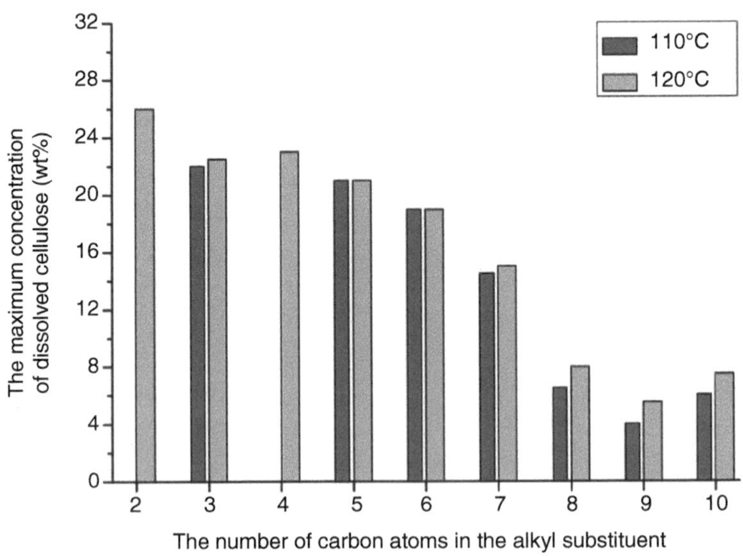

Figure 3.3 ***Dependence of the number of carbon atoms in the alkyl substituent chains on the dissolving power of 1-alkyl-3-methylpyridinium-based ILs with respect to the cellulose alicell super (DP 599) (Sashina et al., 2012b).***

HBA value for imidazolium-based ILs determines their dissolution ability to a large extent, while the HBD value of cellulose dissolving ILs needs to be relatively low (0.29–0.57 (Gericke et al., 2012)). HBA is largely determined by the nature of the anion, using HBA we can describe the solvation of natural polymers by the solvent (Gericke et al., 2012). In general, imidazolium-based ILs with $\beta \geq 0.65$ potentially may dissolve cellulose, but still the most effective solutions have $\beta \geq 0.80$ (Sashina et al., in press). The most effective solvents for cellulose dissolution include 1-butyl-3-methylimidazolium acetate ($\beta = \sim 0.99$–1.20 (Ab Rani et al., 2011; Brandt et al., 2010; Doherty et al., 2010; El Seoud et al., 2011; Ohno and Fukaya, 2009; Wu et al., 2008a)), and 1-butyl-3-methylimidazolium chloride ($\beta = \sim 0.84$–0.95 (El Seoud et al., 2011; Fukaya et al., 2006; Lungwitz and Spange, 2008; Ohno and Fukaya, 2009; Sashina et al., in press; Tamura et al., 2012; Ventura et al., 2009)). Given the many structural similarities of 1-alkyl-3-methylimidazolium- and 1-alkyl-3-methylpyridinium-based ILs, these dissolution trends are equal for both cation types. Thus, the dissolving power of pyridinium-based ILs conforms to the values of the Kamlet–Taft β parameter (see Figure 3.4). We determined these parameters by the maximum absorption value using a spectrophotometer for solutions of the ILs with the dye [Cu(tmen)(acac)](ClO_4) (Sashina et al., in press). It was found that the most effective 1-alkyl-3-methylpyridinium chloride-based ILs (i.e., having alkyl chain lengths from

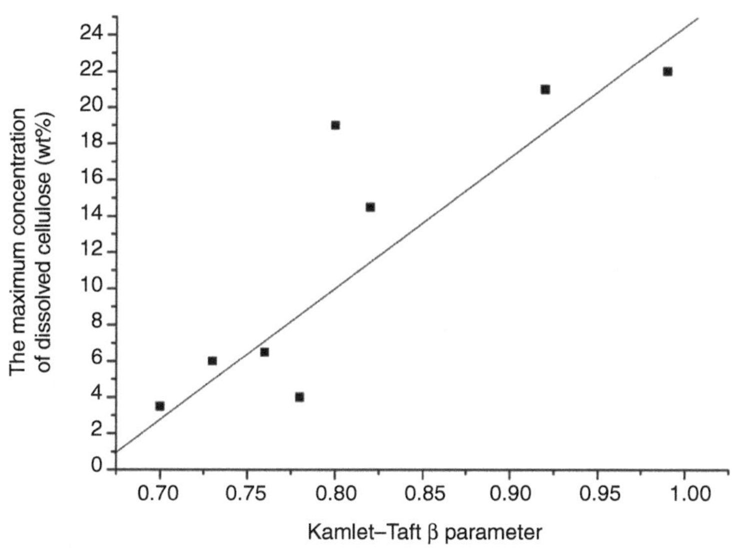

Figure 3.4 ***Correlation between the Kamlet–Taft β parameter values and the dissolving power (with respect to cellulose at 110°C) of 1-alkyl-3-methylpyridinium halide-based ILs (Sashina et al., in press).***

C_3 to C_7) have values for the β parameter between 0.80 and 0.99. As expected, changing from the chloride anion to the bromide anion causes a decrease in the value of the Kamlet–Taft β parameter from 0.99 to 0.70.

It is worth noting, that despite the fact that the hydrogen bond acceptor ability (β) of a solvent tends to be the major parameter defining its cellulose dissolution ability, Hauru et al. (2012) successfully applied the term "net basicity" ($\beta-\alpha$), to better predict the cellulose solubility for a range of different solvents including ILs. It was found that the solvent should have a net basicity within the range of 0.35–0.90 in order to dissolve cellulose; there are some exemptions, however, such as [HOC_2C_1im][(MeO)HPO_2] with quite limited cellulose dissolution ability (Hauru et al., 2012).

Recently, a few non-imidazolium based ILs with net basicities within this "dissolution window" were successfully applied to cellulose dissolution and consequent fiber spinning (Sixta et al., 2015). This new class of ILs are distillable superbases that contain, among others, the cations based on 1,5-diaza-bicyclo[4.3.0]non-5-ene (DBN), 1,8-diazabicyclo-[5.4.0]-undec-7-ene (DBU), and *N,N,N,N*-tetramethylguanidium (TMG). Their dissolution ability is comparable to that of [C_2C_1im][OAc] (currently the favored IL for cellulose dissolution). The researchers showed that the fibers manufactured from 13 wt% cellulose dope with [DBNH][OAc] (which dissolves, at 80°C, in less than 90 min (Michud et al., 2014)) are of higher, or comparable, mechanical strength than current commercial Lyocell fibers (Sixta et al., 2015). The new technology of fiber spinning from these superbase ILs has been named Ioncell-F and has been shown to be extremely promising, with the potential to allow for complete recycling and reuse of the ILs (Hauru et al., 2012; Michud et al., 2015; Sixta et al., 2015). However, the thermal stability of these ILs still needs to be investigated.

Effect of the Structure of ILs on Their Physico-Chemical Properties

The most important characteristics of solvents for natural polymer processing are their thermal properties. When compared to commercially available solvents for cellulose dissolution (i.e., *N*-methylmorpholine-*N*-oxide), ILs have substantially higher thermal stabilities. Dissolution processes are performed at temperatures higher than the solvent's melting point, and the structure of ILs strongly affects their melting point. This is attributed to differences in the strength of the specific interactions between the cations and anions in the ILs, and also to the packing density of the ions. For both 1-alkyl-3-methylimidazolium- and 1-alkyl-3-methylpyridinium-based ILs the elongation

Table 3.2 Chemical shifts of pyridinium ring protons for $[C_2C_1py]Cl$, $[C_2C_1py]Br$, $[C_4C_1py]Cl$, and $[C_4C_1py]Br$ (Sashina et al., 2014)

Proton	$[C_2C_1py]Cl$	$[C_2C_1py]Br$	$[C_4C_1py]Cl$	$[C_4C_1py]Br$
H^2	9.617	9.510	9.633	9.490
H^4	8.182	8.224	8.192	8.236
H^5	7.949	7.972	7.968	7.987
H^6	9.475	9.357	9.450	9.312

of the alkyl-chain substituent usually results in lowering of their melting points (*mp*). For example, $[C_4C_1py]Cl$ has a *mp* of 93°C while $[C_{10}C_1py]Cl$ melts at 55°C (Sashina et al., 2013a). Similar dependence of the elongation of the alkyl-chain substituent to the length of C_8–C_{10} is also observed in the series of 1-alkyl-3-methylimidazolium chlorides (Asikkala, 2008). Changing the anions (with a fixed cation) also affects the melting point. $[C_4C_1py]Cl$ melts at 93°C, while $[C_4C_1py]Br$ remains liquid at room temperature. This is due to the difference in strength of interionic interactions in the IL's structure.

These data may be illustrated by a comparison of the relative strength of H-bonding between the IL anion and cation. It can be traced by changes in the ^1H NMR chemical shift (δ) values. The replacement of the Cl^- anion to the Br^- anion leads to an upfield chemical shift for the more "acidic" H^2 and H^6 protons of 1-alkyl-3-methylpyridinium-based ILs with chain lengths of C_2 and C_4 (see Table 3.2). This behavior is a result of the decrease of hydrogen bond strength. Chemical shifts of H^4 and H^5 protons are practically independent of the elongation of the alkyl-chain substituent from C_2 to C_{10}. But the δ values for bromides are slightly higher than for the corresponding chlorides. Table 3.1 shows that there is a decrease of the potential energy of the IL's cation–anion pair formation (E_p) in the homologous series. As a result, the decrease of the packing density of the ions in the crystal leads to the reduction of the IL's melting point (Sashina et al., in press).

Interactions in the "Cellulose/IL" Systems

The strength of the hydrogen bonds between the ions affects the relative dissolving power of the ILs, and it is vital to define it to clarify the solvation mechanism of the dissolution of natural polymers within ILs.

The anion plays an important role in natural polymer dissolution, as has been shown in many studies (Brandt et al., 2010; Ventura et al., 2009; Liu et al. (2007b,c)). This trend is typical for imidazolium- and pyridinium-based ILs. The principal solvation mechanism, as shown by quantum chemical calculations, is the hydrogen-bond formation between the anion and the cellulose hydroxyls, while the cation does not form hydrogen bonds with the natural polymer (Sashina and Novoselov, 2009; Sashina et al., in press), although other authors (e.g., Zhang et al. (2010a,b)) have a different opinion on the role of the cation in the solvation mechanism of natural polymers with ILs. Figure 3.5 shows the optimized models obtained by quantum chemical calculations for complexes of cellulose (cellotetraose) with 1-butyl-3-methylpyridinium-based ILs, having chloride, bromide, and acetate anions.

The interaction is accompanied by the formation of two hydrogen bonds between the anion of the IL and the primary hydroxyl group, and the neighboring secondary hydroxyl group, of the cellotetraose (see Table 3.3).

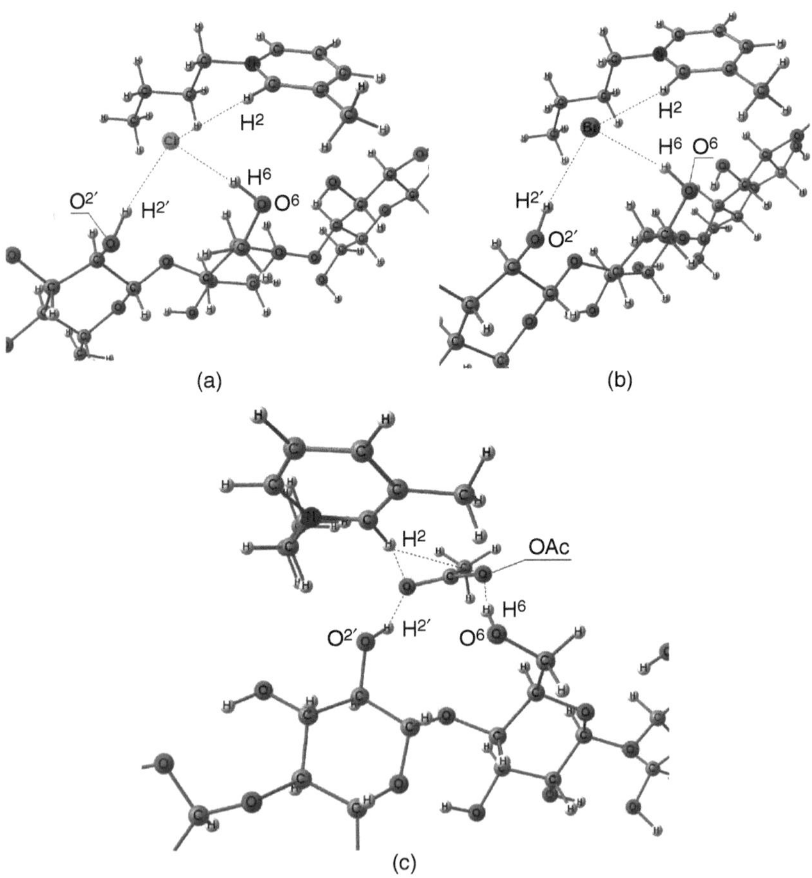

Figure 3.5 ***Optimized solvate complexes of cellotetraoses with 1-butyl-3-methylpyridinium chloride (a), 1-butyl-3-methylpyridinium bromide (b), and 1-butyl-3-methylpyridinium acetate (c) by DFT:B3LYP/6-311 + G(d,p) level (Sashina et al., in press).***

Table 3.3 The main calculated parameters for the solvate complexes of cellotetraose with $[C_4C_1py][OAc]$, $[C_4C_1py]Cl$, and $[C_4C_1py]Br$ (Sashina et al., in press)

ILs	$d_{H^2...X}$ (Å)	Angle $O^{2'}$–$H^{2'}$–X (deg.)	Angle O^6–H^6–X (deg.)	$d_{H^{2'}...X}$ (Å)	$d_{H^6...X}$ (Å)
$[C_4C_1py][OAc]$	2.080, 2.750*	157.56	168.58	1.716	1.737
$[C_4C_1py]Cl$	2.485	167.18	168.50	2.327	2.129
$[C_4C_1py]Br$	2.605	167.16	160.16	2.650	2.374

* From the two oxygen atoms of acetate-anion.

To quantify and characterize solvated complexes of "cellotetraose – 1-butyl-3-methylpyridinium-based ILs" we note the following:

- Distances between the anion and the hydrogen atoms of the cellotetraose hydroxyls ($d_{H6...X}$ and $d_{H2'...X}$) fall between 1.7 Å and 2.7 Å, and valence angles are between 157.6° and 168.6°. These distances increase in the series $[C_4C_1py][OAc] < [C_4C_1py]Cl < [C_4C_1py]Br$. The obtained geometric characteristics satisfy the required criteria for hydrogen bond formation (Steiner, 1998). For example, for hydrogen bond formation with the chloride anion, the distance $C_{sp^3}-OH\cdots Cl^-$ is between ~2 Å and ~2.4 Å, and the valence angle (between atoms O, H, and Cl) is >140°.
- An anion of the solvent is located close to the hydrogen atom of the primary hydroxyl group O^6H^6. It should be noted that the molecular dynamics calculation method for the cellulose interaction with imidazolium-based IL $[C_4C_1im]Cl$ (Liu et al., (2007b,c)) gave similar results; on average, primary OH groups interact more strongly with the chloride anion than secondary OH groups. The interaction of both hydroxyls with the anion is within a range of distances between 2.3 Å and 4 Å (Sashina et al., in press).

Interactions of $[C_4C_1py]Br$ with cellotetraose (see Figure 3.5b), as shown by quantum chemical calculations, are similar to those described earlier; the anion forms hydrogen bonds with the primary and secondary cellotetraose hydroxyls. Distances between bromide anions and cellotetraose hydroxyls are larger than in the case of chloride anions (2.374 Å for $H^{6}\cdots Br$, and 2.650 Å for $H^{2'}\cdots Br$). This seems reasonable if we take into account that the hydrogen bond lengths between halides and H-donors increase when the halide ion volume is larger (Steiner, 1998). Since the hydrogen bond length indirectly characterizes the interaction strength, it is quite understandable that $[C_4C_1py]Br$ dissolves less cellulose than $[C_4C_1py]Cl$ (Sashina et al., 2012b).

The formation of the solvate complex between the solvent and cellotetraose is accompanied by an increase in the distance between the solvent's anion and cation $d_{H2...X}$ with the following values: 0.319 (and 0.118) Å for acetate, 0.372 Å for chloride, and 0.346 Å for bromide. This suggests that the bond between the cation and anion has a tendency to be weakened, and this is confirmed by the results of ^{1}H-NMR studies. We can observe this effect in Table 3.4 by comparing the chemical shifts for the pyridinium ring protons H^2 ($\Delta\delta_{H2} = -0.060$ ppm) and H^6 ($\Delta\delta_{H6} = -0.025$ ppm), whereas the chemical shifts for the H^4 and H^5 atoms are almost unchanged ($\Delta\delta_{H4,5}$ = 0.007–0.009 ppm). This decrease occurs due to the reduced influence of the anion, when the distance between the ions becomes larger (Sashina et al., in press). Similar results were obtained using other calculation method (Sashina and Kashirskii, 2015).

It should be noted that, according to the quantum chemical calculations, the interaction of 1-butyl-3-methylpyridinium acetate with cellotetraose leads to the following properties: the anion is displaced with respect to the plane of the pyridinium ring. Moreover, one oxygen atom of the anion moves slightly away from the cation, while the other oxygen atom comes closer to the cation. This behavior can be explained by the reorientation of the acetate anion in space so that it can form hydrogen bonds with the cellotetraose hydroxyls. The movement of the anion can also be traced by interpreting the ^{1}H-NMR data: the chemical shift of the CH_3-group protons of the anion is significantly displaced downfield (Table 3.4). This strong deshielding can be explained by increased distance between the oxygen atoms and the hydrogen atoms of the pyridinium rings during hydrogen bond formation with cellulose.

Interaction of the anion with the secondary hydroxyls of cellulose glycosidic rings is confirmed by ^{1}H-NMR (see Figure 3.6). To explain the peaks for cellulose, we have used data from the work of Bernet and Vasella (2000), where the chemical shifts of the disaccharide (cellobiose) were investigated. The chemical shifts of the hydrogen atoms in the secondary hydroxyls have the values for cellobiose of δ_{O2H2} = 4.630 ppm, $\delta_{O2'H2'}$ = 5.215 ppm, and are moved downfield ($\Delta\delta_{O2'H2'}$ = 0.171 ppm, $\Delta\delta_{O2H2}$ = 0.067 ppm) when solvated

Table 3.4 Chemical shifts of pyridinium ring and acetate-ion protons for cellulosic solutions with $[C_4C_1py]Cl$ and $[C_4C_1py][OAc]$, and their changing relatively pure ILs in brackets. (Sashina et al., 2014)

Anion of $[C_4C_1py]^+$	H^2	H^4	H^5	H^6	Acetate-ion
Cl^-	9.573 (−0.060)	8.199 (0.007)	7.977 (0.009)	9.425 (−0.025)	
$[OAc]^-$	9.202 (0.004)	8.117 (−0.002)	7.926 (−0.002)	9.202 (0.004)	3.780 (0.240)

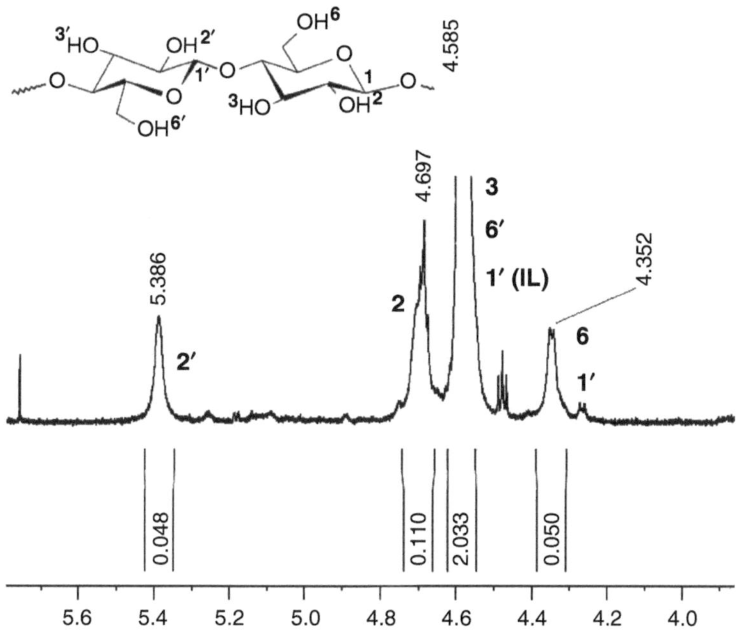

Figure 3.6 ***Partial*** ^{1}***H-NMR Spectrum for Cellulose in Solution With 1-Butyl-3-Methylpyridinium Chloride.*** ^{1}H-NMR spectrum was obtained in DMSO-d_6 (Sashina et al., in press).

by $[C_4C_1py]Cl$. This effect is a result of the interaction of cellulose with the chloride anion. Thus, the intramolecular hydrogen bond in the structure of cellulose is broken during cellulose dissolution in the IL. The hydrogen atom of the secondary OH-group forms the $O^{2'}H^{2'}\cdots Cl^-$ hydrogen bond with the anion. It is difficult to estimate the level of interaction of the anion with the primary hydroxyls by this method since the peaks of the primary hydroxyls of the C^6 (and $C^{6'}$) atom overlap with other peaks, including the peak of the methylene group CH_2 (1′) from $[C_4C_1py]Cl$ (Sashina et al., in press).

During hydrogen bond formation between the anion and the hydroxyl groups of cellulose the interaction of the cation with the anion is simultaneously weakened. This is confirmed by the upfield chemical shifts of the more "acidic" protons $H^{2,6}$ (for 1-butyl-3-methylpyridinium chloride (Sashina et al., 2014)), and $H^{2,4,5}$ (for 1-butyl-3-methylimidazolium chloride (Kuzmina et al., 2014)) when cellulose has been dissolved in these ILs. NMR relaxation data (Remsing et al., 2006), thermochemistry of cellulose/ILs solutions (de Oliveira and Rinaldi, 2015), and MD simulations (Liu et al. (2007b,c)) generally confirm the important role of the anion in hydrogen bond formations with the OH groups of the natural polymer. Nonetheless, some studies (El Seoud et al., 2011; Feng and Chen, 2008) support the assumption that the cation is also involved in hydrogen bonding with cellulose. In this case, the cation has the role of hydrogen bond donor, while the anion acts as a hydrogen bond acceptor. Analysis of the results of molecular dynamics simulations shows that the bulk cation associated with the anion is wedged between adjacent cellulose macromolecules. The ability of the cation to slip between neighboring strands could be a crucial aspect of an IL's ability to dissolve cellulose and might partially explain why aromatic cations are so effective (Rabideau et al., 2013).

Thus, on the basis of all of the available data, we can conclude that the most effective solvents for natural polymer processing are 1-alkyl-3-methylimidazolium- and 1-alkyl-3-methylpyridinium-based ILs with relatively short alkyl substituents on the cation (chain lengths from C_2 to C_7), combined with one of several types of anions, such as Cl^-, $[OAc]^-$, and so on.

During cellulose dissolution with ILs the anion plays the main role. The anion forms strong hydrogen bonds with the hydrogen atoms of the primary hydroxyls of cellulose.

PROPERTIES OF SOLUTIONS OF POLYSACCHARIDES IN IONIC LIQUIDS AND COMPOSITE FILMS PREPARED FROM THESE POLYMERS

Peculiarities of Dissolving Cellulose and Chitin in Ionic Liquids

Cellulose, as well as its derivatives, has found wide application in various fields of human activity, since it is a constantly renewable natural polymer. Cellulose possesses a set of valuable properties and is relatively inexpensive. One of the actual problems in the chemistry of natural polymers is the development of environmentally safe and human-harmless manufacturing methods for processing cellulose into finished products. Natural cellulose dissolves in a limited number of solvents; this complicates manufacturing artificial fibers and films and increases their cost. Until the mid-1980s, hydrated cellulose fibers were produced from cellulose xanthogenate with the use of carbon sulfide, which is a toxic compound. An intensive search for new dissolution systems for cellulose is underway as an alternative to the viscose process. Nonaqueous dissolving systems (tertiary amine *N*-oxides, aprotic solvents with the addition of lithium chloride, dimethylformamide (DMF) combined with N_2O_4, trifluoroacetic acid, and some others) and aqueous systems (solutions of $ZnCl_2$, NaOH, H_3PO_4) have been studied (Liebert, 2010). Most of these solvents have not found practical application for various reasons, such as toxicity and environmental hazard, limited dissolving capacity, difficulties in development of closed technological cycles, and high energy consumption of solvent regeneration, among others.

Chitin is the second most widespread natural polymer, after cellulose. Together with chitosan, chitin has found steadily increasing application in various fields of industry (Varlamov et al., 2011). The use of chitin is limited by the low number of technologically convenient solvents. Chitin dissolves in concentrated solutions of several mineral and organic acids, in the non-aqueous complex solvent dimethylacetamide/lithium chloride (DMAA/LiCl) and in some fluorinated organic solvents (e.g., hexafluoropropanol) (Ohno and Fukaya, 2009). Chitin is transformed to chitosan by saponification; the latter polymer is soluble in acidified aqueous media. However, the very process of transformation of chitin into chitosan involves using concentrated aqueous solutions of acids and alkali, creating problems with environmental pollution and excessive wastewater (Bochek, 2008). Thus, the development of improved manufacturing processes for producing cellulose and chitin fibers and films is still a continuing challenge.

In recent years, considerable attention has been given to ionic liquids, which are capable of dissolving polysaccharides and synthetic polymers (Swatloski et al., 2002; Liebert, 2008; Zhu et al., 2006; Wu et al., 2008; Kadokawa, 2011). The use of this new class of solvents will extend opportunities for processing wood cellulose and chitin and obtaining fibers and films. In contrast to *N*-methylmorpholine-*N*-oxide (NMMO), most of ionic liquids and solutions of cellulose in these liquids are thermally stable (Wendler et al., 2012). The conditions of chitin dissolution in ILs have not been studied in detail, although obtaining chitin films and fibers from solutions of ILs is a promising line of investigation (Wu et al., 2008; Xie et al., 2006).

The polymers considered in the present study include microcrystalline cellulose (MCC) with a degree of polymerization (DP) equal to 100, wood cellulose with a DP of 900, chitin from the North Sea shrimp shells with a DP of 1500, and polyacrylonitrile (PAN) with a DP of 2550 synthesized by a radical process. To study the influence of the degree of deacetylation of chitin and chitosan on their solubility in ILs, a number of samples were prepared in our laboratory, including chitin with degrees of deacetylation (DD) equal to 20 and 36%, as well as chitosan samples with DDs of 84 and 96%.

In order to study the influence of the chemical structure of cellulose, chitin, and chitosan on their solubility in ILs, seven solvents (purchased from Sigma–Aldrich) were used by our research group, see Table 3.5.

Two methods for dissolving polysaccharides were employed. In the first method, a weighed polymer sample was dried at 105°C until a constant weight was reached (i.e., for 3 h); then the sample was placed into the IL to obtain a

Table 3.5 Ionic liquids used for dissolving cellulose and chitin and their melting points T_{mp}

No.	Ionic liquid	Abbreviation	Structural formula	T_{mp} (°C)	Ref.
1	1-Butyl-3-methyl-imidazolium chloride	$[C_4C_1im]Cl$	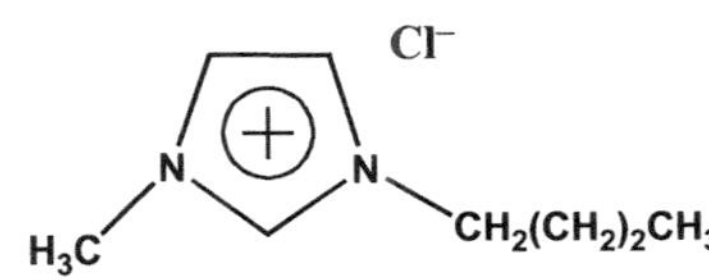	73 66–68	Wendler et al. (2012) Mazza et al. (2009)
2	1-Butyl-3-methyl-imidazolium bromide	$[C_4C_1im]Br$	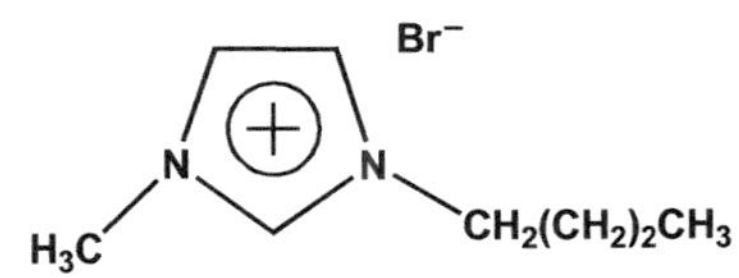		–
3	1-Butyl-3-methyl-imidazolium acetate	$[C_4C_1im][OAc]$	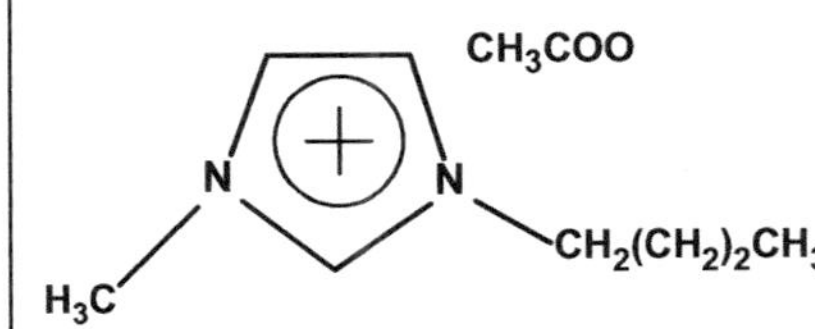	−20	Wendler et al. (2012)
4	1-Ethyl-3-methyl-imidazolium bromide	$[C_2C_1im]Br$	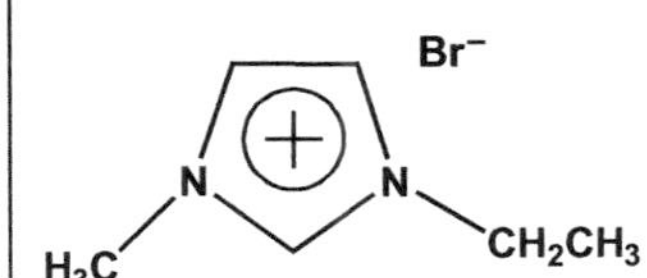	65–79	Zhang et al. (2009)

5	1-Ethyl-3-methyl-imidazolium acetate	$[C_2C_1im][OAc]$	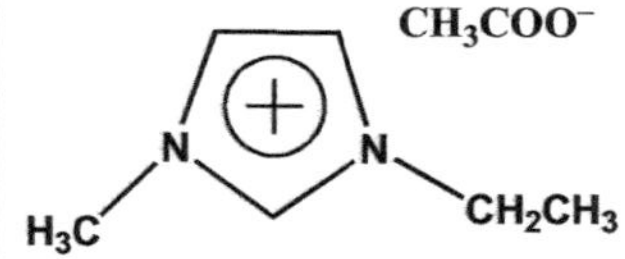	−45	Wendler et al. (2012)
6	1,3-Dimethyl-imidazolium dimethyl phosphate	$[C_1C_1im][PC_1C_1]$	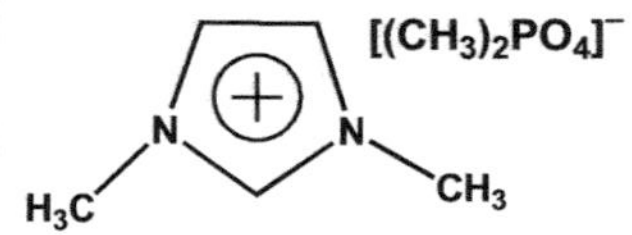	−65	Mazza et al. (2009)
7	1-Benzyl-3-methyl-imidazolium chloride	$[(C_6H_5)^1C_1C_1im]Cl$	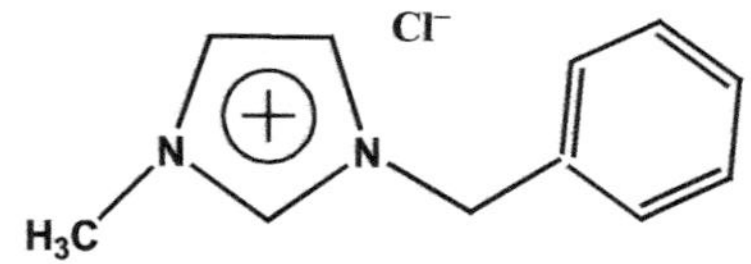		–

1 wt% solution. The mixture was heated to 110°C with stirring in a closed vessel. Dissolution was carried out until a transparent polymer solution was obtained, but for no longer than 11 h. This method was used in the studies of MCC and chitin solubility in seven selected ILs; the time required for preparation of cellulose solutions of various concentrations (1, 5, and 10 wt%) was determined.

The second method involved the preliminary soaking of the polymers in water in order to decrease the time required for dissolution of cellulose and chitin in ILs. In this part of the study, we used cellulose with a DP of 900. This method is close to the technique used in manufacturing of hydrated cellulose fibers by the MMO process (Kosan et al., 2008, 2010). The techniques we used for dissolution of polysaccharides and their solubilities in ILs are described in more detail elsewhere (Bochek et al., 2012).

Table 3.6 presents the results obtained in the studies of MCC and chitin solubility according to the first method. It can be seen that cellulose dissolves in four ionic liquids ([C_4C_1im]Cl, [C_4C_1im][OAc], [C_2C_1im][OAc], and [C_1C_1im][PC_1C_1]). When two other ILs are used, cellulose fibers only swell until a dispersion is formed. It should be noted that MCC dissolves faster in the ILs having acetate anions. With increasing concentration of polymer, differences in dissolving time begin to appear. The dissolving capacity of ionic liquids used for preparation of 10% MCC solutions decreases in the order: [C_2C_1im][OAc] > [C_4C_1im][OAc] > [C_1C_1im][PC_1C_1] > [C_4C_1im]Cl. The differences in time required for dissolution may be caused by different dissolving capacities of the ILs with respect to cellulose as well as differences in solvent viscosities at the dissolution temperature. For instance, [C_4C_1im]Cl melts at 73°C,

Table 3.6 Solubility of chitin (DP = 1500) and microcrystalline cellulose (DP = 100) in ionic liquids; time periods required for preparation of 1 wt% of chitin and 1, 5, and 10 wt% cellulose solutions at 110°C

No.	Ionic liquid	Chitin dissolution time (min)	Cellulose dissolution time (min) 1%	5%	10%
1	[C_4C_1im]Cl	Does not dissolve*	295	300	668
2	[C_4C_1im]Br	Does not dissolve		Does not dissolve	
3	[C_4C_1im][OAc]	185	120	195	330
4	[C_2C_1im]Br	Does not dissolve		Does not dissolve	
5	[C_2C_1im][OAc]	205	70	85	155
6	[C_1C_1im][PC_1C_1]	Does not dissolve	70	180	485
7	[$(C_6H_5)^1C_1C_1$im]Cl	Does not dissolve		Does not dissolve	

The polymers were dissolved in the ILs without preliminary soaking in water (Bochek et al., 2012).
* Sample does not dissolve after exposing for 5 h.

whereas $[C_4C_1im][OAc]$ and $[C_2C_1im][OAc]$ melt at −20 and −45°C, respectively, that is, they are liquids at ambient temperature (Wendler et al., 2012). In the manufacturing of hydrated cellulose fibers, many factors should be taken into account. On the one hand, it is advantageous to dissolve cellulose in ILs giving the shortest dissolution times (i.e., $[C_4C_1im][OAc]$ and $[C_2C_1im][OAc]$). On the other hand, thermal stability of chloride ILs ($[C_4C_1im]Cl$) is significantly higher than that of $[C_4C_1im][OAc]$ and $[C_2C_1im][OAc]$ (Wendler et al., 2012). It is also necessary to account for the energy consumption during the purification of solvents and solvent losses after multiple uses.

Chitin (DD = 0) dissolves in two ionic liquids ($[C_4C_1im][OAc]$ and $[C_2C_1im][OAc]$) and only swells in other solvents. It should be noted that the dissolution of this polymer proceeds faster in $[C_4C_1im][OAc]$ than in $[C_2C_1im][OAc]$. The difference between chitin and cellulose is the presence of an amide group in the C^2 position of the *N*-acetylglucosamine residue of the chitin monomeric unit. It can be assumed that the amide groups of chitin macromolecules contribute to the formation of intermolecular hydrogen bonds present in the natural polymer; as a result, chitin can be dissolved in a limited number of liquids.

We have also studied the solubility of chitin and chitosan samples with different DD values in $[C_4C_1im][OAc]$ and $[C_4C_1im]Cl$. It was found that chitin samples with DDs of 20% and 36% and a chitosan sample with a DD of 84% cannot be dissolved in $[C_4C_1im][OAc]$ or $[C_4C_1im]Cl$. When the first method of solution preparation was used, only chitosan with a DD of 96% was dissolved in $[C_4C_1im][OAc]$, while partial dissolution of the polymers was observed in $[C_4C_1im]Cl$. The scientific literature contains rather contradictory data on the solubility of chitin and chitosan in ILs. For example, it was shown (Xie et al., 2006) that chitin is soluble in $[C_4C_1im]Cl$, while chitosan is not. The opposite result was obtained in another study, that is, chitosan was found to dissolve in $[C_4C_1im]Cl$ and $[C_4C_1im][OAc]$ better than chitin (Wu et al., 2008). These examples demonstrate the necessity of thorough studies of the dissolution processes of chitin and chitosan in ILs.

In further studies of the solubility of cellulose and chitin in ILs, we employed the second method for the dissolution of the polymers (involving preliminary soaking of the samples in water). In this part of the study, we used cellulose with a DP of 900 and chitin with a DP of 1500 to form hydrated cellulose and chitin films from polysaccharide solutions. Since dissolution was carried out under reduced pressure (in order to remove residual water), we examined the influence of pressure and temperature on the duration of dissolution of cellulose in $[C_4C_1im]Cl$ and $[C_4C_1im][OAc]$, and that of chitin in $[C_4C_1im][OAc]$.

Table 3.7 Changes in preparation times for 1% cellulose solutions (DP = 900) in $[C_4C_1im]Cl$ and $[C_4C_1im][OAc]$ and those for chitin solutions in $[C_4C_1im][OAc]$ depending on temperature, pressure, and the amount of added DMSO (Bochek et al., 2012)

No.	Pressure (mbar)	Temperature (°C)	Cellulose dissolving time (min) $[C_4C_1im]Cl$	Cellulose dissolving time (min) $[C_4C_1im][OAc]$	Chitin dissolving time (min)	Amount of DMSO (% of IL mass)
1	20	110	15	15	60	–
2	20	100	15	16	75	–
3	20	90	40	45	180	–
4	20	80	120	30	660	–
5	40	110	18	15	64	–
6	60	110	80	15	168	–
7	82*	110	90	30	190	–
8	82*	110	110	30	220	10
9	82*	110	90	30	440	25
10	82*	110	405	30	–	50

*DMSO starts to evaporate at pressures lower than 82 mbar and 110°C.

Table 3.7 shows that the times required for preparing moderately concentrated (1 wt%) solutions of cellulose in $[C_4C_1im]Cl$ and $[C_4C_1im][OAc]$ in the 100–110°C temperature range under a pressure of 20–40 mbar differ only slightly. Comparison with the data in Table 3.6 demonstrates that preliminary soaking of cellulose in water or mixing ILs with water eliminates this difference in preparation times. This effect is probably caused by the decrease in the viscosity of the aqueous solutions of ILs and by the increasing rate of diffusion of solvent molecules into the polymer matrix (which occurs in the first stage of dissolution). The effect of the polymer concentration is apparent only during the preparation of highly concentrated cellulose solutions (Table 3.7).

Analysis of the data given in Table 3.7 shows a tendency toward decreasing dissolution duration of cellulose solutions prepared in $[C_4C_1im]Cl$ and $[C_4C_1im][OAc]$ with increasing temperature. This is due to an increase in the mobility of solvent molecules and an increased rate of diffusion of solvent molecules into the polymeric matrix as well as a likely change in the thermodynamic affinity of the solvent to the polymer. Similar patterns in the influence exerted by temperature on the dissolution time are observed in the case of dissolution of chitin in $[C_4C_1im][OAc]$. With a temperature increase from 80°C to 110°C the duration of polymer dissolution decreases from 660 min to 60 min.

Some researchers have suggested using aprotic diluents for dissolution of cellulose in tertiary amine *N*-oxides in order to lower consumption of the expensive solvent (Petropavlovskii et al., 1987). This approach has also been applied to ILs (Troshenkova et al., 2008; Kuzmina et al., 2012). As can be seen from Table 3.7, the addition of dimethylsulfoxide (DMSO) to an IL leads, in the cases of cellulose (in [C_4C_1im]Cl) and chitin (in [C_4C_1im][OAc]), to an increased time required for dissolution of the polysaccharides. The only exception is dissolution of cellulose in [C_4C_1im][OAc]; this is probably a result of the better thermodynamic affinity between the polymer and this particular solvent and specific features of the interaction between the solvent ([C_4C_1im][OAc]) and the diluent.

We assume that the increase in the dissolution time of cellulose and chitin in ILs after addition of an aprotic diluent reflects the tendency toward decreasing thermodynamic quality of the mixed solvents (ILs with added DMSO) with respect to the polysaccharides under study.

It is known that cellulose undergoes depolymerization during the course of dissolution in ionic liquids (Bentivoglio et al., 2006; Wendler et al., 2009). Therefore, we examined the depolymerization of cellulose macromolecules in the process of their dissolution in [C_4C_1im][OAc] and [C_4C_1im]Cl and exposure of the resulting solutions at 110°C; depolymerization processes occurring during dissolution of chitin in [C_4C_1im][OAc] were also investigated. It can be seen in Figure 3.7 that the depolymerization of cellulose in [C_4C_1im]Cl occurs

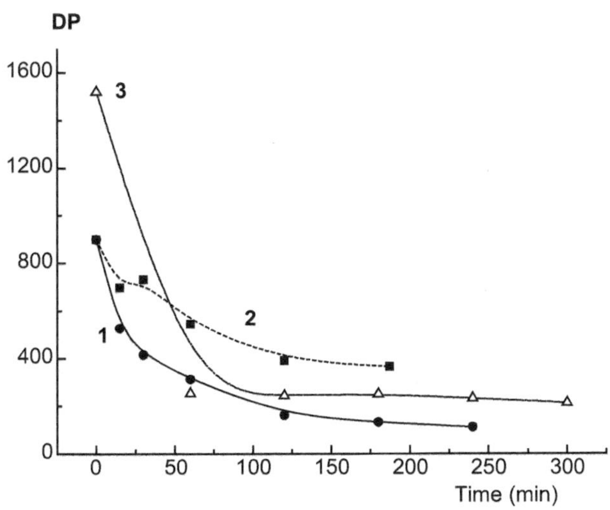

Figure 3.7 ***Changes in degree of polymerization of cellulose in solutions in [C_4C_1im]Cl (1), [C_4C_1im][OAc] (2), and degree of polymerization of chitin in [C_4C_1im][OAc] solution (3) at 110°C (Bochek et al., 2012).***

to a greater extent than that in $[C_4C_1im][OAc]$. To explain the differences in the degree of cellulose depolymerization in two ILs, it is necessary to take into account a number of factors. On the one hand, cellulose was dissolved in two ILs at the same temperature (110°C), and thermal stability of $[C_4C_1im]Cl$ and cellulose solutions in this solvent is higher than that of $[C_4C_1im][OAc]$ (Wendler et al., 2012). It is thus expected that the degree of cellulose depolymerization in $[C_4C_1im]Cl$ should be lower than that in $[C_4C_1im][OAc]$. On the other hand, the presence of water in IL mixtures and elevated temperatures may lead to the formation of certain amounts of hydrochloric acid in $[C_4C_1im]Cl$ aqueous solutions (and acetic acid in $[C_4C_1im][OAc]$ aqueous solutions) during the first stage of polymer dissolution. These acids have different hydrolytic reactivities toward cellulose. Therefore, we presume that the differences in the degree of cellulose depolymerization should be attributed to differences in the reactivity of $[C_4C_1im]Cl$ and $[C_4C_1im][OAc]$ toward cellulose.

Polymer films obtained from solutions of cellulose in $[C_4C_1im]Cl$ and chitin in $[C_4C_1im][OAc]$ by wet spinning were studied by X-ray diffraction and FTIR spectroscopy in order to reveal structural organization. The X-ray diffraction pattern of the initial cellulose sample shows typical reflections characteristic of the Cell-I polymorphic structure, that is, reflections at 2θ angles of 14.5°, 16.5°, 22.4°, and 34.5° associated with the [110], $[1\bar{1}0]$, [020], and [004] planes, respectively (Figure 3.8, curve 1). These results are in good agreement with published data (Bikales and Segal, 1971; Oh et al., 2005). The X-ray diffraction pattern of the hydrated cellulose films prepared from a solution in $[C_4C_1im]Cl$ includes reflections characteristic of Cell-II polymorphic structure at 2θ angles of 12.0°, 20.0°, 22.0°, and 35.0° attributed to the [110], $[1\bar{1}0]$, [020], and [004] planes, respectively (Figure 3.8, curve 2). It is noteworthy that only the reflection at $2\theta = 20.0°$ is distinct. The rest of the reflections are weakly pronounced, which is indicative of a substantially smaller degree of hydrated cellulose crystallinity in the formed film as compared to that of the initial cellulose.

During dissolution of chitin in $[C_4C_1im][OAc]$ according to the second method (involving soaking in water), a sample was taken at the stage of partial dissolution of the polymer. This sample was filtered, the solution was used for the preparation of a film, and insoluble (swollen) particles were washed and dried. The X-ray diffraction pattern of the initial chitin (Figure 3.9, curve 1) demonstrates typical reflections characteristic of the polymorphic structure of highly crystalline α-chitin: reflections at 2θ angles of 9.3°, 12.6°, 19.2°, 22.5°, 26.1°, 29.0°, 32.1°, 35.0°, and 39.2° associated with the [020], [021], [110], [130], [013], [060], [142], [152], and [063] planes, respectively.

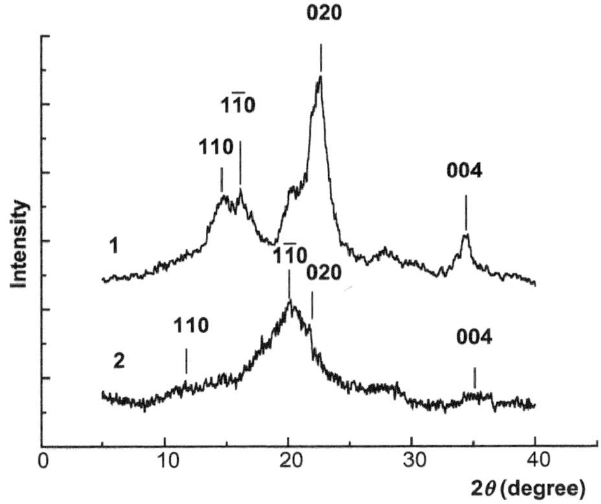

Figure 3.8 ***X-ray diffraction patterns of the initial cellulose (1) and the hydrate cellulose film obtained from 1% cellulose solution in [C_4C_1im]Cl using precipitation with water (2).*** Dissolution of cellulose included preliminary soaking in water (Bochek et al., 2012).

These results are in good agreement with published data (Yen et al., 2009). The X-ray diffraction pattern of chitin particles swollen in [C_4C_1im][OAc] includes similar reflections, with the exception of the [130], [060], and [063] reflections at 2θ angles of 22.5°, 29.0°, and 39.2°, respectively. The lower

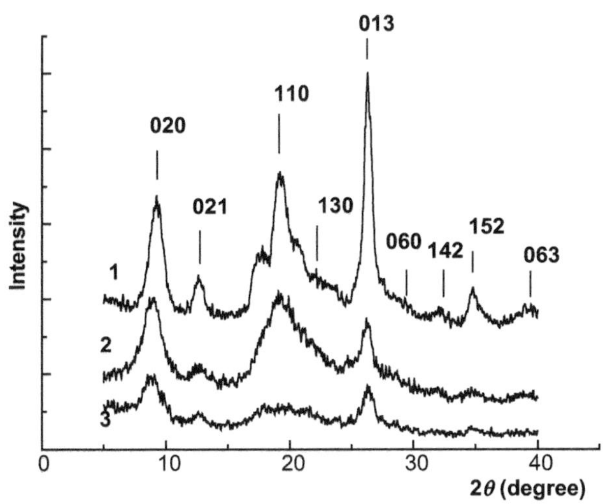

Figure 3.9 ***X-ray diffraction patterns of the initial chitin (1), polymer particles that have not been dissolved in [C_4C_1im][OAc] (2), and the chitin film prepared from the solution (3).*** Dissolution of chitin included preliminary soaking in water.

number of recorded reflections and decrease in their intensity indicate partial disintegration of the initial polymer structure and disruption of long-range order. The X-ray diffraction pattern of the chitin film (Figure 3.9, curve 3) shows only reflections at 2θ angles of 9.3°, 12.6°, 26.1°, and 35.0°, with the reflection at 2θ =19.2° only being observed as an amorphous halo. It can be assumed that chitin is in a nearly amorphous state in the film, that is, the initial structure of chitin was completely disrupted in the course of its dissolution in [C_4C_1im][OAc] with preliminary soaking in water.

It should be noted that the scientific literature contains contradictory data on the completeness of dissolution of chitin in ILs and on the structure of the polymer regenerated from solution. For example, it has been shown (Xie et al., 2006; Prasad et al., 2009) that the X-ray diffraction patterns of chitin precipitated from solutions of ILs are similar to the X-ray diffraction pattern of the initial chitin. Other authors have demonstrated that different X-ray diffraction patterns of the polysaccharides were observed in the cases of partial and complete dissolution of chitin in ILs followed by precipitation with water (Wu et al., 2008).

In the present study, we prepared solutions of chitin in [C_4C_1im][OAc] according to the method described elsewhere (Xie et al., 2006; Wu et al., 2008; Prasad et al., 2009). After that, the polymer was precipitated from solution with water, washed and dried, and its X-ray diffraction pattern was obtained (Figure 3.10). It can be seen that chitin precipitated from

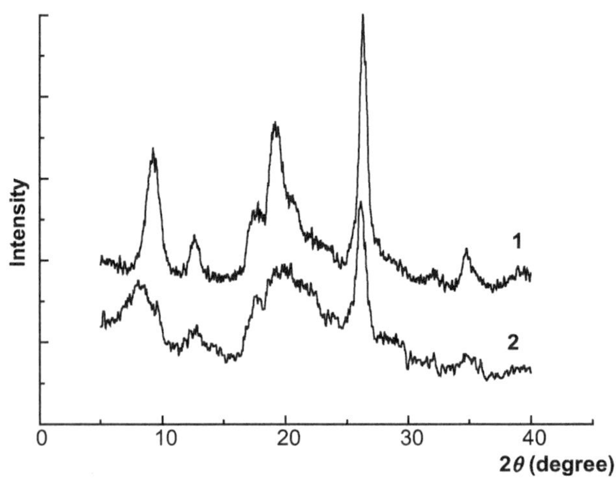

Figure 3.10 ***X-Ray Diffraction Patterns of the Initial Chitin (1) and Chitin Regenerated From the Solution in [C_4C_1im][OAc] (2).*** Dissolution of chitin was performed without preliminary soaking in water.

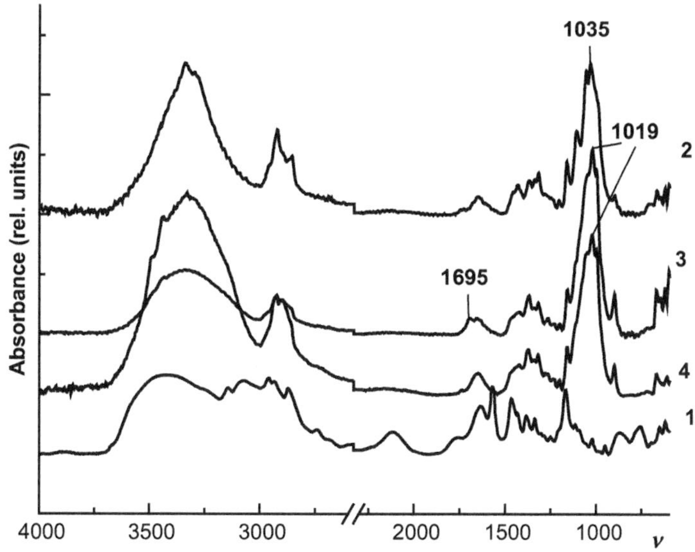

Figure 3.11 ***FTIR spectra of (1) [C_4C_1im]Cl, (2) the initial cellulose, (3) the cellulose regenerated from [C_4C_1im]Cl solution, and (4) commercial hydrated cellulose film (Bochek et al., 2012).***

solution has lower crystallinity than the initial chitin (judging from the less intensive reflections), but the X-ray diffraction patterns are similar. In our opinion, in the given case chitin is only partly dissolved, with the initial structure virtually preserved after precipitation. The solution may contain nanosize crystallites that have not dissolved completely.

Analysis of the IR spectra of cellulose demonstrated that cellulose precipitated from [C_4C_1im]Cl solutions has Cell-II polymorphic modification (Figure 3.11). The shift of the absorption band from 1035 cm^{-1} to 1019 cm^{-1} and the change in the intensities of the set of bands in the 1100–970 cm^{-1} range are indicative of a structural transition of cellulose from the C-I to the C-II polymorphic form. These data are in good agreement with the results of X-ray diffraction analysis (Oh et al., 2005). Upon precipitation of cellulose with water, a new absorption band at 1695–1700 cm^{-1} appears in the IR spectrum of this polymer; it can be identified as a shifted absorption band of the imidazolium ring (1820 cm^{-1}). This shift of the imidazolium ring peak may be explained in two ways. On the one hand, it may be caused by a change in the binding strength of the residual imidazolium cations not washed out from cellulose (Hao et al., 2010). On the other hand, we may suppose that decomposition of the ionic liquid occurs, and residual amounts of decomposition products are present in the film; in addition, chemically bound derivatives of imidazolium cations with aldehyde groups can appear at the ends of cellulose macromolecules (Liebert, 2008).

The IR spectrum of the initial α-chitin shows characteristic absorption bands of CO groups at 1656 and 1625 cm^{-1} (amide I), deformation vibrations of NH groups at 1540 cm^{-1} (amide II), and a band at 1310 cm^{-1} (amide III) (Van de Velde and Kiekens, 2004; Rinaudo, 2006) (Figure 3.12). The IR spectrum of [C_4C_1im][OAc] includes absorption bands at 1570 and 1400 cm^{-1} attributed to the CO groups of acetate ions. The IR spectrum of regenerated chitin also contains all the absorption bands of the amide groups, but their relative intensities are noticeably different. In addition, the absorption band at 1540 cm^{-1} (amide II) is broadened and a band appears at 1400 cm^{-1}; these changes can be explained by the presence of residual IL ([C_4C_1im][OAc]) in the sample. The broadening of the band at 1540 cm^{-1} is caused by the contribution of the vibration band of the ionized carboxyl group (at 1570 cm^{-1}) and the presence of a band at 1400 cm^{-1}. The relative intensities of the bands at 1656 and 1625 cm^{-1} are probably altered because of a change in the system of hydrogen bonds in chitin resulting from the interaction between chitin macromolecules and the remaining IL and/or from changes in the morphological structure of the polymer.

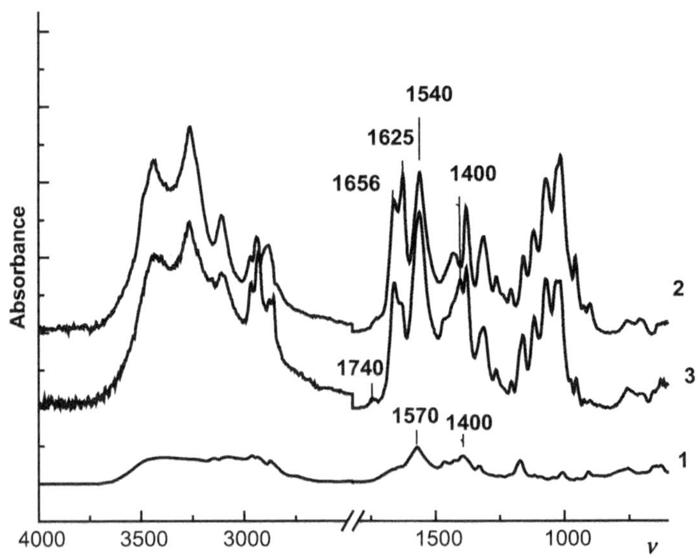

Figure 3.12 ***FTIR Spectra of (1) [C_4C_1im][OAc], (2) the Initial Chitin, and (3) the Regenerated Chitin.***

The increase in the intensity of the absorption band at 1740 cm^{-1} (attributed to acetate groups of polysaccharide esters) may indicate the possible formation of acetate substituents in anhydroglucose units (formation of ester bonds involving end aldehyde groups of polysaccharide macromolecules and $[C_4C_1im][OAc]$) (Liebert, 2008).

Thus, solubility of cellulose and chitin in seven ionic liquids of varied structures was studied. Cellulose was dissolved in four ionic liquids, and chitin was dissolved only in two of the studied ILs. The differences in solubility between cellulose and chitin are caused by the different systems of intra- and intermolecular hydrogen bonds present in the studied polysaccharides. The dissolving capacity of ionic liquids used for preparation of cellulose solutions decreases in the order: 1-ethyl-3-methylimidazolium acetate > 1-butyl-3-methylimidazolium acetate, 1,3-dimethylimidazolium dimethyl phosphate > 1-butyl-3-methylimidazolium chloride.

Preliminary soaking of cellulose and chitin in water or addition of water to an ionic liquid followed by removal of water under vacuum accelerates the dissolution of the polysaccharides. In the course of dissolution of cellulose and chitin, polysaccharide macromolecules undergo depolymerization. Differences in the depolymerization rates of cellulose in solutions of 1-butyl-3-methylimidazolium chloride and 1-butyl-3-methylimidazolium acetate are due to the different reactivities of these solvents with respect to the natural polymer.

Depending on the conditions of chitin dissolution in ionic liquids, considerable differences may be observed in the supramolecular structure of the precipitated polymer. In the case of incomplete dissolution of chitin, the X-ray diffraction pattern of the polymer precipitated from solution is similar to that of the initial chitin. In the case of complete dissolution of chitin and its subsequent precipitation with water, amorphous or partially crystalline polymer structures with a low degree of ordering are formed.

Properties of Mixed Cellulose/Chitin Solutions and Composite Films

One of the most readily available and most convenient ways to impart new functional properties to polymeric materials is through the mixing of polymer solutions in a common solvent or by mixing polymers in a melt. Ionic liquids dissolve many natural and synthetic polymers, including cellulose, chitin, poly(vinyl alcohol) (Swatloski et al., 2002; Liebert, 2008; Zhu et al., 2006; Wu et al., 2008; Kadokawa, 2011), silk fibroin (Mantz et al., 2007), feather keratin (Idris et al., 2013), lignin (Lee et al., 2009; Liebmann et al., 2012), and a number of other polymers (Maki-Arvela et al., 2010).

In order to prepare composite materials based on cellulose and chitin, several authors suggested dissolving both polymers in different ionic liquids and then mixing these solutions (Takegawa et al., 2010). Cellulose/chitosan mixtures in a common solvation system have also been studied (Xiao et al., 2010; Kuzmina et al., 2012). Preparation of composite films and fibers based on cellulose and chitin with the use of a common solvent is of fundamental and practical importance.

Rheological properties of mixed cellulose/chitin solutions in the common solvent have been studied along with the structural organization and thermal stability of composite films. Wood cellulose with a DP of 900 and chitin from North Sea shrimp shells with a DP of 1500 were dissolved in the ionic liquid (1-butyl-3-methylimidazolium acetate ($[C_4C_1im][OAc]$), $T_{mp} = -20°C$ (Wendler et al., 2012)).

Figures 3.13 and 3.14 show flow curves for cellulose and chitin solutions in $[C_4C_1im][OAc]$ at various temperatures. It can be seen that the flow curves for cellulose and chitin in $[C_4C_1im][OAc]$ are the incomplete flow curves of non-Newtonian fluids. These results are in good agreement

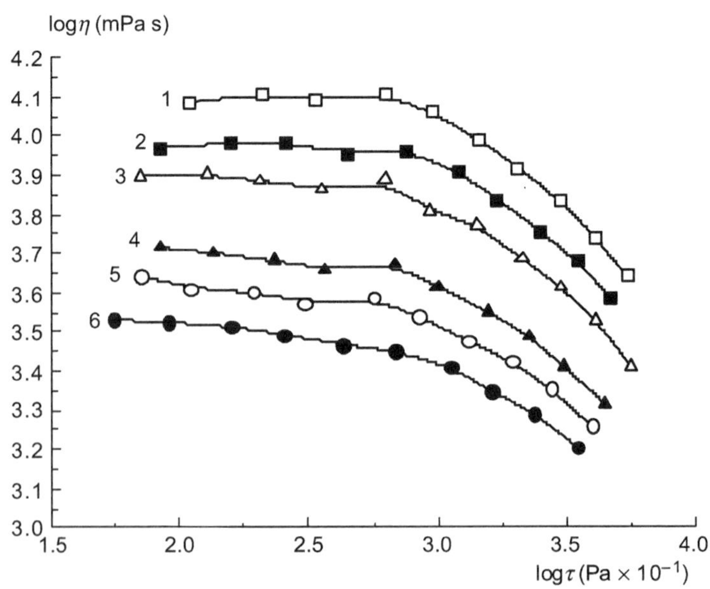

Figure 3.13 ***Flow Curves Obtained for 3% Cellulose Solution in $[C_4C_1im][OAc]$ at (1) 60, (2) 65, (3) 70, (4) 80, (5) 85, and (6) 90°C.***

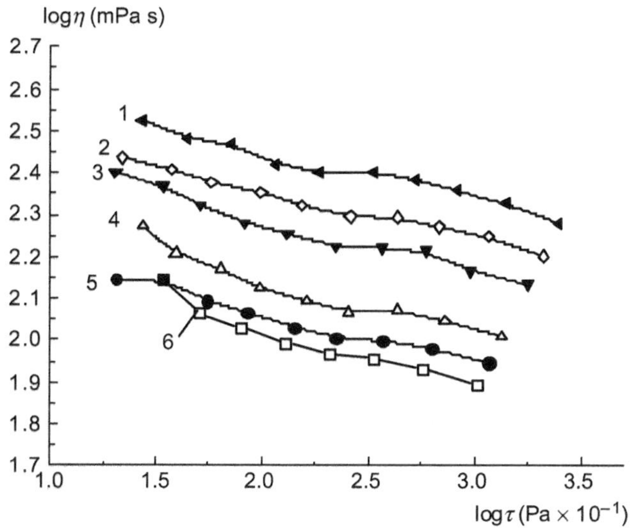

Figure 3.14 ***Flow Curves Obtained for 3% Chitin Solution in [C_4C_1im][OAc] at (1) 60, (2) 65, (3) 70, (4) 80, (5) 85, and (6) 90°C.***

with the data obtained by other authors (Kosan et al., 2008; Sammons et al., 2011; Duan et al., 2011; Lv et al., 2012). The viscosity of concentrated chitin solutions is substantially lower than that of cellulose solutions, although the DP of the initial chitin noticeably exceeds that of cellulose. The observed differences in viscosity between concentrated cellulose and chitin solutions can be explained by two factors. On the one hand, the destruction of polysaccharide molecules occurs in the course of dissolution of cellulose and chitin in [C_4C_1im][OAc], but the destruction of chitin is more pronounced (Bochek et al., 2012). On the other hand, it can be assumed that the thermodynamic quality of [C_4C_1im][OAc] as a solvent is better with respect to chitin as compared to cellulose. It is known that the viscosity of a concentrated solution of the polymer in a solvent with low thermodynamic quality is better with respect to the polymer (Tager, 2007).

With an increasing fraction of chitin in the mixtures, the viscosity of the solutions tends to decrease. From the temperature dependence of the viscosities, we calculated the viscous flow activation heats, E_a. Depending on the composition, concentrated solutions of cellulose/chitin mixtures showed positive or negative deviations from additivity (Figure 3.15, curve 1). The additive values of viscosity were calculated by the formula

$$\log \eta_{calc} = \Sigma W_i \log \eta_i, \tag{3.1}$$

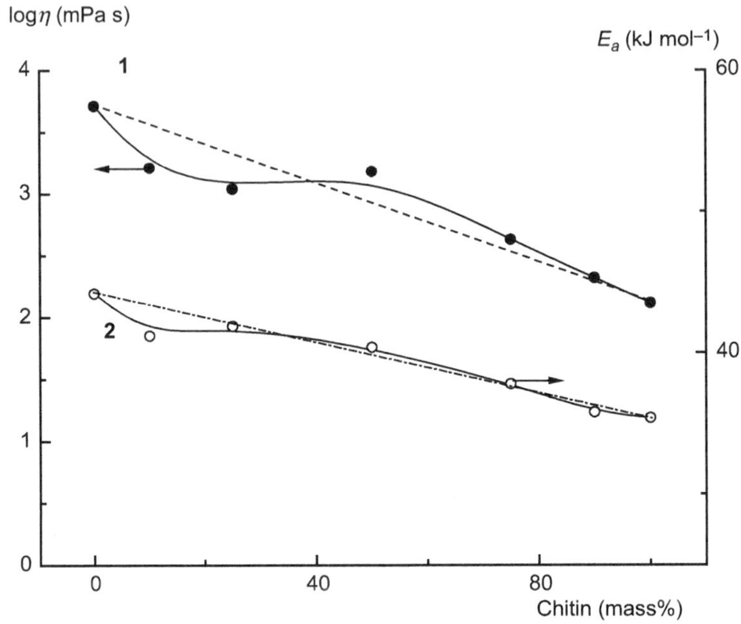

Figure 3.15 ***Dependences of viscosity (1) and viscous flow activation heat E_a, kJ mol^{-1} (2) of 3% solutions of cellulose/chitin mixtures in [C_4C_1im][OAc] at 80°C on mixture composition (Bochek et al., 2013).***

where W_i is the mass fraction of a component in solution, and $\log\eta_i$ is the logarithm of the viscosity of a solution of each of the polymers.

A similar formula was used to calculate the additive values of E_a (Figure 3.15, curve 2). In the range attributed to solutions with chitin content lower than 40 wt%, a negative deviation of the viscosity and E_a from the additive values is observed. At higher contents of chitin, a positive deviation of the viscosity from the additive values is observed, with nearly zero deviation for E_a. A similar behavior was reported for aqueous solutions of mixtures of methyl cellulose with poly(vinyl alcohol) (Bochek et al., 2006). Introduction of chitin into a cellulose solution leads to destruction of the structure, which is characteristic of the cellulose solution. When introducing cellulose into a chitin solution the opposite is true – an increase in the strength of structure of the solution. This may be due to the different compatibilities of the polymers in solution with changing composition of the mixtures.

The viscosity of concentrated solutions of polymer mixtures is affected by numerous factors. New systems of hydrogen bonds being stronger or weaker than those in individual solutions of cellulose and chitin can be formed in mixtures of polymer solutions. Addition of another polymer (chitin) to a cellulose solution may change the thermodynamic affinity of the solvent (IL) with respect to cellulose. A similar behavior should be observed upon the addition of cellulose to a chitin solution. In our case, the addition of up to 40% of chitin to a cellulose solution leads to disintegration of intermolecular entanglement networks in cellulose and the formation of a new mixed system of hydrogen bonds and a weaker intermolecular entanglement network, with the result that the viscosity and strength (E_a) of the structure becomes lower. With the further increase of the chitin content, a slight deviation of the viscosity from the additive values is observed. On the whole, the introduction of a chitin solution into a solution of cellulose in [C_4C_1im][OAc] leads to structural rearrangements in mixed solutions.

The operating characteristics (physico-mechanical and physico-chemical properties) of polymeric composite materials are also affected by numerous factors, including compatibility of the polymers and their degree of ordering, as well as by their structural organization, morphology, thermal stability, and so on. Therefore, it is of fundamental and practical interest to study structural organization and thermal stability of the composites obtained. Figure 3.16 shows X-ray diffraction patterns of films composed of cellulose, chitin, and cellulose/chitin mixtures.

The X-ray diffraction pattern of the cellulose film (Figure 3.16, curve 1) corresponds to the Cell-II polymorphic form (hydrated cellulose); the

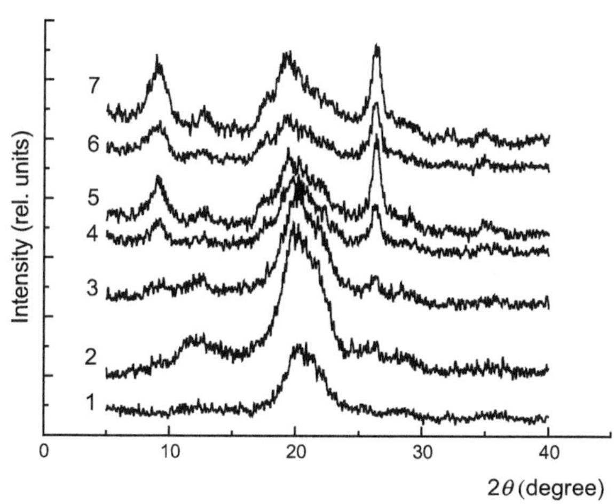

Figure 3.16 ***X-ray diffraction patterns of (1) hydrated cellulose film, (2–6) mixed hydrated cellulose/chitin films, and (7) chitin film prepared from polymer solutions in [C_4C_1im][OAc].*** Chitin content was (2) 10, (3) 25, (4) 50, (5) 75, and (6) 90 wt% (Bochek et al., 2013).

characteristic reflections observed at 2θ angles of 12.0°, 20.0°, and 22.0° are related to the [110], $[1\bar{1}0]$, and [020] planes, respectively. These results are in good agreement with published data (Bikales and Segal, 1971; Oh et al., 2005). It should be noted that the only distinct reflection is that at $2\theta = 20.0°$. The rest of the reflections are weakly pronounced, which indicates a small degree of crystallinity of hydrated cellulose in the regenerated film compared to that of the initial cellulose.

The X-ray diffraction pattern of the chitin film (Figure 3.16, curve 7) shows reflections at 2θ angles of 9.3°, 12.6°, 26.1°, and 35.0°. The obtained results are in a good agreement with published data (Rinaudo, 2006; Wu et al., 2008; Yen et al., 2009). The presence of these reflections indicates that the chitin film is more ordered than the hydrated cellulose film prepared from $[C_4C_1im][OAc]$ solution.

Figure 3.16 (curves 2–6) shows that no new structures are formed in the composite films. If the polymers are compatible and new mixed ordered structures are formed, new reflections characterizing the crystal lattice parameters should appear in the X-ray diffraction patterns. If the polymers in the mixed structures are amorphous, reflections attributed to crystalline structural formations would not be observed in the X-ray diffraction patterns. In the case of partial compatibility of the polymers, X-ray diffraction patterns would show reflections associated with crystallites of one of the polymers, with the absence of reflections related to the other polymer. This pattern was observed in the studies of chitosan/hydroxypropyl cellulose composite films (Yin et al., 2006) and chitosan/glucomannan films (Ye et al., 2006).

Since the reflections of hydrated cellulose and chitin in the 2θ range of 12°–25° are superimposed, the possible formation of ordered domains in hydrated cellulose or chitin can only be judged from the appearance or absence of chitin reflections at 2θ angles of 9.3° and 26.1°, which are not observed in the X-ray diffraction pattern of hydrated cellulose. No chitin reflections were recorded in the composite films containing up to 10 wt% of chitin. This may indirectly indicate that cellulose is partially compatible with chitin. In the case of their incompatibility, chitin may be in the amorphous state. If cellulose is added to chitin, the superposition of chitin and cellulose reflections does not allow us to make an unequivocal conclusion about the compatibility or incompatibility of these two polysaccharides.

Figure 3.17 shows TGA curves and their derivatives for mixtures of cellulose with chitin. The temperature of maximum cellulose decomposition

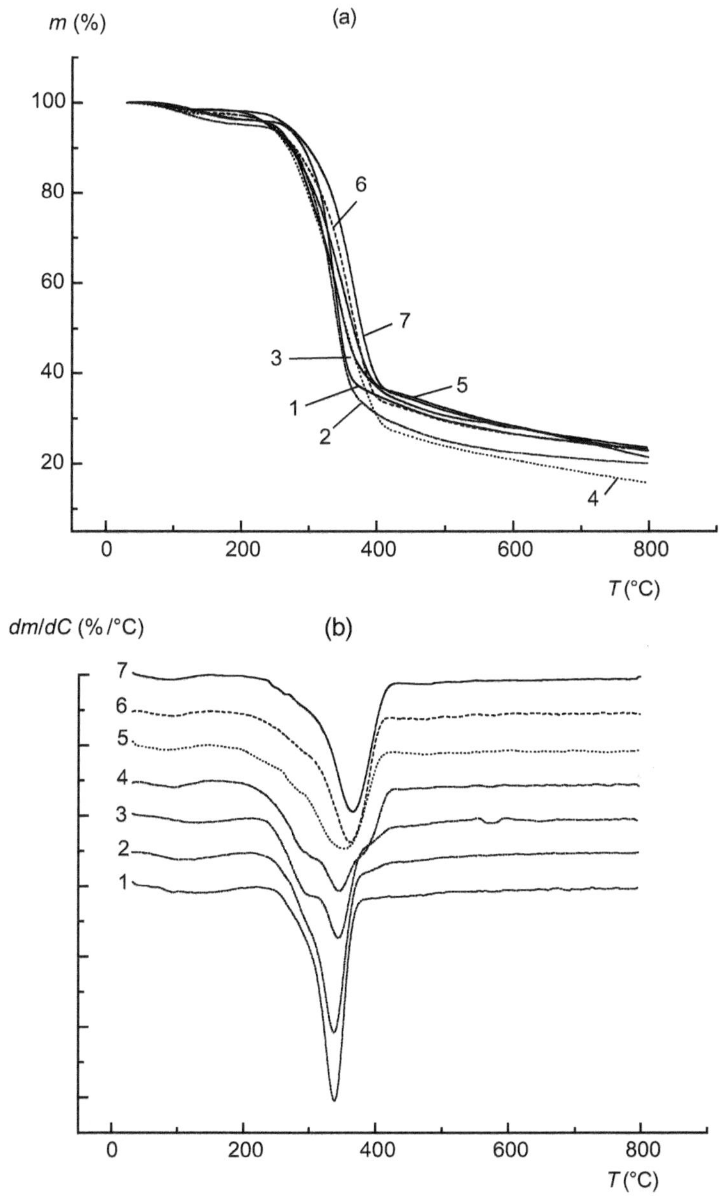

Figure 3.17 ***TGA Curves (a) and Their Derivatives (b) Obtained for (1) Cellulose Films, (2–6) Cellulose/Chitin Mixtures, and (7) Chitin.*** Chitin content was (2) 10, (3) 25, (4) 50, (5) 75, and (6) 90 wt% (Bochek et al., 2013).

(T_{max}) is 338°C, which is lower than that for chitin (370°C). The values of T_{max} obtained for cellulose and chitin are in good agreement with published data (Kim et al., 1994; Mostashari et al., 2009; Nam et al., 2010; Yang et al., 2009). At chitin contents of 25 and 50%, a new destruction process in the mixture is registered (in the form of a shoulder) at temperatures below 338°C. We assume that this decrease in the temperature of mixture destruction may be caused by the presence of products of cellulose and chitin decomposition. Decomposition temperatures (i.e., temperatures of 5% mass loss) for cellulose and chitin are 155–160 and 250°C, respectively. The main peak in the differential TGA curves shifts to higher temperatures with increasing fraction of chitin in the mixtures.

The rheological properties of solutions of cellulose/chitin mixtures in $[C_4C_1im][OAc]$ were also studied. Introduction of chitin into solutions of cellulose in $[C_4C_1im][OAc]$ leads to changes in the structure of the solutions, that is, the intermolecular entanglement network and the system of hydrogen bonds characteristic of cellulose disintegrate, and a new mixed system of hydrogen bonds and a weaker intermolecular entanglement network are formed. As a consequence, the viscosity of the mixed polymer solutions decreased.

Cellulose/chitin composite films were prepared from solutions in $[C_4C_1im][OAc]$, and their structural organization was examined. Partial compatibility of the polymers was observed in composites containing up to 10 wt% of chitin. The thermal stability of the composite films was studied using thermogravimetric analysis. Introduction of chitin into the cellulose matrix is accompanied by an increase in the thermal stability of the composites.

Solutions of Cellulose/Polyacrylonitrile Mixtures and Composite Films Prepared From These Solutions

One of the main applications in the processing of polyacrylonitrile (PAN) fibers is obtaining carbon fibers (CF). Carbon fiber is a high-strength material used mainly as a reinforcing additive to increase the strength of composite materials (carbon-reinforced plastics, carbon–carbon composite materials, etc.) (Chand, 2000). At elevated temperatures, the strength and elastic modulus of CFs are higher compared to fibers prepared from other materials; in addition, they are resistant to mechanical corrosion, whereas fibers prepared from glass and organic polymers degrade in aggressive media. Composite materials based on CFs are ideally suitable when high strength, rigidity, and low weight are required.

They are chemically inert and demonstrate high thermal stability and electrical conductivity and a low coefficient of linear thermal expansion (Feng et al., 2014).

Initially CFs were prepared from natural cellulose materials (cotton and flax fibers) and were used in manufacturing incandescent lamp filaments. In the 1950s, the process for carbonization of viscous fibers into CFs was developed, and these CFs were used in aerospace engineering (Tang and Bacon, 1964). The breakthrough in CF production occurred in the late 1960s after the beginning of commercial production of PAN fibers giving a higher CF yield (50%) than viscous fibers (30%).

The manufacturing procedure for PAN-based carbon fibers is also simpler than the production of CFs from viscous fibers (Fitzer, 1989). Presently, the majority of commercial CFs are manufactured by carbonization of PAN fibers.

It is known that cellulose (Cell) readily dissolves in [C_4C_1im]Cl with added aprotic diluents (Bochek et al., 2012), and polyacrylonitrile dissolves in aprotic DMF. It has been demonstrated (Korobeinyk et al., 2012) that chemical modification of PAN (preparation of polyacrylonitrile/methylmethacrylate (MMA) copolymers with a MMA content of 10 wt%) yields carbon materials of different porosity, and the presence of MMA fragments has an influence on the mechanism and temperature of PAN cyclization. Therefore, preparation of composite materials with new functional properties without chemical modification of PAN (i.e., based on PAN and cellulose) from solutions in a common solvent is of fundamental and practical importance.

Here we describe the study on the preparation of Cell/PAN composite films from solutions of the polymers in binary solvents (IL/DMF) and an investigation of structural organization and thermal stability of these films beside the rheological properties of mixed polymer solutions.

In our experiments, we used wood cellulose with a DP equal to 900 and polyacrylonitrile (DP = 2550) synthesized by radical polymerization. Dissolution of cellulose in [C_4C_1im]Cl was preceded by soaking the polymer in water as described elsewhere (Bochek et al., 2014).

It was demonstrated (Liu et al., 2007a) that PAN dissolves in [C_4C_1im]Cl within 5 h at a temperature of 90°C. This high temperature may lead to the destruction of PAN. Therefore, PAN was dissolved in DMF at 40°C (PAN fully dissolves within 4–5 h) and then mixtures of two solutions (polysaccharide in [C_4C_1im]Cl and PAN in DMF) were prepared in the required proportions.

Since fibers and films of hydrated cellulose as well as cellulose-based polymeric composites are prepared from polymer solutions by wet spinning, it is necessary to understand the behavior of the macromolecules in solutions under shear flow. Figure 3.18a and b and Figure 3.19 show flow curves for cellulose solutions in $[C_4C_1im]Cl$, PAN solutions in DMF, and Cell/PAN mixtures in the $[C_4C_1im]Cl$/DMF binary solvent at various temperatures and ratios between polymer components. All these dependences result in the incomplete flow curves of non-Newtonian fluids. The viscosity of the PAN solution is considerably lower than that of the cellulose solution in $[C_4C_1im]Cl$. It should be noted that the viscosity of the PAN solution in $[C_4C_1im]Cl$ of the same concentration is noticeably higher than that of the DMF solution (Liu et al., 2007a). This phenomenon may be explained by the higher viscosity of the IL solvent ($[C_4C_1im]Cl$) compared to that of DMF and possibly by a better thermodynamic affinity of DMF toward PAN (Tager, 2007). Figure 3.19 demonstrates that combined polymer solutions become less viscous with an increased PAN content in the mixtures.

The temperature dependence of solution viscosities were used in calculating the values of viscous flow activation heat E_a. Figure 3.20 presents the dependences of the viscosities of Cell/PAN solutions and E_a values (curves 1 and 2) on the composition of solutions. It can be seen that over the whole range of compositions studied, negative deviations of the viscosity and E_a from additivity take place; these deviations indicate rearrangements in the structure of mixed solutions with increased PAN content and the formation of some new structures (less stable in comparison to that of cellulose solutions in $[C_4C_1im]Cl$).

Since with increasing PAN fraction in polymer mixtures the composition of the binary solvent changes (as the portion of DMF becomes higher), we have also studied the changes in the viscosity of cellulose solutions in $[C_4C_1im]Cl$ depending on DMF content (Figure 3.20, curve 3). It can be seen that with an increasing portion of DMF in the binary system, a dramatic decrease in viscosity of the cellulose solution occurs. Usually, a deterioration of thermodynamic quality of a solvent (in concentrated solutions) with respect to cellulose and its derivatives is accompanied by an increase in viscosity due to the stronger association (aggregation) of solvated macromolecules (Tager, 2007).

To explain the observed decrease in viscosity of the concentrated cellulose solutions in $[C_4C_1im]Cl$ taking place with an increasing proportion of a diluent (DMF), we should take into account a sharp change in the viscosity of this binary solvent. The melting point of $[C_4C_1im]Cl$ is 73°C, and

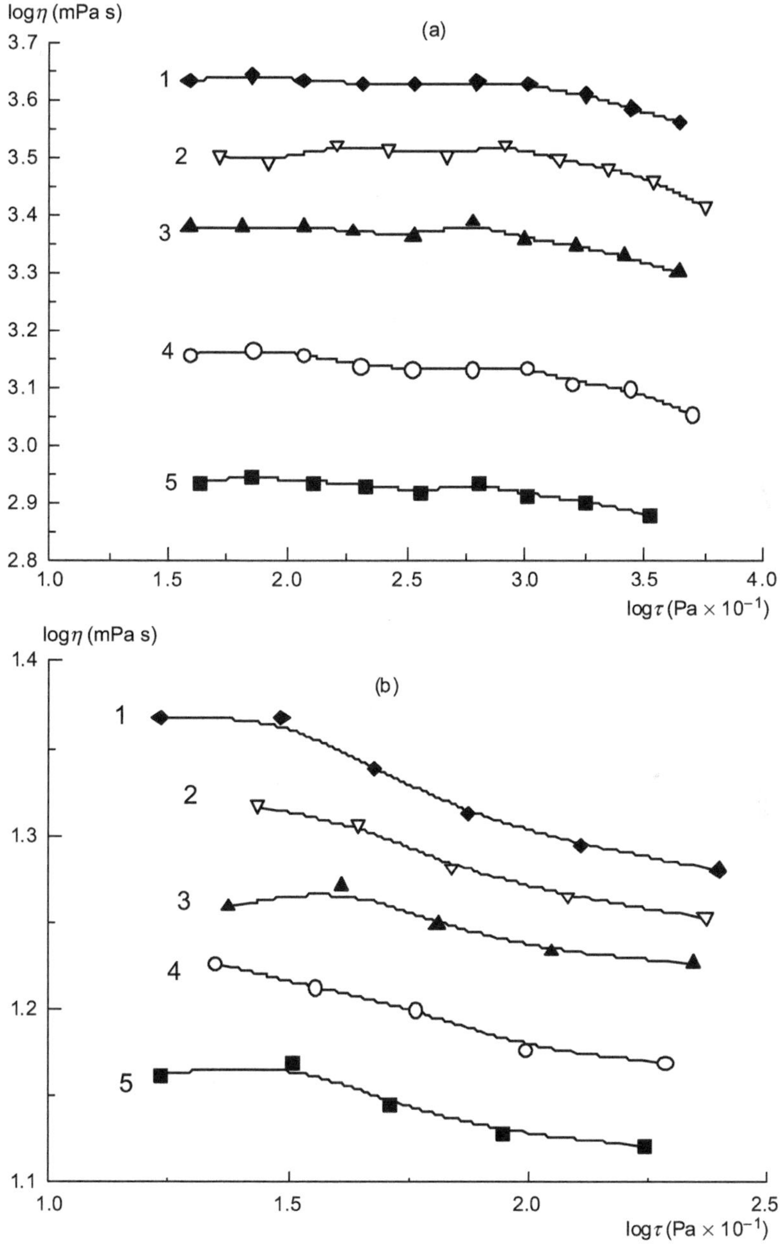

Figure 3.18 ***Flow Curves Obtained for (a) 3% Cellulose Solution in [C_4C_1im]Cl and (b) 3% PAN Solution in DMF at (1) 60, (2) 65, (3) 70, (4) 80, (5) and 90°C.***

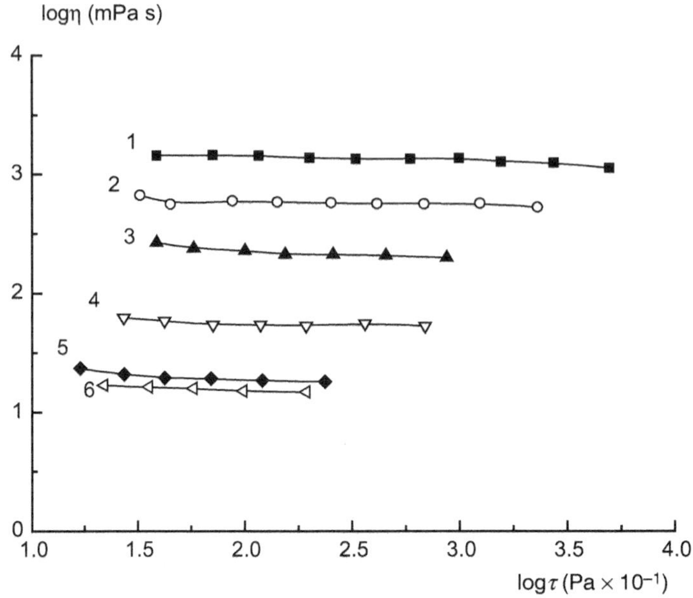

Figure 3.19 ***Flow Curves Obtained at 80°C for 3% Cellulose Solution in [C_4C_1im]Cl (1), Cellulose/PAN Mixtures in [C_4C_1im]Cl/DMF (2–5), and PAN Solution in DMF (6).*** The PAN contents in the solutions were the following: (2) 10, (3) 25, (4) 50, and (5) 75 wt%.

its viscosity is several times greater than that of DMF at this temperature. Rheological characteristics of a polymer solution are influenced by thermodynamic quality and viscosity of a solvent (Tager, 2007). Note that the observed abrupt decrease in the viscosity of Cell/PAN solutions in the binary solvent with increasing the PAN portion should be taken into consideration when solution is intended for use in composite fibers and films.

Figure 3.21 presents X-ray diffractograms of hydrated cellulose films, PAN films, as well as Cell/PAN-based films cast from combined solutions of cellulose in [C_4C_1im]Cl and PAN in DMF. The diffraction pattern of the cellulose film (Figure 3.21, curve 1) is typical of polymorphic modification of hydrated cellulose (Cell-II). PAN powders and films cast from the polymer solution in DMF give one reflection at a 2θ angle of 17°, in good agreement with literature data (Bashir, 1993).

The presence of only one weak reflection indicates that PAN precipitated from DMF solutions acquires an amorphous structure. After introducing PAN into the cellulose matrix, the intensity of cellulose reflections on X-ray diffractograms tends to decrease (Figure 3.21, curves 2–5); this tendency indicates that adding PAN to cellulose solutions inhibits cellulose crystallization during the formation of composite films. When the PAN content exceeds 50%, cellulose reflections are not observed in the diffractograms. It

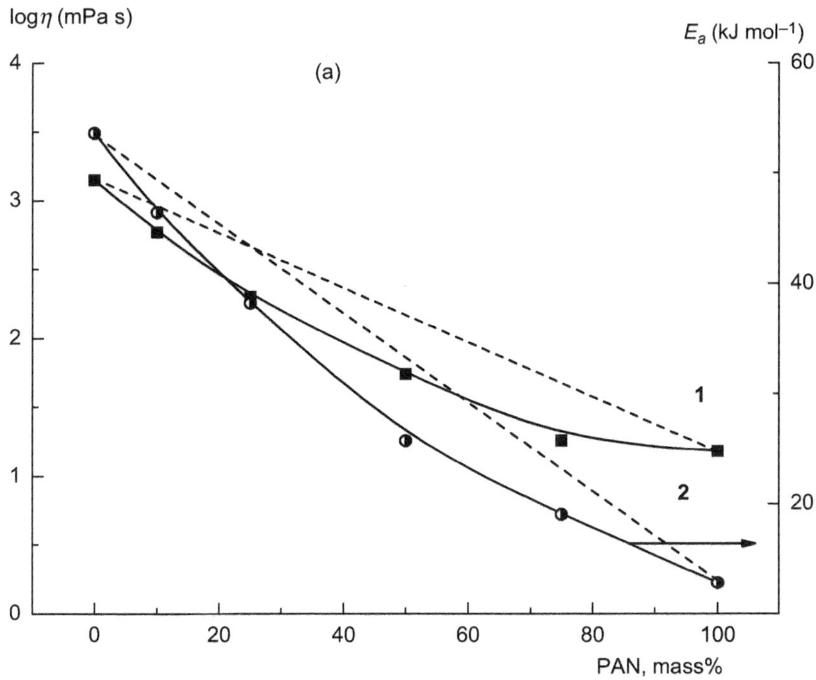

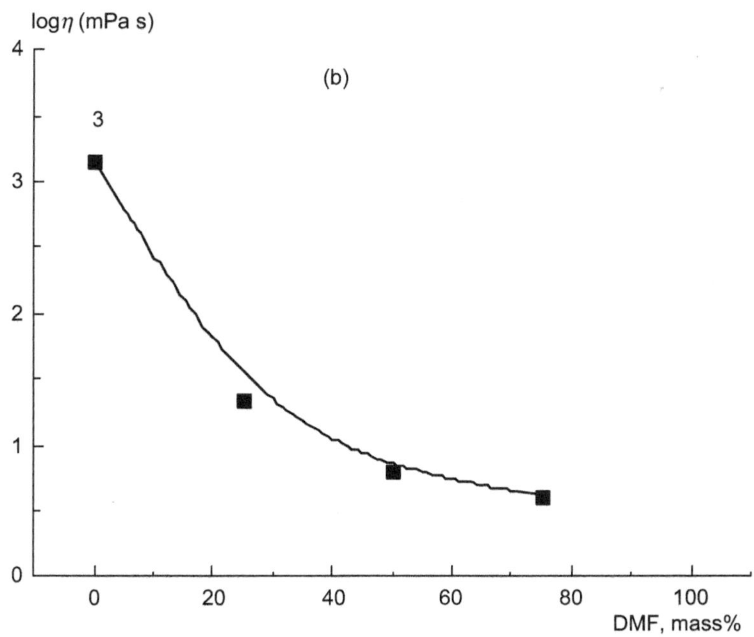

Figure 3.20 ***Dependences of viscosity: a) 3% cellulose/PAN solutions in [C4C1im]Cl/DMF at 80°C (1) and viscous flow activation heat E_a (2) on PAN content (wt%); b) 3% cellulose solutions on DMF content in the binary solvent (3) (Bochek et al., 2014).***

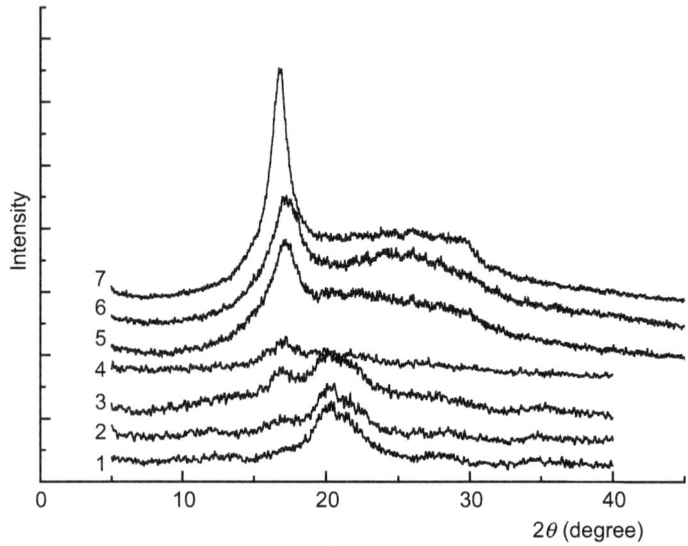

Figure 3.21 ***Diffraction patterns of a hydrated cellulose film (1), mixed hydrated cellulose/PAN films (2–5), a PAN film (6), prepared from solutions in [C_4C_1im]Cl/DMF, and the initial PAN powder (7).*** The PAN contents were (2) 10, (3) 25, (4) 50, and (5) 75 wt%.

can be assumed that partial compatibility of cellulose and polyacrylonitrile occurs in the mixtures containing more than 50% PAN.

Figure 3.22 shows TGA and DTG curves for Cell/PAN mixtures. It can be seen that the carbon residue from PAN films is 52%, while the carbon residue from hydrated cellulose films is 25% (Figure 3.22a). With an increasing PAN fraction in the composite films, the amount of carbon residue increases. These results allow us to conclude that complicated chemical reactions (PAN cyclization followed by thermal destruction of cellulose and PAN) in the composite films lead to chemical interactions between the intermediate products of polymer thermal destruction. The latter processes possibly result in the formation of cross-linked structures with properties differing from those of the initial polymers. A similar trend was observed for chitin/PAN composites prepared from DMF solutions (Sazanov et al., 2005). Overall, the comparison between mass losses of composites in the 600–700°C temperature range demonstrates that introducing PAN into the cellulose matrix leads to an increase in the thermal stability of the composite films.

The differential TGA curves (Figure 3.22b) obtained for PAN include two temperature minima in the 300–310°C and 400–425°C regions. The first area of mass loss is attributed to PAN cyclization, and the next peak corresponds to the temperature of maximum PAN thermal destruction (T_{max}) (Korobeinyk et al., 2012).

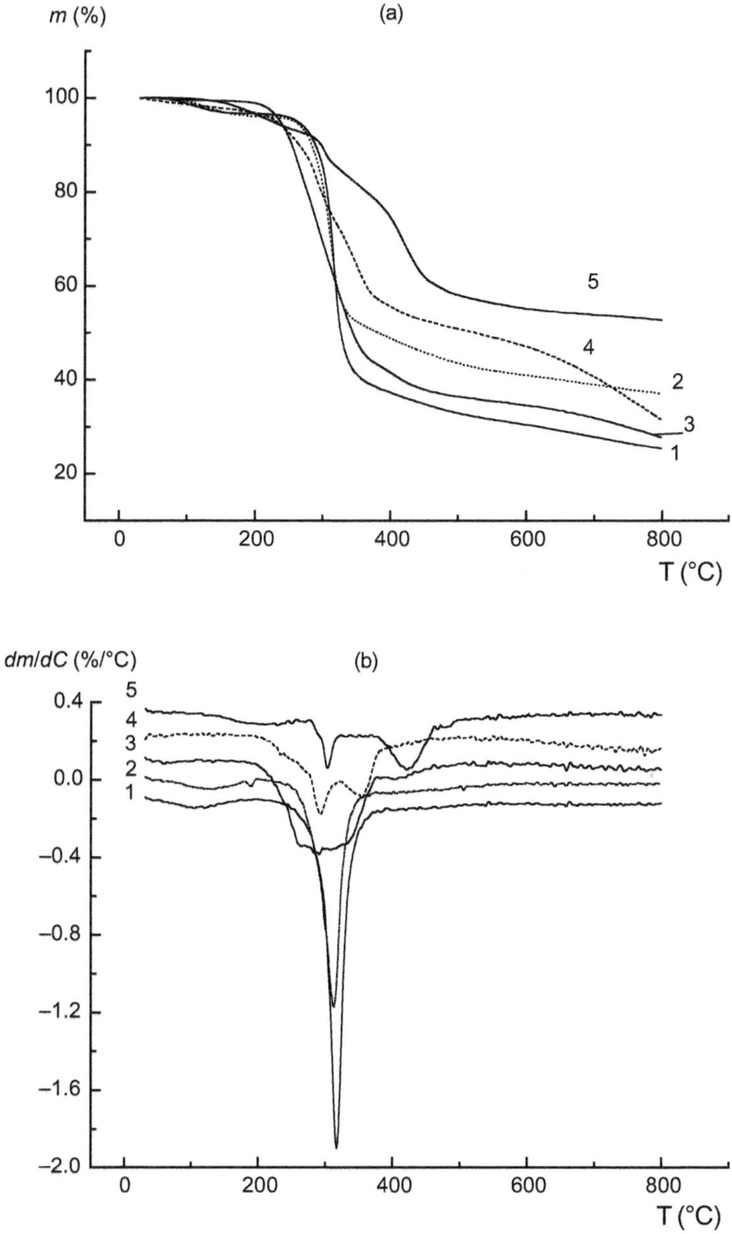

Figure 3.22 ***TGA Curves (a) and Their Derivatives (b) Obtained for Cellulose Films (1), Cellulose/PAN Mixtures (2–4), and PAN (5).*** The PAN content was (2) 25, (3) 50, and (4) 75 wt% (Bochek et al., 2014).

The maximum temperature of cellulose thermal destruction (T_{max}) was 338°C. It can be seen that with introducing PAN into the cellulose matrix, the intensity of the cellulose T_{max} peak decreases, thus indicating that PAN acts as a stabilizer in the thermal decomposition of the polysaccharide. When the PAN content in composites reaches 50%, peaks of PAN cyclization and PAN thermal destruction shift closer to each other, and the peak of cellulose destruction practically cannot be isolated, that is, the composite behaves as a single entity. This behavior implies that the two polymers are compatible. The TGA results are consistent with the previous data of X-ray structural analysis revealing that the polymers are compatible in composites containing more than 50% PAN.

Note that introduction of the cellulose into the PAN matrix leads to a decrease in the temperature of PAN thermal destruction and the initial temperature of carbonization. This phenomenon would allow preparation of CFs with lower energy expenditures (i.e., at lower temperatures), and the presence of PAN in the cellulose matrix results in an increased thermal stability of these composites as compared to that of pure cellulose.

Thus, we have obtained Cell/PAN composite films from polymer solutions in the binary solvent [C_4C_1im]Cl/DMF and studied the structural organization of these materials. In composites containing more than 50 wt% PAN, partial compatibility of the polymers was observed.

Introducing the solution of PAN in DMF into the solution of cellulose in [C_4C_1im]Cl leads to changes in solution structure, namely, disintegration of the intermolecular entanglement network and the system of hydrogen bonds inherent to cellulose. Instead, a new mixed system of weaker hydrogen bonds and a similarly weaker intermolecular entanglement network is formed. Introduction of cellulose into PAN leads to a lowering of the initial temperature of PAN cyclization, while the presence of polyacrylonitrile increases the thermal stability of the composites. The observed changes in thermal properties of the polymers with varying content of cellulose and polyacrylonitrile in their composites can be used in developing manufacturing procedures for the production of carbon fibers.

PROPERTIES OF SOLUTIONS OF POLYPEPTIDES IN IONIC LIQUIDS

Solubility of Fibroin in ILs

Ionic liquids (ILs) as solvents can also be applied for the conversion and modification of natural polypeptides. These polymers and their mixtures can be used to create new functional materials. Some types of ILs are efficient solvents not only for polysaccharides, but also for fiber- and film-forming polypeptides (keratin from hair, feathers, and wool, as well as silk and spider fibroin). Fibroin is a natural fiber-forming polypeptide. It is composed of 16 different amino acids, among them the main ones are glycine, alanine, and serine. Up to now, renewable resources of fibroin are only partially used (mainly for textile fibers). For utilization of short fibers with lower quality and silk industry waste, one can use dissolution processes in order to produce nontoxic, biocompatible, and biodegradable materials for medicine and biotechnology (films, gels, sponges) (Cheng et al., 2008; Hofmann et al., 2006; Sashina and Novoselov, 2005b; Sashina et al., 2005, 2006a, 2010).

Fibroin is a sparingly soluble scleroprotein. Its molecular structure is stabilized by hydrogen bonding, ionic bonding, and van der Waals interactions between active groups. Concentrated solutions of salts, acids, and alkali can be used as fibroin solvents; however, these solvents have no practical applications due to several reasons (Sashina et al., 2006). In acid solutions polymers have a tendency to degrade, and salt solutions must be subjected to a long and energy-intensive dialysis. Using *N*-methylmorpholine-*N*-oxide is not widely accepted (Novoselov and Sashina, 2003) because the thermal stability of this solvent is not sufficient to create a sustainable process. Taking into account their physical and chemical properties, ILs can be very promising compared to alternative solvent systems.

There are reports (Kuzmina et al., 2009; Phillips et al., 2004, 2005; Xie et al., 2006a) that fibroin can be dissolved to obtain highly concentrated solutions with 1-butyl-3-methylimidazolium chloride and 1-butyl-3-methylimidazolium acetate. However, in this dissolution process, the dissolved state of the fibroin and the possibility to regulate the properties of solutions and materials recovered from them are still poorly understood. Solubility of fibroin with ionic liquids can be characterized by enthalpies of solution. These thermodynamic characteristics can be useful to estimate the dissolving efficacy of the solvents. The enthalpy of dilute fibroin solutions does not depend on the polymer concentration, but on the nature of the solvent that is typical for cellulose solutions (Novoselov et al., 1999). Figure 3.23 shows the dependence of the enthalpies of solutions for fibroin (washed silk fibers) with $[C_4C_1im]Cl$ (Figure 3.23, line 1) and $[C_4C_1im][OAc]$ (Figure 3.23, line 2) on the concentration of the polymer in solution.

The dissolution of fibroin with $[C_4C_1im][OAc]$ is accompanied by an exothermic effect of more than 190 J g^{-1}. This value is higher than the enthalpy of solution for cellulose with *N*-methylmorpholine-*N*-oxide at 85°C (≈130 J g^{-1}; Novoselov et al. (1999)), and it is more than three times

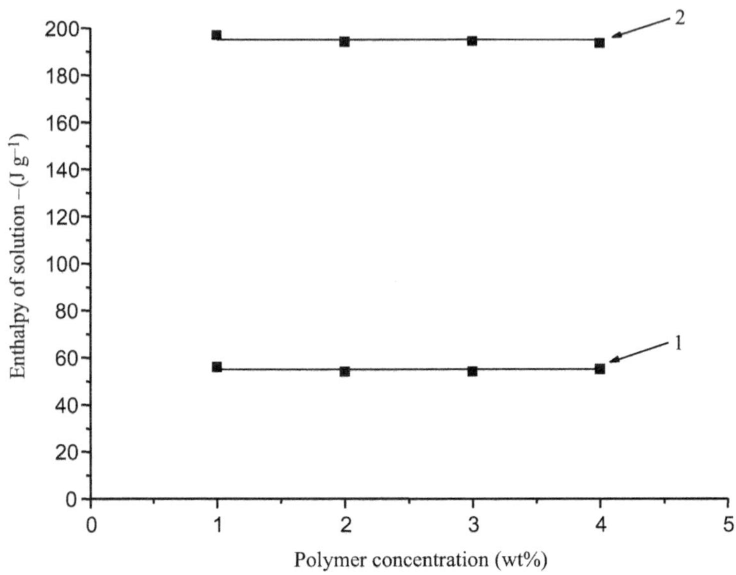

Figure 3.23 ***Enthalpies of Solutions for Silk Fibroin With ILs Solutions Prepared at 90°C.*** (1) 1-Butyl-3-methylimidazolium chloride and (2) 1-butyl-3-methylimidazolium acetate (Sashina et al., 2012).

the enthalpy of solution for fibroin with $[C_4C_1im]Cl$. This significant difference between the ΔH_{dis} values of fibroin solutions with chloride- and acetate-based ILs can be explained as follows. The integral enthalpy of solution is the following sum of the individual components:

$$\Delta H_{dis} = \Delta H_{sol-sol} + \Delta H_{pol-pol} + \Delta H_{pol-sol} + \Delta H_{dil} \quad (3.2)$$

where the first two components of the sum characterize the endothermic effect of breaking the bonds between similar molecules, $\Delta H_{pol-sol}$ is the heat of an exothermic reaction of the polymer with the solvent, and ΔH_{dil} is the enthalpy of dilution.

It is obvious that the endothermic effect of the destroying of the structure ($\Delta H_{sol-sol}$) is less significant for $[C_4C_1im][OAc]$ than for $[C_4C_1im]Cl$ because acetate IL is a liquid at room temperature but chloride has a melting point above 60°C. It is also clear that the interaction energy of the polymer macromolecules ($\Delta H_{pol-sol}$) with $[C_4C_1im][OAc]$ is higher than with $[C_4C_1im]Cl$. Therefore, the acetate IL is a better solvent for fibroin than the chloride IL. This fact is also confirmed by other data related to the dissolution of cellulose. In particular, the calculation of the interaction energy for cellulose with the solvent gives values of -100.5 kJ mol^{-1} ($[C_4C_1im]Cl$) and -129.8 kJ mol^{-1} ($[C_4C_1im][OAc]$) (Sashina and Novoselov, 2009).

The contribution of the fracture energy of the crystal structure of the polymer ($\Delta H_{pol-pol}$) to the enthalpy of solution was estimated (Sashina et al., 2012). A comparison of the enthalpy of solution for native silk fibers with the enthalpy of solution for films that have been reprecipitated from a solution of fibroin with lithium chloride allowed us to calculate that the destruction of the highly ordered β-structure of fibroin at dissolution is accompanied by the absorption of ~21 J g^{-1} of energy by the polymer.

Thus, ionic liquids can effectively destabilize the highly ordered supramolecular structure of fibroin, and translate this polymer to the dissolved state.

Properties of Fibroin Solutions With Ionic Liquids

The properties of solutions and the methods of processing the solutions to create polymeric materials are affected by such characteristics of the dissolved state as the conformation of macromolecules, the nature of their interactions, the rheological properties, and the particle size before dissolution.

It has been reported that fibroin films obtained from solutions with different solvents have unsatisfactory physical and mechanical properties and, moreover, have high brittleness. Such strong differences in the properties of regenerated and native silk fibroin can be attributed to the strong degradation of the polymer during dissolution (Holland et al., 2007), or to the characteristics of the conformational state of fibroin macromolecules. For example, fibroin has a predominantly α-helix conformation, and forms random coils with a predominance of intramolecular bonds in aqueous-salt solutions and in hexafluoroisopropanol (Chen et al., 2001; Sashina et al., 2005a). The absence of the interactions between these coils prevents film-forming from these solutions. These important properties determine the ability of fibroin solutions to be processed for continuous fiber or film formation. Moreover, little is known regarding fiber- and film-forming from fibroin solutions with ionic liquids.

Fibroin may be dissolved in either $[C_4C_1im]Cl$ or $[C_4C_1im][OAc]$ at 90°C to significant concentrations (up to 15 wt% and 20 wt%, respectively). However, the high dissolution temperature leads to a substantial degradation of the polymer. During dissolution of native fibroin with ILs at 90°C, its molecular weight reduces to 30–90 kDa (native silk fibroin after washing has a molecular weight of up to 250 kDa).

The degradation of the polymer during the dissolution process at high temperature likely explains the relatively low viscosity of fibroin solutions with ILs. This can be illustrated by the viscosity curves of 5, 10, and 15 wt% fibroin solutions with $[C_4C_1im]Cl$ and $[C_4C_1im][OAc]$ at 90°C (Figure 3.24).

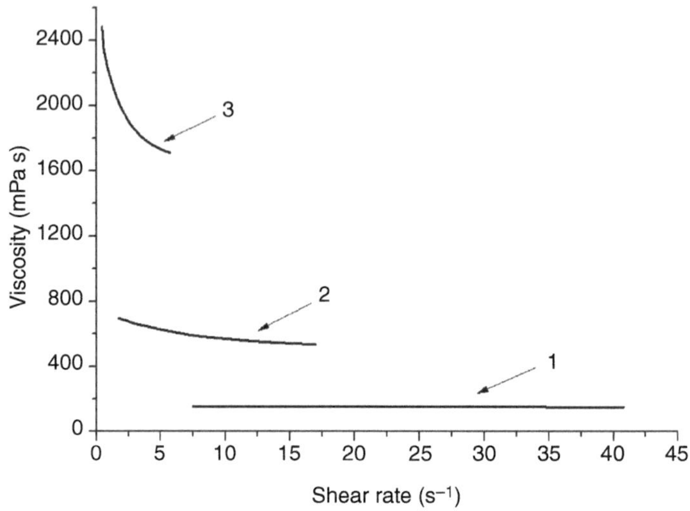

Figure 3.24 ***Dependence of the Viscosity of Silk Fibroin Solutions With [C_4C_1im][OAc] on the Shear Rate at 90°C.*** (1) 5, (2) 10, and (3) 15 wt%.

Fibroin solutions of 5 wt% concentration behave as Newtonian fluids; it can be seen that structure formation is absent in these solutions. It is probable that fibroin macromolecules in solution form compact globules. The viscosity of the solutions increases with an increase in polymer concentration, leading to increased interactions between the macromolecules in solution. At the greater gradient of the shear rate the flow behavior is changed (indicated by a change in tilt angle on the viscosity curves).

The ability of the solution structure to restore to a prior state after removal of a disruptive load shear is important for the continuous processes of polymer solution moldings. We can estimate the ability of the fibroin solutions to restore their intermolecular interaction networks after load removal by the following data. We measured the viscosities of the solutions, then allowed for 30-min relaxation after the initial viscosity data were taken (they are shown in Figures 3.24 and 3.25), then these measurements were repeated at the minimum shear rate (corresponding to the initial measurement for each of the viscosity curves). It was found that the viscosity of 5–15 wt% fibroin solutions with [C_4C_1im]Cl (after removal of strain shear) is practically unchanged. Therefore, the solution structures are confirmed to be destroyed. In contrast, the viscosity of 10–15 wt% fibroin solutions with [C_4C_1im][OAc] (after removal of strain shear and relaxation) restored 94% and 96% of their relative initial values. This fact means that the 10–15 wt% of fibroin solutions with [C_4C_1im][OAc] can be considered as thixotropic

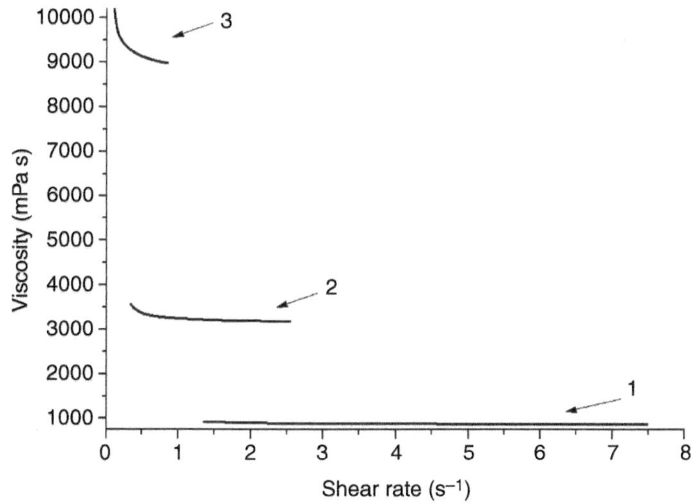

Figure 3.25 ***Dependence of the Viscosity of Silk Fibroin Solutions With [C_4C_1im]Cl on the Shear Rate at 90°C.*** (1) 5, (2) 10, and (3) 15 wt%.

fluids, in which the structural network destroyed by a mechanical action is able to recover itself after removal of the acting force. However, their incomplete recovery indicates the existence of fibroin denaturation under the action of load shear.

Restructuring in the structural networks of the dissolved fibroin occurs as well under temperature changes, and during this process there are apparent activation energy values of viscous flow. Figures 3.26 and 3.27 show the dependence of the viscosity for 15 wt% of fibroin solution with ionic liquids on temperature at different shear rates. Table 3.8 gives the activation energy values for each temperature range.

Comparing the activation energy values for fibroin solutions with [C_4C_1im]Cl and [C_4C_1im][OAc], we should take into account that 1-butyl-3-methylimidazolium chloride is a solid up to 65°C while 1-butyl-3-methylimidazolium acetate is a liquid even at room temperature. The activation energy of viscous flow for fibroin solutions with [C_4C_1im][OAc] is 80 kJ mol^{-1} at 25–40°C. To compare, aqueous collagen solutions are characterized by activation energies of 22–63 kJ mol^{-1} at 23–27°C (Vasil'ev, 2004). The observed decrease in the apparent activation energy with increased temperature is typical for solutions with synthetic polymers.

The rheological behavior of the silk fibroin solutions with ionic liquids characterizes them as globular structures, which interact strongly with one another (depending on the concentration and the temperature). The

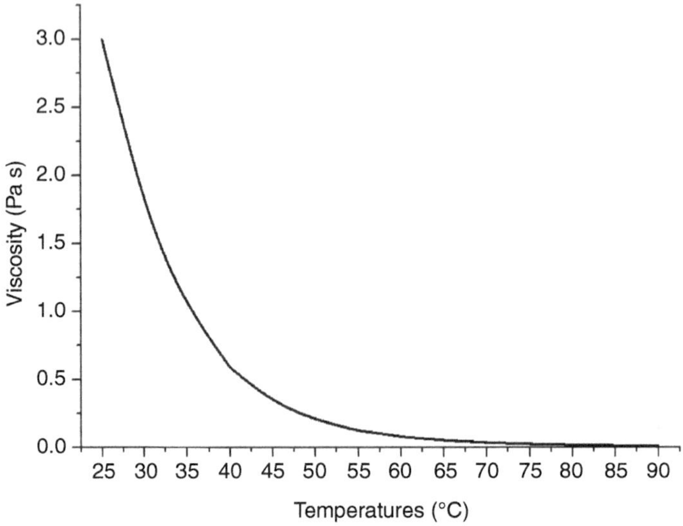

Figure 3.26 ***The Temperature Dependence of the Viscosity of 15 wt% Fibroin Solution With [C_4C_1im][OAc].***

addition of water to the solution (or another liquid that is not the fibroin solvent) leads to polymer precipitation. On the phase diagram of fibroin/IL/water systems (see Figure 3.28), regions of existence of homogeneous

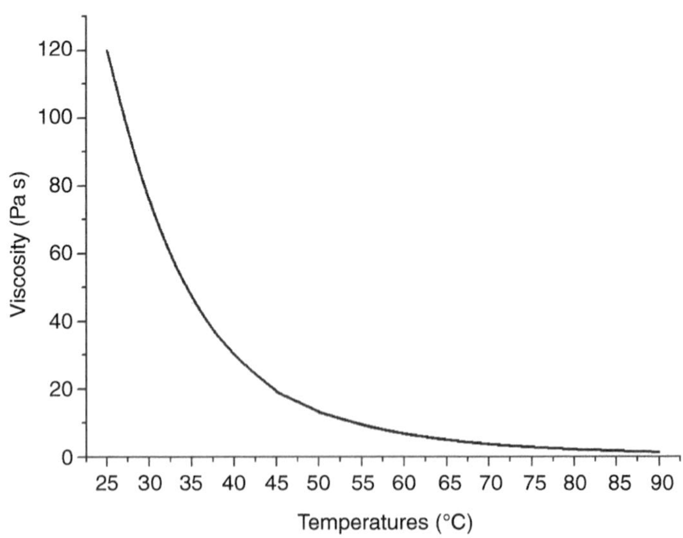

Figure 3.27 ***The Temperature Dependence of the Viscosity of 15 wt% Fibroin Solution With [C_4C_1im]Cl.***

Table 3.8 Activation energy of the viscous flow for 15 wt% fibroin solutions with ILs

Shear rate (s^{-1})	Temp. diapason (°C)	Activation energy values at the viscous flow (kJ mol^{-1})
[C_4C_1im][OAc]		
0.068	25–45	71.9
0.102	40–50	65.5
0.204	55–65	57.2
0.34	55–70	58.4
0.408	50–75	57.7
0.476	70–80	54.4
0.68	80–85	46.9
0.85	85–90	42.8
[C_4C_1im]Cl		
0.0034	25–40	82.7
0.0102	40–55	87.2
0.017	50–60	83.2
0.0238	60–65	76.6
0.0306	65–70	72.6
0.034	65–70	76.1
0.068	70–80	72.5
0.102	75–85	69.6
0.136	75–90	68.8
0.17	85–90	66.7

fibroin solutions with [C_4C_1im]Cl and [C_4C_1im][OAc] are indicated by the curves.

When we use [C_4C_1im][OAc] as a solvent, the region of unlimited component mixing for the three-phase systems is larger than for [C_4C_1im]Cl. This fact can be explained by a higher dissolving power of the acetate-containing ILs. The critical water content in the system, after which there is a delamination of the solution, is 8 wt% for [C_4C_1im]Cl, and 12.5 wt% for [C_4C_1im][OAc].

However, the mechanical properties of the fibroin films (and fibers) are extremely poor; these films are brittle and short-lived materials. Therefore, to improve the formation properties of the solutions we need methods that control the conformational state of the polymer. The transition from a globular conformation to an extended one of macromolecules can be performed by introducing specific additives into the solution, or by altering the precipitation process.

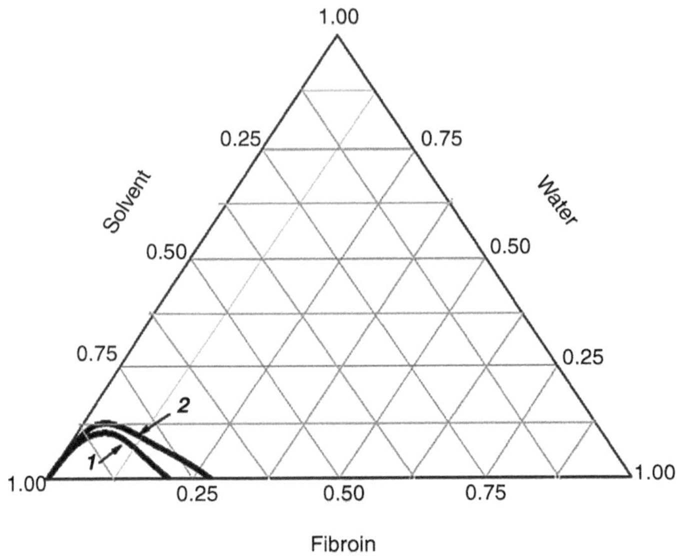

Figure 3.28 ***Ternary Phase Diagram for Fibroin/Ionic Liquid/Water Systems With Following ILs: (1) 1-butyl-3-methylimidazolium chloride and (2) 1-butyl-3-methylimidazolium acetate (Golubikhin et al., 2013).***

Regulation of the Conformational State of Fibroin Solutions in Ionic Liquids

Solutions of globular polymers do not show good fiber-forming properties, so an important challenge is to transform macromolecules, compactly folded in the solution of fibroin, into an elongated state. Fibroin macromolecules can proceed through reversible conformational transitions:

α-Helix ⟷ random coil ⟷ β-sheet

↕

Globule

Conformational changes in macromolecules can occur as a result of interactions with solvents of different natures. In aqueous-salt solutions the fibroin macromolecules have the conformation of a random coil and helix (Sashina et al., 2005a), which is preserved by dissolving it in hexafluoroisopropanol (Sashina et al., 2003). The same effect can be observed when fibroin is dissolved in $[C_4C_1im]Cl$, and $[C_4C_1im][OAc]$, as evidenced by the X-ray of cooled fibroin solutions in these ILs (see Figure 3.29).

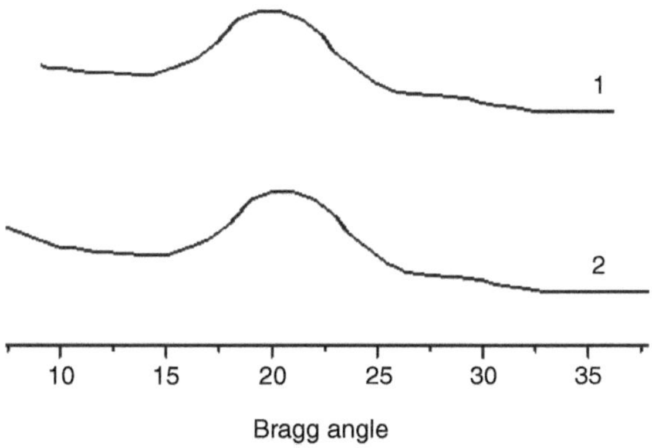

Figure 3.29 ***X-Ray Images of 12 wt% Fibroin Solutions With 1-Butyl-3-Methylimidazolium Chloride (1), and 1-Butyl-3-Methylimidazolium Acetate (2).***

Two X-ray images show the absence of reflectances caused by a β-conformation; there is only the amorphous halo ~20°.

It is known that the addition of nonpolar compounds (benzene, *o*- and *p*-xylene, chloroform, olive oil and castor oil, and carbon tetrachloride) to liquid silk (soluble fibroin, secreted by the gland of the silkworm) leads to an initiation of α-helix formation. Adding polar compounds (alcohols, acetone, and acetic acid) leads to initiation of β-sheet formation (Magoshi et al., 1985). Thus, the interaction of macromolecules with different compounds leads to different forms of spatial organization; it is associated with the effect of the hydrophobic allylic groups of the polymer macromolecules. Nonpolar solvents are surrounded by hydrophobic allylic chains, which are located in the external fibroin area of the coil or helix. Also, peptide groups interact with each other inside the conformation. In polar solvents the β-conformation is thermodynamically more stable due to the repulsion of the polar solvent molecules and the hydrophobic side chains of fibroin. Neighboring fibroin macromolecules have an antiparallel stretching, with the hydrophobic groups facing each other; such a structure is typical for the elongated sheet conformation.

It has been shown (Magoshi et al., 1985; Sashina et al., 2005a) that in polar solvents the formation of the β-sheet crystal structures of fibroin during methanol precipitation is the most intensive process. Dioxane and acetone are less effective in inducing the β-conformation in comparison to some alcohols. The effectiveness of alcohols to form polypeptides' β-conformation weakens as the hydrophobic chain length of the molecules increases, that is, the HBD properties of the molecules. The number of groups of C═O, and NH in the polypeptide structure is approximately equal, but the C═O group is a strong HBA, and at the same time, the NH group is a weak donor. Alcohols are strong HBD, and they compete with the NH groups of the polypeptides in the process of coil unwrapping. Dioxane and acetone are not stronger HBAs compared to the CO groups of the peptides, and therefore they cannot compete with the CO groups. Therefore, the interaction of the peptide bonds inside the helix of the rolled macromolecule is maintained, and the helical conformation is maintained.

Fibroin Precipitation From Solution in ILs

Taking into account the above-mentioned facts, we conclude that the impact of the precipitation bath on fibroin solution processing lies not only in the desolvation of the macromolecules, but also in some other conformational effects. It is important that the process imparts satisfactory physical and mechanical properties to the final product.

The amount and the speed of polymer desolvation during the precipitation process depends on the physico-chemical characteristics of the "solvent/precipitant" system. Thermochemical studies of "solvent/precipitant" systems (Sashina et al., 2012, 2013; Sashina and Golubikhin, 2012a; Golubikhin et al., 2013) showed that there are significant differences in the values of the enthalpy of solution for aprotic and protic precipitants with ILs. Aprotic precipitants are less intensive in destroying the "fibroin/solvent" solvate complexes; when we use these precipitants coagulation of fibroin is difficult. The use of protic precipitants, including water, results in a more complete and rapid precipitation. This may be due to a strong exothermic interaction between the protic precipitant (e.g., alcohol) and the ionic liquid. Figures 3.30 and 3.31 show concentration dependences of the enthalpy of solution for $[C_4C_1im]Cl$ and $[C_4C_1im][OAc]$ with water and alcohols.

Enthalpies of solution are comprised of the thermal effects of the interaction between solvent and solute. In general, the values of the enthalpies of solution and the strength of the solvation complexes formed by polyhydric alcohols with ILs is higher than the values for the polymers with water. However, not only the energy, but also the mobility of solvent/precipitant solvate complexes affects the precipitation process and its rate. For ionic liquids, which are ionic systems in the molten state, we can estimate the mobility of solvate complexes from the electrical conductivity values. Figures 3.32 and 3.33 show the concentration dependences of the electrical conductivity of IL/precipitant mixtures.

The IL/water solvate complexes have the highest electrical conductivities, and hence the highest mobility. At the same time, polyols that interact with the solvent in some cases more vigorously than water (we can estimate this effect by the values of the enthalpies of solution), can produce films of

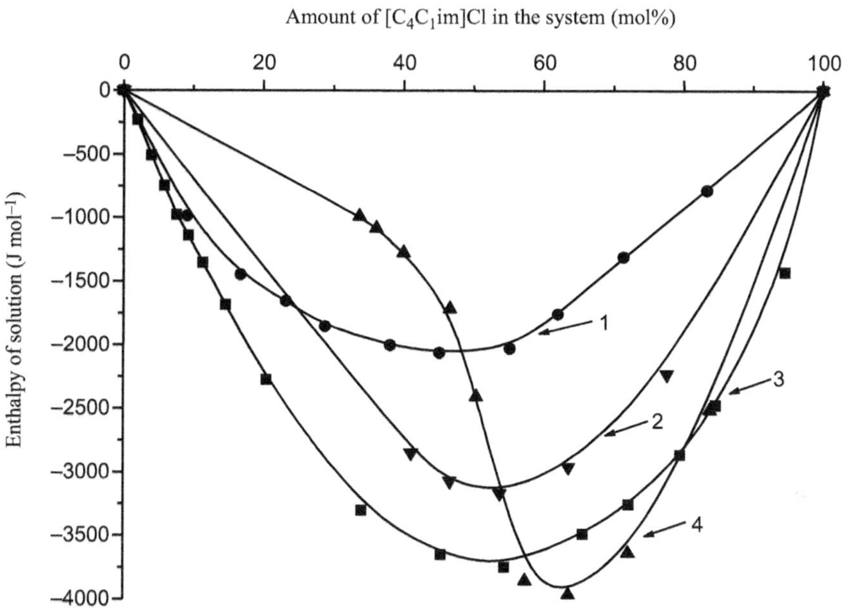

Figure 3.30 ***The Concentration Dependence of the Enthalpy of Solution for [C_4C_1im]Cl with Protic Precipitants at 90°C.*** (1) Water, (2) ethylene glycol, (3) *n*-pentanol, and (4) glycerol (Susanin et al., 2014).

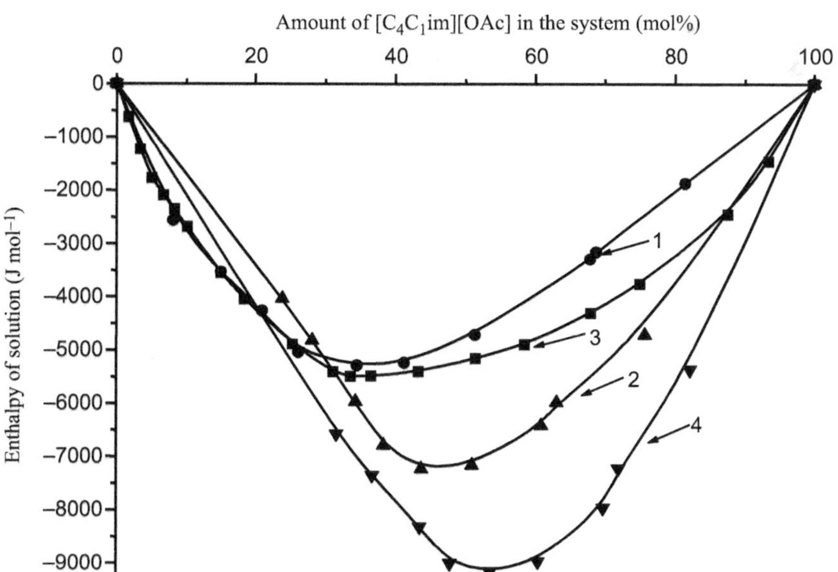

Figure 3.31 ***The Concentration Dependences of the Enthalpy of Solution for [C_4C_1im][OAc] With Protic Precipitants at 90°C.*** (1) Water, (2) ethylene glycol, (3) *n*-pentanol, and (4) glycerol (Susanin et al., 2014).

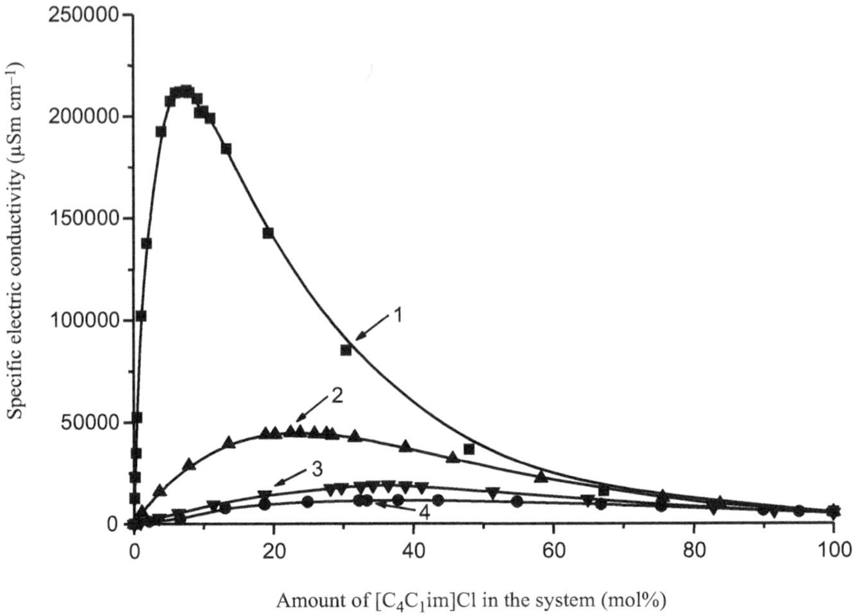

Figure 3.32 ***The Concentration Dependences of the Specific Electric Conductivity for [C_4C_1im]Cl With Precipitants.*** (1) Water, (2) ethylene glycol, (3) *n*-pentanol, and (4) glycerol (Susanin et al., 2014).

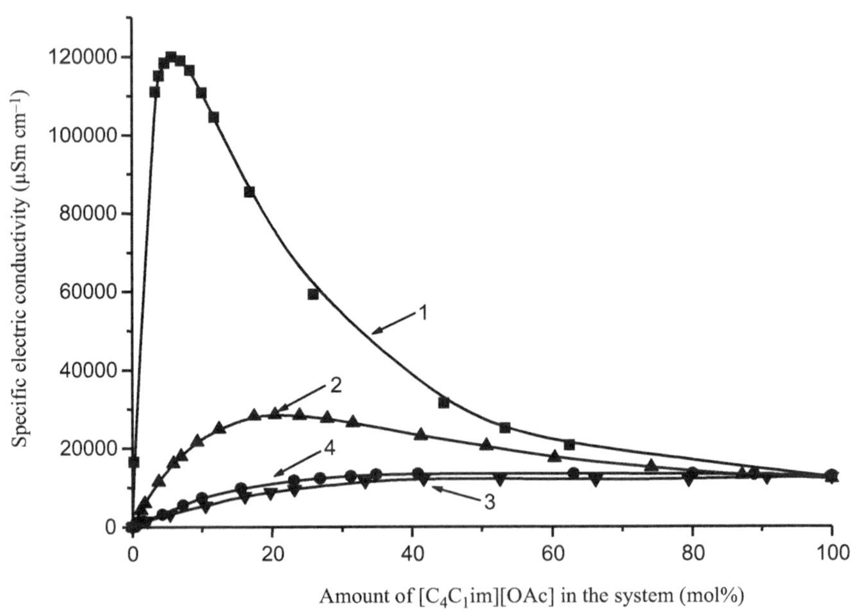

Figure 3.33 ***The Concentration Dependence of the Specific Electric Conductivity for [C_4C_1im][OAc] With Precipitants.*** (1) Water, (2) ethylene glycol, (3) *n*-pentanol, and (4) glycerol (Susanin et al., 2014).

relatively higher quality. This can be explained by a lower mobility of the solvent/precipitant solvate complexes (this mobility can be estimated from the electrical conductivity of the mixtures). When the solvent is removed slowly, the polymer surface layer stays flexible, and it is deformed by internal stresses. In this case internal stresses in the fiber are weakened. As a result, we have a more homogeneous structure of the fiber or film material.

Pyridinium- and imidazolium-based ionic liquids can give solutions of cellulose and fibroin with a high concentration for the subsequent preparation of materials for various purposes. These solvents are also ideal for the creation of blends of the natural polymers. However, a number of issues remain unresolved, but it is safe to assume that effective methods will be developed for the processing of natural polymers such, as polysaccharides and polypeptides, using ionic solvents.

ACKNOWLEDGMENTS

This work was supported by the Ministry of Education and Science of the Russian Federation (State Task No. 2014/186). The work was conducted at the Center for Collective Use at the Research Institute "Chemical Technology and Ecology" (St. Petersburg State University of Technology and Design). The authors thank the staff of the scientists of the Federal State Budgetary Institution of Science Institute of Macromolecular Compounds of Russian Academy of Sciences, St. Petersburg, for providing scientific support in the study of mixtures of cellulose.

REFERENCES

Ab Rani, M.A., Brant, A., Crowhurst, L., Dolan, A., Lui, M., Hassan, N.H., Hallett, J.P., Hunt, P.A., Niedermeyer, H., 2011. Understanding the polarity of ionic liquids. Phys. Chem. Chem. Phys. 13, 16831–16840.

Asikkala, J., 2008. Application of ionic liquids and microwave activation in selected organic reactions. Academic dissertation. The University of Oulu. Available from: http://herkules.oulu.fi/isbn9789514287190/isbn9789514287190.pdf.

Basa, M.L.T.N., 2010. Ionic liquids: solvation characteristics and cellulose dissolution. Academic dissertation. The University of Toledo. Available from: http://utdr.utoledo.edu/theses-dissertations/791/.

Bashir, Z., 1993. The effect of solvent on the X-ray scattering from polyacrylonitrile. J. Mater. Sci. Lett. 12, 1526–1528.

Bentivoglio, G., Röder, T., Fasching, M., Buchberger, M., Schottenberger, H., Sixta, H., 2006. Cellulose processing with chloride-based ionic liquids. Lenzinger Ber. 86, 154–161.

Bernet, B., Vasella, A., 2000. Intra- and intermolecular H-bonds of alcohols in DMSO, 1H-NMR analysis of inter-residue H-bonds in selected oligosaccharides: cellobiose, lactose, *N,N'*-diacetylchitobiose, maltose, sucrose, agarose, and hyaluronates. Helv. Chim. Acta 9 (83), 2055–2071.

Bikales, N., Segal, L., 1971. Cellulose and Cellulose Derivatives. Wiley, New York, Mir, Moscow, 1974.

Bochek, A.M., 2008. Prospects for use of polysaccharides of different origin and environmental problems in processing them. Fibre Chem. 40 (3), 192–197.
Bochek, A.M., Shevchuk, I.L., Kalyuzhnaya, L.M., 2006. Properties of aqueous solutions of methyl cellulose-polyvinyl alcohol blends. Russ. J. Appl. Chem. 79 (12), 2007–2012.
Bochek, A.M., Murav'ev, A.A., Novoselov, N.P., Zaborski, M., Zabivalova, N.M., Petrova, V.A., Vlasova, E.N., Volchek, B.Z., Lavrent'ev, V.K., 2012. Specific features of cellulose and chitin dissolution in ionic liquids of varied structure and the structural organization of regenerated polysaccharides. Russ. J. Appl. Chem. 85 (11), 1718–1725.
Bochek, A.M., Murav'ev, A.A., Novoselov, N.P., Zabivalova, N.M., Petrova, V.A., Yudin, V.E., Popova, E.N., Lavrent'ev, V.K., 2013. Rheological properties of mixtures of cellulose with chitin in 1-butyl-3methylimidazolium acetate of the composite films obtained. Russ. J. Appl. Chem. 86 (12), 1913–1917.
Bochek, A.M., Murav'ev, A.A., Novoselov, N.P., Popova, E.N., Sazanov, Yu.N., Lavrent'ev, V.K., 2014. Composite cellulose-polyacrylonitrile films prepared from solutions in a mixed solvent, 1-butyl-3-methylimidazolium chloride-dimethylformamide. Russ. J. Appl. Chem. 87 (5), 634–639.
Brandt, A., Hallett, J.P., Leak, D.J., Murphy, R.J., Welton, T., 2010. The effect of the ionic liquid anion in the pretreatment of pine wood chips. Green Chem. 12, 672–679.
Chand, S., 2000. Review: carbon fibers for composites. J. Mater. Sci. 35, 1303–1313.
Chen, X., Knight, D.P., Shao, Z., Vollrath, F., 2001. Regenerated *Bombyx* silk solutions studied with rheometry and FTIR. Polymer 42, 9969–9974.
Cheng, C., Shao, Z.Z., Volrath, F., 2008. Silk fibroin-regulated crystallization of calcium carbonate. Adv. Funct. Mater. 18 (15), 2172–2179.
de Oliveira, H.F.N., Rinaldi, R., 2015. Understanding cellulose dissolution: energetics of interactions of ionic liquids and cellobiose revealed by solution microcalorimetry. ChemSusChem 8 (9), 1577–1584.
Doherty, T.V., Mora-Pale, M., Foley, S.E., Linhardt, R.J., Dordick, J.S., 2010. Ionic liquid solvent properties as predictors of lignocelluloses pretreatment efficacy. Green Chem. 12, 1967–1975.
Duan, X., Xu, J., He, B., Li, J., Sun, Y., 2011. Preparation and rheological properties of cellulose/chitosan homogeneous solution in ionic liquid. BioResources 6 (4), 4640–4651.
El Seoud, O.A., da Silva, V.C., Possidonio, S., Casarano, S.R., Areas, E.P.G., Gimenes, P., 2011. Microwave-assisted derivatization of cellulose. 2 – the surprising effect of the structure of ionic liquids on the dissolution and acylation of the biopolymer. Macromol. Chem. Phys. 23 (212), 2541–2550.
Erdmenger, T., Haensch, C., Hoogenboom, R., Schubert, U.S., 2007. Homogeneous tritylation of cellulose in 1-butyl-3-methylimidazolium chloride. Macromol. Biosci. 4 (7), 440–445.
Feng, L., Chen, Z., 2008. Research progress on dissolution and functional modification of cellulose in ionic liquids. J. Mol. Liq. 142 (1–3), 1–5.
Feng, L., Xie, N., Zhong, J., 2014. Carbon nanofibers and their composites: a review of synthesizing, properties and applications. Materials 7, 3919–3945.
Fitzer, E., 1989. Pan-based carbon fibers – present state and trend of the technology from the viewpoint of possibilities and limits to influence and to control the fiber properties by the process parameters. Carbon 27 (5), 621–645.
Fukaya, Y., Sugimoto, A., Ohno, H., 2006. Superior solubility of polysaccharides in low viscosity, polar, and halogen-free 1,3-dialkylimidazolium formates. Biomacromolecules 7 (12), 3295–3297.
Gericke, M., Fardim, P., Heinze, T., 2012. Ionic liquids – promising but challenging solvents for homogeneous derivatization of cellulose. Molecules 17 (6), 7458–7502.
Golubikhin, A.Yu., Susanin, A.I., Sashina, E.S., 2013. Phase diagrams of fibroin – ionic liquid – precipitator systems [in Russian]. Mater. Design Technol. 5 (30), 40–42.

Hao, Y., Peng, J., Hu, Sh., Li, J., Zhai, M., 2010. Thermal decomposition of allyl-imidazolium-based ionic liquid studied by TGA–MS analysis and DFT calculations. Thermochim. Acta 501 (1–2), 78–83.

Hauru, L.K.J., Hummel, M., Michud, A., Sixta, H., 2014. Dry jet-wet spinning of strong cellulose filaments from ionic liquid solution. Cellulose 21 (6), 4471–4481, http://doi.org/10.1007/s10570-014-0414-0.

Hofmann, S., Wong Po Foo, C.T., Rossetti, F., Textor, M., Vunjak-Novakovic, G., Kaplan, D.L., Merkle, H.P., Meinel, L., 2006. Silk fibroin as an organic polymer for controlled drug delivery. J. Controlled Release 111 (1–2), 219–227.

Holland, C., Terry, A.E., Porter, D., Vollrath, F., 2007. Natural and unnatural silks. Polymer 48 (12), 3388–3392.

Idris, A., Vijayaraghavan, R., Ali Rana, U., Fredericks, D., Patti, A.F., MacFarlane, D.R., 2013. Dissolution of feather keratin in ionic liquids. Green Chem. 15 (2), 525–534.

Kadokawa, J., 2011. Preparation of polysaccharide-based materials compatibilized with ionic liquids. In: Kokorin, A. (Ed.), Ionic Liquids: Application and Perspectives. InTech, Rijeka, Croatia, pp. 95-114/674 p.

Kamlet, M.J., Taft, R.W., 1976. The solvatochromic comparison method. 1. The β scale of solvent hydrogen-bond acceptor (HBA) basicities. J. Am. Chem. Soc. 98 (2), 377–383.

Kamlet, M.J., Abboud, J.L., Taft, R.W., 1977. The solvatochromic comparison method. 6. The π* scale of solvent polarities. J. Am. Chem. Soc. 99 (18), 6027–6038.

Kim, S.S., Kim, S.J., Moon, Y.D., Lee, Y.M., 1994. Thermal characteristics of chitin and hydroxypropyl chitin. Polymer 35 (15), 3212–3216.

Korobeinyk, A.V., Whitby, R.L.D., Mikhalovsky, S.V., 2012. High temperature oxidative resistance of polyacrylonitrile-methylmethacrylate copolymer powder converting to a carbonized monolith. Eur. Polym. J. 48, 97–104.

Kosan, B., Michels, C., Meister, F., 2008. Dissolution and forming of cellulose with ionic liquids. Cellulose 15 (1), 59–66.

Kosan, B., Schwikal, K., Meister, F., 2010. Solution states of cellulose in selected direct dissolution agents. Cellulose 17 (3), 495–506.

Kuzmina, O.G., Sashina, E.S., Novoselov, N.P., Zaborski, M., 2009. Blends of cellulose and silk fibroin in 1-butyl-3-methylimidazolium chloride-based solutions. Fibres Text. East. Eur. 77 (6), 36–39, http://www.fibtex.lodz.pl/2009/6/36.pdf.

Kuzmina, O., Heinze, T., Wawro, D., 2012. Blending of cellulose and chitosan in alkyl imidazolium ionic liquids. Polym. Sci., 1–9, Article ID 251950.

Kuzmina, O., Jankowski, S., Fabiańska, A., Sashina, E., Wawro, D., 2014. Preswelling of cellulose pulp for dissolution in ionic liquid. Cellul. Chem. Technol. 48 (1–2), 45–51.

Lee, S.H., Doherty, T.V., Linhardt, R.J., Dordick, J.S., 2009. Ionic liquid-mediated selective extraction of lignin from wood leading to enhanced enzymatic cellulose hydrolysis. Biotechnol. Bioeng. 102 (5), 1368–1376.

Liebert, T., 2008. Innovative concepts for the shaping and modification of cellulose. Macromol. Symp. 262 (1), 28–38.

Liebert, T., 2010. Cellulose solvents – remarkable history, bright future. In: Liebert, T., Heinze, T., Edgar, K. (Eds.), Cellulose Solvents: For Analysis, Shaping and Chemical Modification. ACS Symposium Ser. 1033, 3–54.

Liebmann, B., Friedl, A., Rodrigues, J.F.C., 2012. Lignocellulosic biomass dissolution and fractioning using ionic liquids as a solvent. Chem. Eng. Trans. 29, 553–558.

Liu, W., Cheng, L., Zhang, H., Zhang, Y., Wang, H., Yu, M., 2007a. Rheological behaviors of polyacrylonitrile/1-butyl-3-methylimidazolium chloride concentrated solutions. Int. J. Mol. Sci. 8, 180–188.

Liu, Z., Remsing, R.C., Moore, P.B., Moyna, G., 2007b. Molecular dynamics study of the mechanism of cellulose dissolution in the ionic liquid 1-*n*-butyl-3-methylimidazolium chloride. ACS Symp. Ser. 975 (23), 335–350.

Liu, Z., Remsing, R.C., Moore, P.B., Moyna, G., 2007c. Ionic Liquids IV. Am. Chem. Soc. 975 (23), 253–255.
Lungwitz, R., Spange, S., 2008. A hydrogen bond accepting (HBA) scale for anions, including room temperature ionic liquids. New J. Chem. 32, 392–394.
Lv, Y., Wu, J., Zhang, J., Niu, Y., Liu, Ch.-Y., He, J., Zhang, J., 2012. Rheological properties of cellulose/ionic liquid/dimethylsulfoxide (DMSO) solutions. Polymer 53 (12), 2524–2531.
Magoshi, J., Magoshi, Y., Nakamura, S., 1985. Crystallization, liquid crystal, and fiber formation of silk fibroin. J. Appl. Polym. Sci. Appl. Polym. Symp. 41, 187–204.
Maki-Arvela, P., Anugwom, I., Virtanen, P., Sjoholm, R., Mikkola, J.P., 2010. Dissolution of lignocellulosic materials and its constituents using ionic liquids – a review. Ind. Crops. Prod. 32 (3), 175–201.
Mantz, R.A., Fox, D.M., Marshall Green, III, J., Fylstra, P.A., De Long, H.C., Trulove, P.C., 2007. Dissolution of biopolymers using ionic liquids. Zeitschrift fur Naturforschung A 62, 275–280.
Mazza, M., Catana, D.-A., Vaca-Garcia, C., Cecutti, C., 2009. Influence of water on the dissolution of cellulose in selected ionic liquids. Cellulose 16 (2), 207–215.
Michud, A., King, A., Parviainen, A., Sixta, H., Hauru, L., Hummel, M., Kilpeläinen, I., 2014. Process for the production of shaped cellulose articles.
Michud, A., Tanttu, M., Asaadi, S., Ma, Y., Netti, E., Kaariainen, P., Sixta, H., 2015. Ioncell-F: ionic liquid-based cellulosic textile fibers as an alternative to viscose and Lyocell. Textile Res. J., http://doi.org/10.1177/0040517515591774.
Mostashari, S.M., Baghi, O., Mostashari, S.Z., 2009. Thermogravimetric analysis of a cellulosic fabric treated with nickel sulfate hexahydrate as a flame-retardant. Cellul. Chem. Technol. 43 (1–3), 95–98.
Nam, Y.S., Park, W.H., Ihm, D., Hudson, S.M., 2010. Effect of the degree of deacetylation on the thermal decomposition of chitin and chitosan nanofibers. Carbohydr. Polym. 80 (1), 291–295.
Novoselov, N.P., Sashina, E.S., 2003. Rheological properties of dilute solutions of natural silk fibroin in *N*-methylmorpholine oxide. Russ. J. Phys. Chem. 77 (5), 761–763.
Novoselov, N.P., Sashina, E.S., Kozlov, I.L., Kurlykin, M.P., 1999. Polythermic study of enthalpies of solution of cellulose with *N*-methylmorpholine-*N*-oxide. Zh. Prikl. Khim. 72 (7), 1192–1194, [in Russian].
Oh, S.Y., Yoo, D.I., Shin, Y., Kim, H.C., Kim, H.Y., Chung, Y.S., Park, W.H., Youk, J.H., 2005. Crystalline structure analysis of cellulose treated with sodium hydroxide and carbon dioxide by means of X-ray diffraction and FTIR spectroscopy. Carbohydr. Res. 340 (15), 2376–2391.
Ohira, K., Abe, Y., Kawatsura, K., Suzuki, K., Mizuno, M., Amano, Y., Itoh, T., 2012. Design of cellulose dissolving ionic liquids inspired by nature. ChemSusChem 2 (5), 388–391.
Ohno, H., Fukaya, Y., 2009. Task specific ionic liquids for cellulose technology. Chem. Lett. 38 (1), 2–7.
Petropavlovskii, G.A., Bochek, A.M., Shek, V.M., 1987. Properties and application amine *N*-oxides as solvents for cellulose. Khim. Drev. 2, 3–27, [in Russian].
Phillips, D.M., Drummy, L.F., Conrady, D.G., Fox, D.M., Naik, R.R., Stone, M.O., Trulove, P.C., De Long, H.C., Mantz, R.A., 2004. Dissolution and regeneration of *Bombyx mori* silk fibroin using ionic liquids. J. Am. Chem. Soc. 126 (44), 14350–14351.
Phillips, D.M., Drummy, L.F., Naik, R.R., De Long, H.C., Fox, D.M., Trulove, P.C., Mantz, R.A., 2005. Regenerated silk fiber wet spinning from an ionic liquid solution. J. Mater. Chem. 15, 4206–4208.
Prasad, K., Murakami, M., Kaneko, A., Takada, A., Nakamura, Y., Kadokawa, J., 2009. Weak gel of chitin with ionic liquid, 1-allyl-3-methylimidazolium bromide. Int. J. Biol. Macromol. 45 (3), 221–225.
Rabideau, B.D., Agarwal, A., Ismail, A.E., 2013. Observed mechanism for the breakup of small bundles of cellulose Iα and Iβ in ionic liquids from molecular dynamics simulations. J. Phys. Chem. B 117 (13), 3469–3479.

Remsing, R.C., Swatloski, R.P., Rogers, R.D., Moyna, G., 2006. Mechanism of cellulose dissolution in the ionic liquid 1-*n*-butyl-3-methylimidazolium chloride: a ^{13}C and $^{35/37}Cl$ NMR relaxation study on model systems. Chem. Commun. 12, 1271–1273.
Rinaudo, M., 2006. Chitin and chitosan: properties and applications. Prog. Polym. Sci. 31 (7), 603–632.
Sammons, R.J., Collier, J.R., Rials, T.G., Petrovan, S., 2011. Rheology of 1-butyl-3-methylimidazolium chloride cellulose solutions. I. Shear rheology. J. Appl. Polym. Sci. 110, 1175–1181.
Sashina, E.S., Golubikhin, A.Y., 2012a. Thermochemistry of mixing imidazole-based ionic liquids with water and organic solvents. Russ. J. Gen. Chem. 82 (10), 1731–1733.
Sashina, E.S., Novoselov, N.P., 2005b. Polyelectrolyte complexes of fibroin with chitosan. Russ. J. Appl. Chem. 78 (3), 487–491.
Sashina, E.S., Novoselov, N.P., 2009. Effect of structure of ionic liquids on their dissolving power toward natural polymers. Russ. J. Gen. Chem. 79 (6), 1057–1062.
Sashina et al., 2003. A method of silk dissolution. Patent DE 10313877 (16.10.2003), from request No. 10313877.3. March 17, 2003.
Sashina, E.S., Novoselov, N.P., Heinemann, K., 2005. Properties of solutions and films of fibroin-poly-3-hydroxybutyrate polymer blends. Russ. J. Appl. Chem. 78 (1), 153–158.
Sashina, E.S., Novoselov, N.P., Vorbach, D., Meister, F., 2005a. Conformational changes in fibroin upon its dissolution in hexafluoroisopropanol. Polym. Sci. A 47 (10), 1096–1103.
Sashina, E.S., Bochek, A.M., Novoselov, N.P., Kirichenko, D.A., 2006. Structure and solubility of natural silk fibroin. Russ. J. Appl. Chem. 79 (6), 869–876.
Sashina, E.S., Vnuchkin, A.V., Novoselov, N.P., 2006a. Properties of films prepared from solutions of fibroin-cellulose blends in *N*-methylmorpholine *N*-oxide. Russ. J. Appl. Chem. 79 (5), 806–810.
Sashina, E.S., Golubikhin, A.Yu., Novoselov, N.P., 2010. The method of obtaining the silk fibroin-based film for produce contact lenses. Patent RUS 2443805. August 20, 2010.
Sashina, E.S., Golubikhin, A.Yu., Novoselov, N.P., 2012. Thermochemical study on the dissolution and regeneration of fibroin from solutions in imidazole-based ionic liquids. Russ. J. Gen. Chem. 82 (8), 1440–1443.
Sashina, E.S., Kashirskii, D.A., Zaborski, M., Jankowski, S., 2012b. Synthesis and dissolving power of 1-alkyl-3-methylpyridinium-based ionic liquids. Russ. J. Gen. Chem. 82 (12), 1994–1998.
Sashina, E.S., Golubikhin, A.Yu., Susanin, A.I., 2013. The use of ionic liquids for the utilization of lignocellulosic biomass. Izvestiya VUZ (Technol. Light Industry) 22 (4), 38–40, [in Russian].
Sashina, E.S., Kashirskii, D.A., Janowska, G., Zaborski, M., 2013a. Thermal properties of 1-alkyl-3-methylpyridinium halide-based ionic liquids. Thermochim. Acta 568, 185–188.
Sashina, E.S., Kashirskii, D.A., Jankowski, S., 2014. PMR study of structural features of ionic liquids based on 1-alkyl-3-methylpyridinium and mechanism of their interaction with cellulose. Fibre Chem. 45 (5), 268–273.
Sashina, E.S., Kashirskii, D.A., 2015. Pyridinium-based ionic liquids – application for cellulose processing. In: Handy, S. (Ed.), Ionic Liquids – Current State of the Art. InTech. Rijeka, Croatia, pp. 389–417.
Sashina, E.S., Kashirskii, D.A., Busygin, K.N., (in Press). Dissolution of cellulose with pyridinium-based ionic liquids: effect of chemical structure and interaction mechanism. Cellul. Chem. Technol. 2.
Sazanov, Yu.N., Novoselova, A.V., Nud'ga, L.A., Ugolkov, V.L., Fedorova, G.N., Kulikova, E.M., Kostycheva, D.M., Gribanov, A.V., 2005. Cocarbinization of blends of polyacrylonitrile with chitin and chitosan. Russ. J. Appl. Chem. 78 (8), 1320–1324.
Sixta, H., Michud, A., Hauru, L., Asaadi, S., Ma, Y., King, A.W.T., Beadle, C., 2015. Ioncell-F: a high-strength regenerated cellulose fibre. Nordic Pulp Paper Res. J. 30 (1), 43–57.
Steiner, T., 1998. Hydrogen-bond distances to halide ions in organic and organometallic crystal structures: up-to-date database study. Acta Crystallogr. B 54, 456–463.

Susanin, A.I., Golubikhin, A.Yu., Sashina, E.S., 2014. The problem of the selection ionic solvent – precipitant systems for optimization of processes at the obtained of biopolymer materials [in Russian]. Mater. Design Technol. 5, 64–68.

Swatloski, R.P., Spear, S.K., Holbrey, J.D., Rogers, R.D., 2002. Dissolution of cellose with ionic liquids. J. Am. Chem. Soc. 124 (18), 4974–4975.

Taft, R.W., Kamlet, M.J., 1976. The solvatochromic comparison method. 2. The α-scale of solvent hydrogen-bond donor (HBD) acidities. J. Am. Chem. Soc. 98 (10), 2886–2894.

Tager, A.A., 2007. Physical Chemistry of Polymers. In: Askadskii, A.A. (Ed.). Nauchnyi Mir, Moscow [in Russian].

Takegawa, A., Murakami, M.-aki., Kaneko, Y., Kadokawa, J.-ichi., 2010. Preparation of chitin/cellulose composite gels and films with ionic liquids. Carbohydr. Polym. 79 (1), 85–90.

Tamura, K., Nakamura, N., Ohno, H., 2012. Cytochrome c dissolved in 1-allyl-3-methylimidazolium chloride type ionic liquid undergoes a quasi-reversible redox reaction up to 140°C. Biotechnol. Bioeng. 109 (3), 729–735.

Tang, M.M., Bacon, R., 1964. Carbonization of cellulose fibers – I. Low temperature pyrolysis. Carbon 2 (3), 211–214.

Troshenkova, S.V., Sashina, E.S., Novoselov, N.P., Petrenko, V.E., 2008. A semiempirical investigtion of a model for the complex formed by fibroin with 1,1,1,3,3,3-hexafluoropropan-2-ol. Fibre Chem. 40 (3), 376–380.

Van de Velde, K., Kiekens, P., 2004. Structure analysis and degree of substitution of chitin, chitosan and dibutyrylchitin by FT-IR spectroscopy and solid state ^{13}C NMR. Carbohydr. Polym. 58 (4), 409–416.

Varlamov, V., Bratskaya, Sv., Yakovleva, I., Senel, S., 2011. Advances in chitin science. Proceedings of the Tenth International Conference of the European Chitin Society, S.- Petersburg, May 20–24, 2011, 372 p.

Vasil'ev, M.P., 2004. Collagen Filament. Fiber- and Film-Based Materials. Saint-Petersburg State University of Technology and Design [in Russian], Saint-Petersburg.

Ventura, S.P.M., Neves, C.M.S.S., Freire, M.G., Marrucho, I.M., Oliveira, J., Coutinho, J.A.P., 2009. Evaluation of anion influence on the formation and extraction capacity of ionic-liquid-based aqueous biphasic systems. J. Phys. Chem. B 113 (27), 9304–9310.

Wendler, F., Kosan, B., Krieg, M., Meister, F., 2009. Cellulosic shapes from ionic liquids modified by activated charcoals and nanosilver particles. Lenzinger Ber. 87, 106–116.

Wendler, F., Todi, L.-N., Meister, F., 2012. Thermostability of imidazolium ionic liquids as direct solvents for cellulose. Thermochim. Acta 528 (1), 76–84.

Wu, Y., Sasaki, T., Irie, S., Sakurai, K., 2008. A novel biomass-ionic liquid platform for the utilization of native chitin. Polymer 49 (9), 2321–2327.

Wu, Y., Sasaki, T., Kazushi, K., Seo, T., Sakurai, K., 2008a. Interactions between spiropyrans and room-temperature ionic liquids: photochromism and solvatochromism. J. Phys. Chem. B 112 (25), 7530–7536.

Xiao, W., Chen, Q., Wu, Y., Wu, T., Dai, L., 2010. Dissolution and blending of chitosan using 1,3-dimethylimidazolium chloride and 1-*H*-3-methylimidazolium chloride binary ionic liquid solvent. Carbohydr. Polym. 83 (1), 233–238.

Xie, H., Zhang, S., Li, S., 2006. Chitin and chitosan dissolved in ionic liquids as reversible sorbents of CO_2. Green Chem. 8, 630–633.

Xie, H., Li, S., Zhang, S., 2006a. Ionic liquids as novel solvents for the dissolution and blending of wool keratin fibers. Green Chem. 7, 606–608.

Yang, H., Yan, R., Chen, H., Lee, D.H., Zheng, C., 2009. Characteristics of hemicellulose, cellulose and lignin pyrolysis. Fuel 86, 1781–1788.

Ye, X., Kennedy, J.F., Li, B., Xie, B.J., 2006. Condensed state structure and biocompatibility of the konjac glucomannan/chitosan blend films. Carbohydr. Polym. 64 (4), 532–538.

Yen, M.-T., Yang, J.-H., Mau, J.-L., 2009. Physicochemical characterization of chitin and chitosan from crab shells. Carbohydr. Polym. 75 (1), 15–21.

Yin, J., Luo, K., Chen, X., Khutoryanskiy, V.V., 2006. Miscibility studies of the blends of chitosan with some cellulose ethers. Carbohydr. Polym. 63 (2), 238–244.
Zhang, S., Lu, X., Zhou, Q., Li, X., Zhang, X., Li, S., 2009. Ionic Liquids: Physicochemical Properties, first ed. Elsevier Science Ltd., Amsterdam-Oxford, 520 p.
Zhang, J., Zhang, H., Wu, J., Zhang, J., He, J., Xiang, J., 2010a. NMR spectroscopic studies of cellobiose solvation in EmimAc aimed to understand the dissolution mechanism of cellulose in ionic liquids. Phys. Chem. Chem. Phys. 12 (8), 1941–1947.
Zhang, J., Zhang, H., Wu, J., Zhang, J., He, J., Xiang, J., 2010b. Reply to "Comment on 'NMR spectroscopic studies of cellobiose solvation in EmimAc aimed to understand the dissolution mechanism of cellulose in ionic liquids' by R.C. Remsing, I.D. Petrik, Z. Liu and G. Moyna, Phys. Chem. Chem. Phys., 2010, 12, DOI: 10.1039/c004203j. Phys. Chem. Chem. Phys. 12 (44), 14829–14830.
Zherenkova, L.V., Khalatur, P.G., 2010. Nanostructure self-organization of ionic liquids. Russ. J. Phys. Chem. A 6 (84), 987–993.
Zhu, S., Wu, Y., Chen, Q., Yu, Z., Wang, C., Jin, S., Ding, Y., Wu, G., 2006. Dissolution of cellulose with ionic liquids and its application: a mini-review. Green Chem. 8, 325–327.

CHAPTER 4

Formation of Metal–Ligand Complexes in Ionic Liquids and Perspective of ILs Application in Purification Technologies

Konstantin Popov, Vyacheslav Chistov, Elena Esipova, Julia Dikareva, Andrei Vendilo
R&D Fine Chemicals Centre, Moscow, Russia

FORMATION OF COMPLEXES IN ILS

Room temperature ionic liquids (RTILs) are gaining increasing interest as a unique medium for complex formation and development of luminescent, nano, and other novel inorganic materials (Richter et al., 2013, 2015; Steinrueck and Wasserscheid, 2015; Dai et al., 2014; Wimberg et al., 2014; Takao et al., 2013; Yan et al., 2010; Murding and Tang, 2010; Billard and Gaillard, 2009; Nockemann et al., 2008, 2009; Cocalia et al., 2006a,b; Billard et al., 2003). Among the most promising fields are RTIL-based lithium batteries (Yim et al., 2015; Zhao and Bostrom, 2015; Cecchetto et al., 2012; Lewandowski and Swiderska-Mocek, 2009) and recent applications of ionic liquids (ILs) in separation technology (Li et al., 2010; Vander Hoogerstraete et al., 2015; Pospiech and Kujawski, 2015; Perales et al., 2014; Shkrob et al., 2014; Billard et al., 2013; Sun et al., 2012; Dietz, 2006).

Most of these applications require knowledge of the thermodynamics of complex formation in RTILs. This chapter is intended as a review of data on the formation of complexes of cations, in hydrophilic and hydrophobic RTILs (Table 4.1), with chelating agents (Table 4.2) of relevance to inorganic cation extraction and separation processes. It represents an update of our previous review (Popov et al., 2011) since the number of papers on the stability constants of complexes in RTILs has doubled during the last 5 years.

Application, Purification, and Recovery of Ionic Liquids
http://dx.doi.org/10.1016/B978-0-444-63302-6.00004-2

Table 4.1 Chemical formulae and abbreviations of RTILs commonly used in this chapter

Chemical formula	Abbreviation	Name, Comment
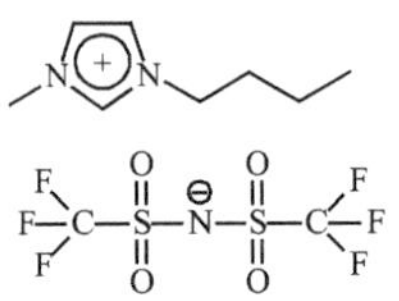	$[C_4C_1im][NTf_2]$	1-Butyl-3-methylimidazolium bis[(trifluoromethyl)sulfonyl]imide, hydrophobic RTIL
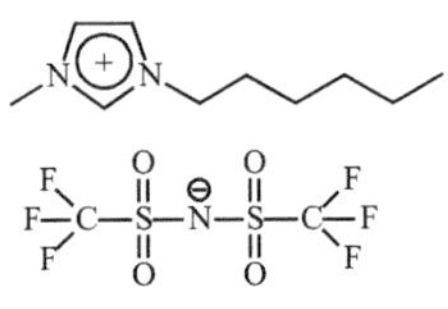	$[C_6C_1im][NTf_2]$	1-Hexyl-3-methylimidazolium bis[(trifluoromethyl)sulfonyl]imide, hydrophobic RTIL
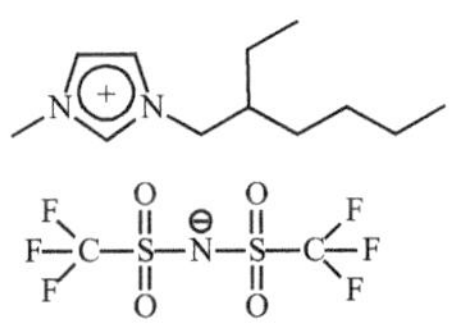	$[(C_2)^2C_6C_1im][NTf_2]$	1-(2-Ethylhexyl)-3-methylimidazolium bis[(trifluoromethyl)sulfonyl]imide, hydrophobic RTIL
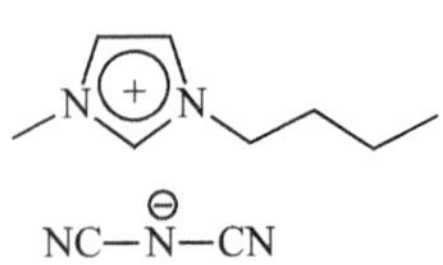	$[C_4C_1im][N(CN)_2]$	1-Butyl-3-methylimidazolium dicyanamide, hydrophilic RTIL
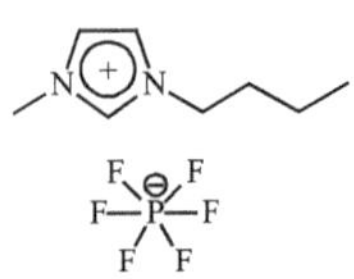	$[C_4C_1im][PF_6]$	1-Butyl-3-methylimidazolium hexafluorophosphate, hydrophobic RTIL
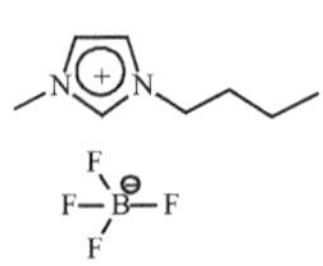	$[C_4C_1im][BF_4]$	1-Butyl-3-methylimidazolium tetrafluoroborate, hydrophilic RTIL
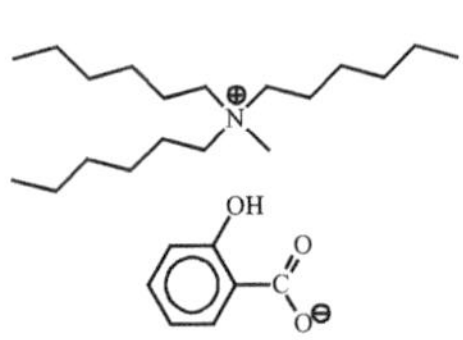	$[N_{8881}][Sal]$	Trioctylmethylammonium salicylate, task-specific hydrophobic RTIL

Table 4.1 Chemical formulae and abbreviations of RTILs commonly used in this chapter *(cont.)*

Chemical formula	Abbreviation	Name, Comment
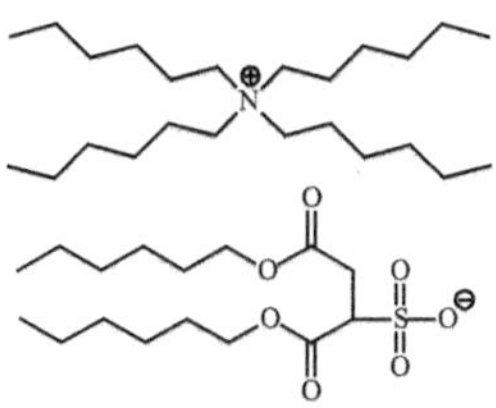	$[N_{6666}][DHSS]$	Tetrahexylammonium dihexylsulfosuccinate, task-specific hydrophobic RTIL
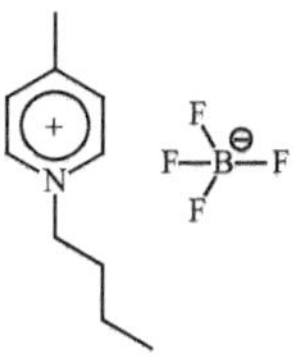	$[C_4C_1py][BF_4]$	*N*-Butyl-4-methyl-pyridinium tetrafluoroborate, hydrophilic RTIL

RTILs are intensively studied in solvent extraction processes due to such important advantages over conventional organic solvents as:

- negligible vapor pressures;
- low flammability;
- high moisture stability;
- high temperature stability;
- wide temperature range of liquid phase existence;
- relatively high radiation stability;
- wide electrochemical window;
- easy tunability (e.g., a possibility to change physicochemical properties through the appropriate combination of cations and anions);
- possibility to eliminate aqueous phase acidification during extraction.

Inorganic salts usually have poor solubility in both hydrophobic and hydrophilic RTILs at a level of 0.05–0.005 mol dm^{-3} (Pereiro et al., 2012; Yan et al., 2010; Seddon et al., 2000; Djigailo, 2010). However it may exceed 0.5 mol dm^{-3} at 70°C for $NiCl_2$ and $Cu(NO_3)_2$ (Abebe et al., 2013) and even 4 mol dm^{-3} at ambient temperature for $LiClO_4$ in $[C_4C_1im][SCN]$ (Rosol et al., 2009).

Thus, similar to the case of conventional organic solvents, some complexing agents are required for successful extraction of cations from an aqueous solution into a hydrophobic RTIL (Sun et al., 2012; Cocalia et al., 2006a; Visser et al., 2000). It has been demonstrated that the extraction efficacy of RTILs can be modulated by addition of a chelating agent. Dai et al. (1999),

Table 4.2 Chemical formulae and abbreviations of chelating agents commonly used in this chapter

Chemical formula	Abbreviation	Name
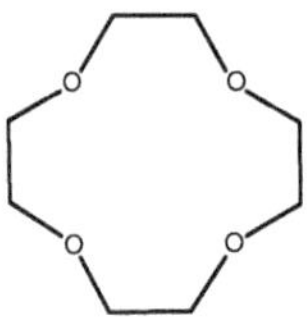	12C4	12-crown-4
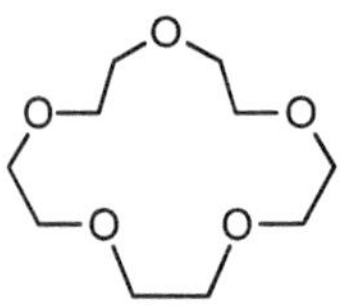	15C5	15-crown-5
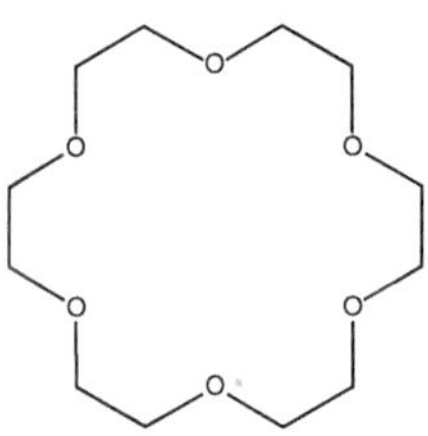	18C6	18-crown-6
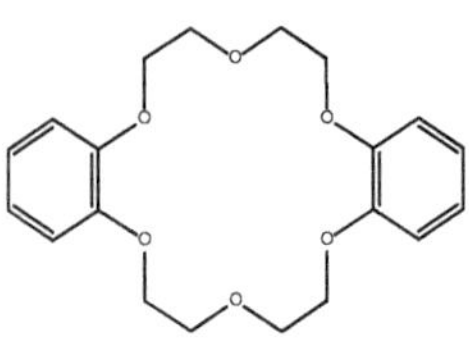	DB18C6	Dibenzo-18-crown-6
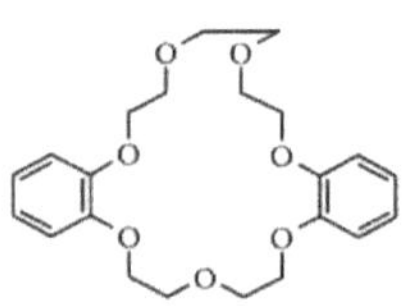	DB21C7	Dibenzo-21-crown-7
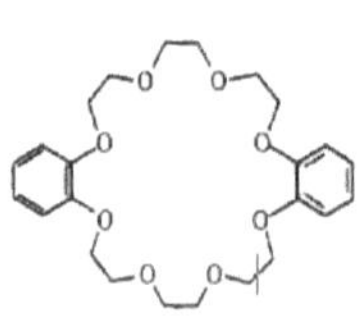	DB24C8	Dibenzo-24-crown-8
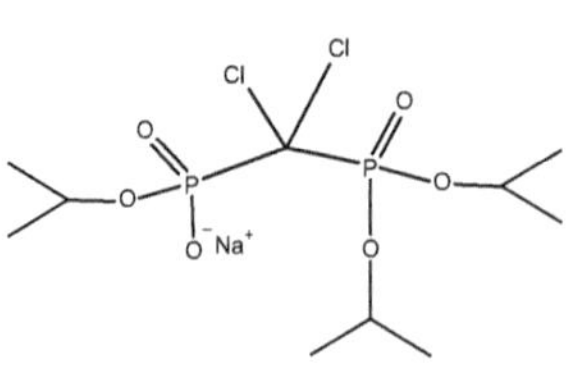	DCPTE	Dichloromethylene bis(phosphonic acid) triisopropyl ether sodium salt

for example, first discovered that highly efficient extraction of strontium ions can be achieved when dicyclohexane-18-crown-6 (DC18C6) is combined with 1-ethyl-3-methylimidazolium bis[(trifluoromethyl)sulfonyl]imide. At the same time, it has been shown that polydentate chelating agent tetraglyme when used as an additive in an *N*-methyl-*N*-propylpyrrolidinium bis(trifluoromethylsulfonyl) imide $[C_3C_1pyrr][NTf_2]/0.65\ mol\ dm^{-3}\ LiNTf_2$ dramatically enhances the transport properties of this electrolyte (Bayley et al., 2010). Recently the chelating agents are either added to the RTIL or are synthetically included into the RTIL's anion or cation structure (task-specific RTILs or TSILs). This in turn requires knowledge of the particular complexes participating in the extraction equilibria and the chemical species involved.

However, the number of reports on complex composition, equilibrium constants, and complex formation thermodynamics is surprisingly small compared to that of conventional molecular solvents (Table 4.3). Indeed, in the most comprehensive International Union of Pure and Applied Chemistry (IUPAC) database with more than 400,000 equilibrium constants the corresponding data for RTILs are few and far between (IUPAC, 2007). It should be noted that the data on $\Delta H°$ and $\Delta S°$ of complex formation (Keypour et al., 2014; Banić et al., 2014; Vendilo et al., 2014; Popov et al., 2011, 2013) are even less common.

Besides the pure RTILs listed in Table 4.3, the stability of some complexes was studied in mixed RTIL/molecular solvent systems:

- cesium and sodium complexes with DB18C6 in $[C_4C_1py][BF_4]$/water-mixed solvents by nuclear magnetic resonance (NMR) ^{133}Cs, ^{23}Na at 25°C (Popov et al., 2007a,b);
- lithium complexes with 12-crown-4, 15-crown-5, 18-crown-6, Benzo-15-crown-5 in a binary molten salt mixture: 55/45 mol% 1-methyl-3-ethylimidazolium chloride and $AlCl_3$ by $^7Li^+$ NMR, 22°C (Eyring et al., 1993);
- lithium complexes with 12-crown-4, 15-crown-5, cryptands C222, C221, C211 in 50% by weight mixtures of $[C_2C_1im][PF_6]/CH_3CN$ and $[C_2C_1im][BF_4]/CH_3CN$ at 25°C by $^7Li^+$ NMR (Samsipur and Irandoust, 2012);
- Li^+, Mn^{2+}, Zn^{2+}, Cd^{2+} complexes with 3,6-bis(2-pyridylmethyl)-5-methyl-3,6-diazahexane-1-amine and 4,7-bis(2-pyridylmethyl)-6-methyl-4,7-diazaheptane-1-amine in 50 and 25% mixtures $[C_2C_1im][BF_4]/CH_3CN$ at 25–60°C by $^7Li^+$ NMR (Keypour et al., 2014);
- complexes of dioxovanadium (V) with *D*-(–)-quinic acid in water/$[C_4C_1im][BF_4]$ (Majlesi and Rezaienejad, 2013).

Table 4.3 Complexes in RTIL, characterized by stability constants (updated from Popov et al., 2011)

Cation	Ligand	RTIL	Method	Ref.
UO_2^{2+}	NO_3^-	$[C_4C_1im][NTf_2]$	UV–Vis, 18.5°C	Georg et al. (2010)
Nd^{3+}	NO_3^-	$[C_4C_1im][NTf_2]$	UV–Vis, 20–70°C	Liu et al. (2013)
Nd^{3+}	NO_3^- Cl^-	$[C_4C_1im][NTf_2]$	UV–Vis, room temperature	Ansari et al. (2015a)
Eu^{3+}	Calix[4]arene derivatives	$[C_4C_1im][NTf_2]$	UV–Vis, 25–60°C	Ansari et al. (2015b)
Am^{3+}	TODGA††	$[C_6C_1im][PF_6]$	Cyclic voltammetry; room temperature	Sengupta et al. (2013a)
Ag^+	Cryptand 222	$[C_2C_1im][BF_4]$, $[C_4C_1im][BF_4]$, $[C_2C_1im][CF_3SO_3]$ $[C_2C_1im][NTf_2]$ $[C_4C_1im][NTf_2]$ $[C_3C_1pyrr][NTf_2]$ $[C_4C_1pyrr][NTf_2]$	Potentiometry; Ag electrode, 25°C	Lewandowski et al. (2005)
Li^+, K^+	18C6	$[C_2C_1im][NTf_2]$	^{13}C-NMR, 25°C	Umecky et al. (2014)
H^+, Li^+, Na^+K^+, Rb^+, Cs^+, NH_4^+	Valinomycin	[TDMA][TFPB]★	Cyclic voltammetry; 25°C	Langmaier and Samec (2009)
Li^+, Na^+, K^+, Rb^+	DB18C6	[ODIQ][TFBP]★★	Cyclic voltammetry; 56°C	Nishi et al. (2006)
Li^+	OH^-	$[N_{6666}][DHSS]$	^{7}Li-NMR, 25°C	Djigailo (2010)
Cs^+	12C4, 15C5	$[C_4C_1im][NTf_2]$	^{133}Cs-NMR, 25–50°C in a water-saturated RTIL	Vendilo et al. (2014)

Cs^+	18C6	$[C_4C_1im][BF_4]$, $[N_{6666}][DHSS]$ $[C_4C_1im][NTf_2]$, $[C_4C_1py][BF_4]$ $[C_4C_1im][PF_6]$, $[N_{8881}][Sal]$ $[C_4C_1im][N(CN)_2]$ $[C_4C_1im][CH_3SO_4]$ $[C_6C_1im][NTf_2]$ $[(C_2)^2C_6C_1im][NTf_2]$	^{133}Cs-NMR, 25–50°C in water saturated RTILs	Popov et al. (2007) Vendilo et al. (2009b, 2010c)
Cs^+	DB18C6	$[C_4C_1im][NTf_2]$ $[N_{6666}][DHSS]$, $[N_{8881}][Sal]$	NMR ^{133}Cs, 25–50°C	Vendilo et al. (2011)
Cs^+	DB21C7, DB24C8	$[C_4C_1im][NTf_2]$	NMR ^{133}Cs, 25–50°C	Vendilo et al. (2012)
Li^+	15C5, 18C6	$[C_4C_1im][NTf_2]$	NMR ^{7}Li, 25–50°C	Popov et al. (2013)
Sr^{2+}	DCPTE	$[C_4C_1im][N(CN)_2]$	NMR ^{31}P, 25°C	Vendilo et al. (2010d)
Ni^{2+}, Pb^{2+}	2-aminothiophenol	$[C_4C_1im][PF_6]$	25°C, extraction	Lertlapwasin et al. (2010)
Co^{2+}	Cl^-, NO_3^-	$[HORC_1im]\ NO_3$†	UV–Vis, 25–55°C	Banić et al. (2014)
Cu^{2+}	Ethylenediamine	$[C_4py][NO_3]$	UV–Vis, 25–45°C	Song et al. (2005)

* Tridodecylmethylammonium tetrakis[3,5-bis(trifluoromethyl)phenyl]borate.
** *N*-octadecylisoquinolinium tetrakis[3,5-bis(trifluoromethyl)phenyl]borate.
† 1-(2-Hydroxyethyl)-3-methylimidazolium nitrate.
†† *N,N,N′,N′*-tetraoctyl digycolamide.

All of the constants mentioned in the current chapter represent so-called "conditional" equilibrium constants defined as:

$$\beta_n = [ML_n][M]^{-1}[L]^{-n} \tag{4.1}$$

where [M], [L], and $[ML_n]$ correspond to the concentrations of cation M, ligand L, and complex species ML_n (electrostatic charges are omitted for simplicity). In the same way:

$$K_n = [ML_n][ML_{n-1}]^{-1}[L]^{-1} \tag{4.2}$$

Complex formation studies in aqueous solutions requires ionic strength control (Critical Surveys, 1975; Popov and Wanner, 2005), which is normally provided by an inert supporting electrolyte added in amounts that exceed by several orders of magnitude the concentration of the reagents participating in the equilibrium. Therefore, correct comparison of equilibrium data is possible only when they are referred to at equal ionic strength values.

In the case of RTILs, the ionic strength is provided by the solvent itself. Thus the ionic strength I depends on the density and molar mass of an RTIL and varies over a rather wide range. For example, $I = 5.35$ mol dm^{-3} for $[C_4C_1im][BF_4]$, 4.83 for $[C_4C_1im][PF_6]$, and 2.91 for $[C_4C_1im][NTf_2]$. In our review we have used the "conditional" constants in the form that they have been published, although these can be transformed into concentration constants and recalculated to some uniform ionic strength value if required.

The reliability of equilibrium complex formation data is a serious concern for IUPAC (Critical Surveys, 1975). Thus, the experimental data are given along with the values reported previously for RTILs and molecular solvents, and are compared with IUPAC recommended (*R*) or provisional (*P*) values where possible.

Unfortunately, up to now, there have been no RTIL systems studied by at least two independent research groups under the same experimental conditions. At the same time, the stability constants published evidently have different quality. For example, all constants presented in Eyring et al. (1993) have too high standard deviations to be treated by IUPAC even as provisional. Indeed, log K_1 values for Li^+ complexes with 12-crown-4, 15-crown-5, benzo-15-crown-5, and 18-crown-6 in a mixture of RTILs are reported to be 3.1 ± 2.3, 4.1 ± 3.5, 3.5 ± 2.9, and 1.8 ± 0.6, respectively

(Eyring et al., 1993). Sengupta et al. (2013a) did not provide such important experimental details as metal-to-ligand ratio and standard deviations for stability constants. In the report by Ansari et al. (2015a) temperature is not specified.

Some constants (Lertlapwasin et al., 2010) are not actually stability constants, but "effective" ones. Those published for valinomycin (Langmaier and Samec, 2009) seem to be overestimated, insofar as they exceed any known constants for this ligand in any solvent by several orders of magnitude. However, the data that are reported for Ag^+ seem to be reliable, as the value presented by this group for dimethyl sulfoxide (DMSO) fits in very well with the corresponding data of several independent research groups.

Some of the log *K* values for RTILs are presented in Tables 4.4–4.6 along with the stability of the same complexes in some molecular solvents (IUPAC, 2007; Arnaud-Neu et al., 2003; Kikuchi and Sakamoto, 2000).

As can be seen from Table 4.4 the stability of such different species as $[UO_2(NO_3)]^+$, $[Nd(NO_3)]^{2+}$, $[Ag(Cryptand\ 2.2.2)]^+$, and $[Cs(18C6)]^+$ in a "classical" RTIL is generally higher than in water, but lower than in less polar molecular solvents. The most drastic increase related to water is observed for nitrate complexes (Liu et al., 2013; Georg et al., 2010). This is the same result we recently obtained for complexes of Sr^{2+} with dichloromethylene bis(phosphonic acid) triisopropyl ether: $\log K_1 = 3.5$ in $[C_4C_1im][N(CN)_2]$, while in water $\log K_1 = -0.37$ (Vendilo et al., 2010d). Banić et al. (2014) for the $[CoCl_4]^{2-}$ complex in 1-(2-hydroxyethyl)-3-methylimidazolium nitrate reported higher stability than in water and nearly the same as in 1,2-dichloroethane (1,2-DCE) (IUPAC, 2007).

This trend can be observed clearly for $[Cs(18\text{-crown-}6)]^+$ stability in a number of hydrophobic and hydrophilic RTILs (Table 4.6). The only exception registered up to now is in the case of a task-specific RTIL, $[N_{6666}]$[DHSS]. Here the anion of the IL competes with 18-crown-6 for Cs^+ and therefore this diminishes the $\log K_1$ of the $[Cs(18\text{-crown-}6]^+$ complex. Thus, this complex is less stable in this RTIL, and that should be considered at extraction of cations from aqueous solutions.

It has been noted that the $\log K_1$ values for RTILs falls in between the range of those for the molecular solvents acetonitrile and water (Vendilo et al., 2010a; Popov et al., 2013). At the same time, the $\log K_1$ values for an alkali metal cation with crown ethers in molecular solvents is correlated (Mei et al., 1977) with the donor number (DN) of the solvent

Table 4.4 Stability constants (log K_1) of complexes in RTILs and in molecular solvents at 25°C

Complex	$[C_4C_1im][NTf_2]$	H_2O*	CH_3CN*	DMSO*	References
$[UO_2NO_3]^+$	4.8	0.3	7.9		Georg et al. (2010)
$[NdNO_3]^{2+}$, dry RTIL	7.88	−0.19			Liu et al. (2013)
$[NdNO_3]^{2+}$, humid RTIL	3.23				Ansari et al. (2015a)
[Eu(calix[4]arene)]$^{3+\dagger}$, humid RTIL	7.47				Ansari et al. (2015b)
$[Ag(cryptand\ 222)]^+$	10.0	9.6	9.01	8.4	Lewandowski et al. (2005)
$[Li(18C6)]^+$, humid RTIL	2.0	~0	2.32	~0	Popov et al. (2013)
$[Li(18C6)]^+$, dry RTIL**	2.6				Umecky et al. (2014)
$[Li(18C6)]^+$, humid RTIL**	2.5				Umecky et al. (2014)
$[Cs(15C5)]^+$, humid RTIL	1.75	0.8	2.9–3.2		Vendilo et al. (2014)

* Data for molecular solvents are taken from IUPAC (2007) and Arnaud-Neu et al. (2003) for ionic strength 0–0.2 mol dm^{-3}.
** In $[C_2C_1im][NTf_2]$.
† Calix[4]arene is appended with four pyrazine pendent arms containing diphenyl phosphoryl substituent.

Table 4.5 Stability constants (log K_1) of alkali metal complexes in RTILs and molecular solvents* at 25°C. Updated from Popov et al. (2011)

	DB18C6			Valinomycin		15C5	
M^+	[TDMA][TFBP], Langmaier and Samec (2009)	Water	Methanol	[ODIQ][TFBP], Nishi et al. (2006)	Methanol	$[C_4C_1im][NTf_2]$, Vendilo et al. (2014); Popov et al. (2013)	Water
Li^+	5.0	–	–	11.1	–	1.66	~0
Na^+	7.0	1.16	4.4	12.8	0.67–11	–	
K^+	8.2	1.67	5.05	17.2	4.48–4.90	–	
Rb^+	7.3	1.08	4.23	15.7	4.81	–	
Cs^+	–	–	–	15.1	3.9	1.6	0.8

* Data for molecular solvents are taken from IUPAC (2007) and Arnaud-Neu et al. (2003) for ionic strength 0–0.2 mol dm^{-3}; –, no data available.

Table 4.6 Stability constants (log K_1) and thermodynamic quantities ΔH°_1, ΔS°_1 of cesium complexes with 18C6 in RTILs and molecular solvents at 25°C

Solvent	DN*	log K_1	ΔH°_1 (kJ · mol^{-1})	ΔS°_1 (J · mol^{-1} · K^{-1})	References
1,2-Dichloroethane	0	7.98			Kikuchi and Sakamoto (2000)
Acetonitrile	14.1	4.8	−17	35	Arnaud-Neu et al. (2003)**
Propylene carbonate	15.1	4.50	−43	−18	Arnaud-Neu et al. (2003)
Acetone	17.0	4.51	−53	−27	Arnaud-Neu et al. (2003)
$[C_6C_1im][NTf_2]$		4.4	−10	53	Vendilo et al. (2010c)
DMFA	26.6	3.64	−49	−28	Arnaud-Neu et al. (2003)
$[(C_2)^2C_6C_1im][NTf_2]$		3.4	9	100	Vendilo et al. (2010c)
$[C_4C_1im][NTf_2]$		3.4	−6.8	42	Vendilo et al. (2010c)
DMSO	29.8	3.04			Arnaud-Neu et al. (2003)
MeOH	31.5	4.6	−47	−71	Arnaud-Neu et al. (2003)
$[C_4C_1im][N(CN)_2]$		3.03	−47	−30	Popov et al. (2007a)
$[C_4C_1im][BF_4]$		2.8	−80	−65	Popov et al. (2007a)
$[C_4C_1py][BF_4]$		2.6	−47	−32	Popov et al. (2007a)
$[C_4C_1im][PF_6]$		2.4	−21	−25	Vendilo et al. (2010c)
$[N_{8881}][Sal]$		1.43	3.9	40	Vendilo et al. (2010c)
$[C_4py][MeSO_4]$		1.20			Vendilo et al. (2008)
Water	33	0.96	−17	−39	Arnaud-Neu et al. (2003)
$[N_{6666}][DHSS]$		0.76	2.1	22	Vendilo et al. (2010c)

* Gutmann and Vichera (1966).
** Log K_1 values presented by Arnaud-Neu et al. (2003) indicate IUPAC selection and correspond to $I = 0–0.1$ mol dm^{-3}.

(Table 4.6). There are some exceptions for $[Cs(18C6)]^+$ with pyridine: DN 33.1; log K_1 = 5.7 (Mei et al., 1977) and with MeOH (Table 4.6). In this sense, the log K_1 values presented in Table 4.6 suggest that the hydrophobic RTILs $[C_6C_1im][NTf_2]$, $[C_4C_1im][NTf_2]$, $[C_4C_1im][PF_6]$, $[N_{8881}]$ [Sal] as well as the hydrophilic RTILs $[C_4C_1im][N(CN)_2]$, $[C_4C_1py][BF_4]$, $[C_4pyr][MeSO_4]$ have a DN between 33 (water) and 14 (acetonitrile), and span the same range that polar molecular solvents do, but are rather far removed from such nonpolar solvents as 1,2-DCE (DN 0; Kikuchi and Sakamoto, 2000). This observation gives much wider frontiers to the data reported earlier by Nishi et al. (2006) for Li^+, Na^+, K^+, and Rb^+ complexes with dibenzo-18-crown-6 in a hydrophobic RTIL *N*-octadecylisoquinolinium tetrakis[3,5-bis(trifluoromethyl)phenyl]borate, which is demonstrated to have a DN between 4.4 (nitrobenzene) and 0 (1,2-DCE). Indeed, for cesium complexes with 15C5 and 18C6 in $[C_4C_1im][NTf_2]$ one can get a rather good fit for this RTIL within 25 < DN < 28 (Figure 4.1).

There is also some qualitative agreement between the log K_1 values and solvent polarity derived from the E^N_T scale – an empirical scale based on the solvatochromic pyridinium *N*-phenolate betaine dye (Reichardt, 2005), and with Fourier transform infrared spectroscopy (FTIR)-based estimations (Köddermann et al., 2006). An examination of Table 4.6 reveals that the polarity of RTILs with a range of 0.60–0.65 fits well into the existing empirical solvent polarity scale for molecular solvents characterizing them not as "superpolar," but as quite normal polar solvents. In this respect,

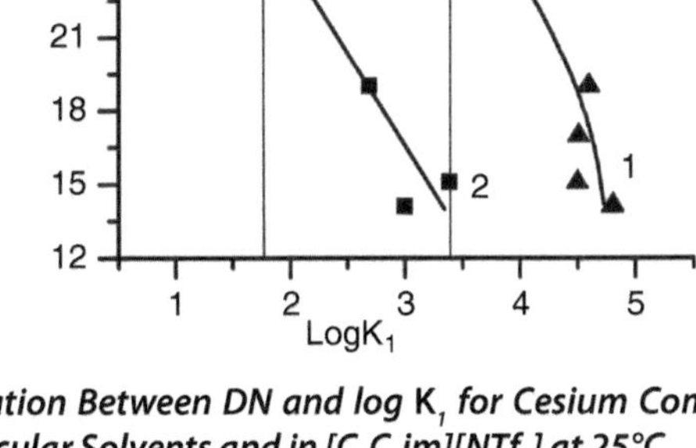

Figure 4.1 ***A Correlation Between DN and log K$_1$ for Cesium Complexes with 18C6 (1) and 15C5 (2) in Molecular Solvents and in $[C_4C_1im][NTf_2]$ at 25°C.***

it is interesting to note that the water content in RTILs strongly affects both the stability constants (Ansari et al., 2015; Banić et al., 2014; Popov et al., 2007a,b) and the E^N_T values (Reichardt, 2005). All log K_1 values in Table 4.6 correspond to water-saturated samples; this might be an explanation for the nonideal correlation between E^N_T and stability constants.

A tendency to form complexes with two ligands (ML_2) decreases symbaticly with DN, or antibaticly with E^N_T. Therefore, Cs^+ forms only *ML* complexes with 18-crown-6 in water (E^N_T 10), DMSO (E^N_T 0.45), $[C_4C_1im][PF_6]$, $[N_{8881}][Sal]$, and $[C_4C_1im][BF_4]$ (E^N_T 0.66–0.67), while in weakly coordinating solvents 1,2-DCE (E^N_T ~0.3), acetone (E^N_T 0.35), DMFA (E^N_T 0.4), $[C_6C_1im][NTf_2]$ (E^N_T 0.654), and $[C_4C_1im][NTf_2]$ (E^N_T 0.596), species of ML_2 are also registered. The log K_2 values for $[(C_2)^2C_6C_1im][NTf_2]$, $[C_6C_1im][NTf_2]$, and $[C_4C_1im][NTf_2]$ appeared to be 2–3 log units lower than log K_1 (Popov et al., 2011). These values correspond well to those found for molecular solvents, and for UO_2^{2+} complexes with NO_3^- in $[C_4C_1im][NTf_2]$ (Georg et al., 2010).

Indeed, the bis[(trifluoromethyl)sulfonyl]imide anion $[NTf_2]^-$ has very weak solvating ability, being easily substituted in the coordination sphere of UO_2^{2+} and Nd^{3+} even by a nitrate ion, which demonstrates a very high stability of complexes in $[C_4C_1im][NTf_2]$. It was long believed that $[NTf_2]^-$ does not form complexes. However, the crystal structures of $[C_3C_1im][Eu(NTf_2)_4]$, $[C_4C_1im][Eu(NTf_2)_4]$, $[C_4C_1py]_2[Eu(NTf_2)_5]$ (Tang et al., 2008), and $[C_3C_1pyrr]_2[Yb(NTf_2)_5]$, where C_3C_1pyrr indicates 1-methyl-1-propylpyrrolidinium (Murding et al., 2005), have demonstrated that it can coordinate as a bidentate ligand via oxygen atoms forming either 6-membered or 4-membered rings:

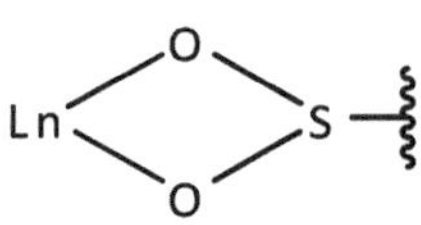

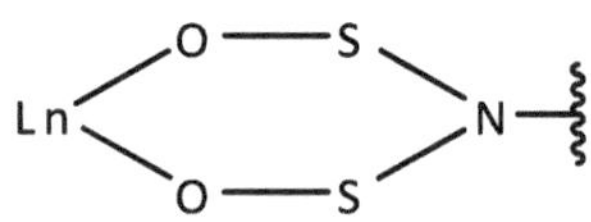

Similar structures have been found for alkaline earth metals in $[C_3C_1pyrr]_2[Ca(NTf_2)_4]$, $[C_3C_1pyrr]_2[Sr(NTf_2)_4]$, and $[C_3C_1pyrr][Ba(NTf_2)_3]$ (Babai and Murding, 2006). Moreover, even the alkali ions are found to form weak bonds with $[NTf_2]^-$. An X-ray structure of $[K(18C6)][NTf_2]$ (Yan et al., 2010) revealed that the $[NTf_2]^-$ anion forms an asymmetric bridge between two $[K(18C6)]^+$ units. One unit has a monodentate bond with an oxygen atom of one Tf fragment of the NTf_2^- anion, while another Tf arm is coordinated via two oxygen atoms resulting in the formation of a chelate ring with K^+ to the second $[K(18C6)]^+$ unit. Potassium ions therefore have the coordination numbers of 7 and 8.

Saturation of RTILs with water generally leads to a decrease in stability constant values for alkali metal complexes with crown ethers. For $[Cs(18C6)]^+$ in $[C_4C_1py][BF_4]$/water-mixed solvents a linear dependence of log K_1 on RTIL content was observed (Popov et al., 2007a,b). This behavior is similar to that in water/ethanol mixtures (Arnaud-Neu et al., 2003).

With the exception of $[N_{6666}]$[DHSS] and $[N_{8881}]$[Sal], cesium nitrate solubility in RTILs is much less than in water. The metal–solvent interaction is likely to be stronger in water than in the RTILs; that is, less energy is needed to break the metal–(RTIL anion) bonds. Thus, the differences in metal–solvent interactions are expected to make complex formation more exothermic in RTILs than in water. Indeed, ΔH°_1 of $[Cs(18C6)]^+$ formation is negative for all hydrophilic RTILs, for hydrophobic $[C_6C_1im][NTf_2]$, $[C_4C_1im][NTf_2]$, and $[C_4C_1im][PF_6]$, and for almost all polar molecular solvents (Table 4.6).

Thus, it can be seen that the enthalpy change promotes complex formation in hydrophilic RTILs, whereas the corresponding change of entropy is negative and provides the decomposition of $[Cs(18C6)]^+$. However, regarding ΔS°_1 this is not the case for hydrophobic RTILs $[C_6C_1im][NTf_2]$, $[C_4C_1im][NTf_2]$, $[N_{8881}]$[Sal], and $[N_{6666}]$[DHSS]. These reveal a positive entropy change like that for acetonitrile. Moreover, the entropy change gives the dominating contribution to $[Cs(18C6)]^+$ stability in $[C_4C_1im][NTf_2]$, $[C_4C_1im][NTf_2]$, and $[N_{8881}]$[Sal]. For as yet unclear reasons only one hydrophobic RTIL ($[C_4C_1im][PF_6]$) demonstrates the same behavior as hydrophilic RTILs and polar molecular solvents (Table 4.6).

Thus, the thermodynamic quantities clearly indicate that the contributions to the overall stability of $[Cs(18C6)]^+$ complexes may differ rather significantly. The complexation of Cs^+ is the most exothermic in $[C_4C_1im][BF_4]$. Moreover, the observed ΔH^0_1 value for $[Cs(18C6)]^+$ is highest in both molecular solvents and RTILs. At the same time, the corresponding entropy change for this solvent is also highest, diminishing the enthalpy contribution to log K_1. The data represented in Table 4.6 indicate that both the cation and anion of the RTIL affect the stability of complex formation and the change of thermodynamic function.

This is not simply explained in terms of the solvation of the cesium ion and 18C6. The tentative scheme of complex formation in RTILs shown in (4.3) is more complicated than that in molecular solvents (Ohtsu and Ozutsumi, 2003).

In general, both ions forming the IL (its cation Z^+ and anion X^-) may interact with cesium complex constituents. Z^+ is competing with cesium for the ligand, while X^- solvates cesium, thereby resisting complex formation:

$$CsX_n^{\,1-n} + Z_mL^{m+} \rightleftarrows CsLX_{n-p}^{\,1-n+p} + mZ^+ + pX^- \tag{4.3}$$

In this respect ^{133}Cs NMR data can be considered as a very sensitive probe for such species as $CsLX_{n-p}^{\,1-n+p}$ formation. The NMR chemical shift data (Popov et al., 2011) indicate that the crown ether does not substitute all RTIL anions X^- in the coordination sphere of cesium in $[Cs(18C6)]^+$ complexes, even in triflimide-based RTILs (Figure 4.2a). This observation agrees well with X-ray structural data for cesium complexes with 18-crown-6 in the solid state (Gjikaj and Adam, 2006; Ellerman et al., 1998) and in an aqueous solution (Ozutsumi et al., 1989). In all of these structures cesium is located above the mean oxygen plane of the crown ether ring since its size is larger than the cavity size of 18C6 (170 and 130 pm, respectively (Arnaud-Neu et al., 2003)). Thus, the exposed part of Cs^+ may interact with RTIL anions making a coordination number equal to 8 or 9, which are observed crystallographically for complexes in molecular solvents. This supposition is also in good agreement with structural data for $[K(18C6)][NTf_2]$ (Yan et al., 2010).

It is interesting to note that ^{133}Cs NMR chemical shifts of $[Cs(18C6)_2]^+$ are essentially independent of the solvent's nature, providing almost the same values for RTILs and for molecular solvents. This observation indicates that in a "sandwich"-type complex the two-crown ether molecules effectively shield the cesium ion from interaction with the solvent (Figure 4.2b). By contrast, the chemical shifts of $[Cs(18C6)]^+$ are strongly solvent dependent both in RTILs and in molecular solvents.

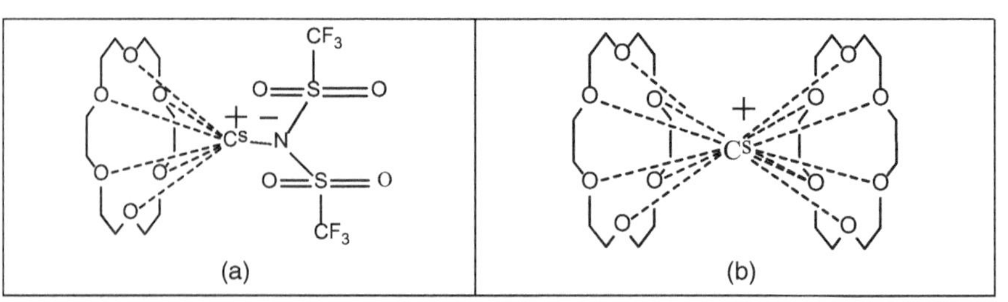

Figure 4.2 ***Tentative Structures of $[Cs(18C6)NTf_2]^0$ (a) and $[Cs(18C6)_2]^+$ (b) in $[C_4C_1im][NTf_2]$.***

On the other hand, the strong influence of the IL cation, Z^+, on log K_1, $\Delta H°_1$ and $\Delta S°_1$ values of complex formation (4.3) is observed when $[C_4C_1im][BF_4]$ and $[C_4C_1py][BF_4]$ are compared, indicating Z^+ (crown ether) interactions of various intensity. Such interactions have analogs among molecular solvents. For example, for acetonitrile (AN), nitromethane (NM), and even dichloromethane, 18-crown-6·2AN, 18-crown-6·2NM, and 18-crown-6·2CH_2Cl_2 solvates have been isolated and their structures have been determined by X-ray crystallography (de Boer et al., 1982; Rogers et al., 1988; Garrell et al., 1988; Jones et al., 1994). Hence, both the cation and the anion of RTILs have an impact on the resultant stability constant and thermodynamic quantities of chelated compounds.

For $[Cs(18C6)_2]^+$ complexes in hydrophobic RTILs the situation is significantly different from that of $[Cs(18C6)]^+$ complexes: $\Delta H°_2$ is more negative than $\Delta H°_1$. At the same time, $\Delta S°_2$ is negative, while $\Delta S°_1$ is positive (Popov et al., 2011). For the second crown ether molecule coordination in $[C_4C_1im][NTf_2]$, $[C_6C_1im][NTf_2]$, and $[(C_2)^2C_6C_1im][NTf_2]$ (4.4), a possible explanation could be associated with a very weak X^- bonding in the $CsLX_n^{\ 1-n}$ species:

$$CsLX_n^{\ 1-n} + Z_mL^{m+} \rightleftarrows CsL_2^{\ +} + mZ^+ + nX^- \tag{4.4}$$

An extreme case, when $n = 0$, is also possible. This is consistent with the poor coordinating ability of the NTf_2^- anion toward alkali and alkaline earth cations (Jensen et al., 2002). Alternatively, in $[C_4C_1im][PF_6]$, $[C_4C_1im][BF_4]$, $[C_4C_1py][BF_4]$, and $[C_4C_1im][N(CN)_2]$, the solvent's anion X^- could be bound to cesium more tightly and thus prevented $[Cs(18C6)_2]^+$ complex formation. Such an interpretation is supported by the much better extraction of $CsNO_3$ from water into RTILs for $[N_{8881}][Sal]$, $[N_{6666}][DHSS]$, and $[C_4C_1im][PF_6]$ relative to $[C_4C_1im][NTf_2]$, $[C_6C_1im][NTf_2]$, and $[(C_2)^2C_6C_1im][NTf_2]$, in the absence of 18C6. This further suggests that complex formation plays an important role in extraction processes.

METAL ION EXTRACTION INTO RTILS

Solvent extraction represents a significant separation technique for the recovery or removal of both metal ions and organic compounds, with numerous advantages over competing techniques. However, volatile organic compounds (VOCs)-based liquid–liquid extraction usually employs water-immiscible organic solvents, many of which are toxic, flammable, and volatile (Sun et al., 2012; Dietz, 2006). ILs exhibit some properties that make them attractive alternatives to conventional organic solvents (Dietz, 2006). Indeed, for the separation of gases, organic molecules, and biomolecules (e.g., amino acids) the distribution coefficients *D* have been found to exceed those of conventional solvents. However, the situation with inorganic cation extraction is not so clear.

One of the first communications on Sr^{2+} extraction from water into a series of dialkylimidazolium-based RTILs in the presence of dicyclohexane-18-crown-6 reported log $D = 4.04$ (Dai et al., 1999), while most molecular solvents revealed very low distribution coefficients under the same conditions. Moreover, this work has demonstrated that the extraction degree is tunable by varying either the RTIL anion or cation. Since this report, an intensive study of RTIL extraction ability, selectivity, and extraction mechanism for inorganic cation separation has been undertaken worldwide (Vander Hoogerstraete et al., 2015; Janssen et al., 2014; Shkrob et al., 2014; Sun et al., 2012; Murding and Tang, 2010; Vendilo et al., 2009a; Yakshin et al., 2012; Cocalia et al., 2006a,b; Dietz, 2006; Visser and Rogers, 2003; Chun et al., 2001; Visser et al., 2000).

Table 4.7 clearly demonstrates the advantages of mutual RTIL and crown ether application in Cs^{+} extraction relative to conventional organic solvents. The distribution coefficients D_{Cs} in some cases increase by 2–4 orders of magnitude. At the same time, acidification of the feeding aqueous solution generally diminishes RTIL efficiency, while that for molecular solvents increases.

Luo et al. (2006b) using DC18C6 in $[C_2C_1im][BETI]$ (BETI: bis(perfluoroethanesulfonyl)imide) managed to exceed the value reported by Dai et al. (1999) for strontium: log $D_{Sr} = 4.9$.

The extraction of silver ions was found to be greatly enhanced by combined application of RTIL and calyx[4]arene compared to that of chloroform (Shimojo and Goto, 2004). It is reported that octyl(phenyl)-*N*,*N*-diisobutylcarbamoylmethylphosphine oxide, tri-*n*-octylphosphine oxide, and relative complexing agents dissolved in RTILs drastically enhance the extractability of lanthanides and actinides, in comparison with conventional organic solvents (Yang et al., 2013; Visser and Rogers, 2003). For Am^{3+} Pu^{4+}, Th^{4+}, and UO_2^{2+} chelating agent–assisted distribution ratios in RTIL are reported to exceed those for VOCs over one order of magnitude (Sun et al., 2012; Visser and Rogers, 2003). It was demonstrated

Table 4.7 Crown ether–assisted cesium extraction efficiency from aqueous phase into RTIL and conventional organic solvents at 20–25°C

Aqueous phase	Organic phase	Crown ether	log D_{Cs}	References
$CsNO_3$; pH ~ 7	DCE	DC18C6, 0.4 mol dm^{-3}	−1.1	Yakshin et al. (2012)
$CsNO_3$ in 1 mol dm^{-3} HNO_3	$CHCl_3$	DC18C6, 0.05 mol dm^{-3}	−1.54	Yakshin et al. (2006)
$CsNO_3$ in 1 mol dm^{-3} HNO_3	$CHCl_3$	DB18C6, 0.05 mol dm^{-3}	−2.0	Yakshin et al. (2006)
$CsNO_3$ in 1 mol dm^{-3} HNO_3	$CHCl_3$	18C6, 0.05 mol dm^{-3}	−1.89	Yakshin et al. (2006)
$CsNO_3$ in 1 mol dm^{-3} HNO_3	$CHCl_3$	DB21C7, 0.1 mol dm^{-3}	−1.96	Yakshin et al. (1992)
$CsNO_3$ in 1 mol dm^{-3} HNO_3	$C_6H_5NO_2$	DB21C7, 0.1 mol dm^{-3}	2.03	Yakshin et al. (1992)
$CsNO_3$; pH ~ 7	Octanol	DC18C6, 0.05 mol dm^{-3}	−0.89	Ali et al. (2014)
$CsNO_3$; pH ~ 7	$[C_4C_1im][NTf_2]$	DC18C6, 0.01 mol dm^{-3}	1.9	Yakshin et al. (2012)
$CsNO_3$; pH ~ 7	$[C_4C_1im][NTf_2]$	DC18C6, 0.1 mol dm^{-3}	2.59	Luo et al. (2006a)
$CsNO_3$; pH ~ 7	$[C_2C_1im][BETI]$	DC18C6, 0.1 mol dm^{-3}	3.06	Luo et al. (2006b)
$CsNO_3$; pH ~ 7	$[C_4C_1im][NTf_2]$	DC18C6, 0.05 mol dm^{-3}	1.97	Ali et al. (2014)
$CsNO_3$; pH ~ 7	$[C_4C_1im][PF_6]$	DC18C6, 0.1 mol dm^{-3}	1.76	Luo et al. (2006)
$CsNO_3$; pH ~ 7	$[C_4C_1im][NTf_2]$	18C6, 0.15 mol dm^{-3}	1.56	Vendilo et al. (2009a)
$CsNO_3$; pH 1.27	$[C_4C_1im][NTf_2]$	18C6, 0.15 mol dm^{-3}	−0.23	Vendilo et al. (2009a)
$CsNO_3$; pH ~ 7	$[C_4C_1im][NTf_2]$	DB18C6, 0.15 mol dm^{-3}	2.28	Vendilo et al. (2009a)
$CsNO_3$; pH 1.92	$[C_4C_1im][NTf_2]$	DB18C6, 0.15 mol dm^{-3}	−0.47	Vendilo et al. (2009a)
$CsNO_3$; pH ~ 7	$[C_4C_1im][NTf_2]$	DB21C7, 0.15 mol dm^{-3}	1.67	Vendilo et al. (2012)
$CsNO_3$; pH ~ 7	$[C_4C_1im][NTf_2]$	DB24C8, 0.15 mol dm^{-3}	0.94	Vendilo et al. (2012)
$CsNO_3$; pH ~ 7	$[N_{8881}][Sal]$	No crown added	1.21	Vendilo et al. (2009a)
$CsNO_3$; pH ~ 7	$[N_{6666}][DHSS]$	No crown added	0.69	Vendilo et al. (2009a)

BETI, bis(perfluoroethanesulfonyl)imide.

that Cs^{+} extraction efficiency (Vendilo et al., 2010a, 2015) and Am^{3+}/Eu^{3+} extraction selectivity (Sengupta et al., 2013a) correlate well with complex stability constant values in the RTIL phase. It is worth mentioning that the high extraction efficiency would sufficiently increase cost efficiency for industrial separation applications.

Along with ordinary RTILs, task-specific RTILs with chelating moieties built into the cation or anion have been intensively developed (Li et al., 2015; Sengupta et al., 2013b; Sun et al., 2012; Visser et al., 2001a, 2002; Harjani et al., 2008).

Recently, two distinct directions of improvements in RTIL-based extraction applications have been identified:

1. The first is associated with attempts to vary the combination of an RTIL with negligible complexing ability and a complexing agent dissolved in the IL to enhance the target cation's solubility in order to obtain either a maximal D value or higher extraction selectivity (Sun et al., 2012). This research area may be subdivided into: the use of extractant systems employing neutral extractants (e.g., crown ethers, calixarenes, tri-*n*-butylphosphate), and acidic or anionic ones (e.g., organophosphorus acids, pseudohalides, 2-thenoyltrifluoroacetone).
2. Another path aimed at achieving the same objective through modification of either the cation or anion of RTILs in such a way that they gain the ability to form complexes with extracted substrates (i.e., TSILs) (Li et al., 2015; Dietz, 2006; Luo et al., 2006a). This second area may also be subdivided into synthetic implementation of a complexing agent fragment into a RTIL cation or into its anion. In some cases an anion itself can initially possess sufficient complexing ability (e.g., $[DHSS]^{-}$, $[Sal]^{-}$).

Recently, the use of neutral extractants has received the most attention as they have played an important role in cation separation based on conventional solvents, and therefore direct comparison is possible (Sun et al., 2012; Dietz, 2006). Over the past decade some important peculiarities, advantages, and drawbacks of RTILs relative to conventional solvents have been found. Higher extraction efficiency of RTILs over conventional solvents was definitively observed for a series of cations in the presence of neutral complexing agents (Sun et al., 2012; Murding and Tang, 2010).

It should be noted that hydrophobic RTILs and organic solvents have generally different extraction mechanisms. For conventional solvents, the extraction of neutral species of a cation M^{n+} is required; for example, if M is present in an aqueous solution in the form of nitrate, then it's partitioning into the organic phase is promoted by a neutral (nonionized) complexing agent *L*. This can be presented by a general equation:

$$M^{n+}_{aq} + nNO^{-}_{3\ aq} + L_{o} \rightleftarrows [M(NO_3)_n L]_o \qquad (4.5)$$

where indexes "aq" and "o" indicate aqueous and organic phases, respectively.

Quite a different and more complicated situation of ion exchange was observed for the extraction of metal ion nitrates from water into common RTILs; for example, $[C_4C_1im][PF_6]$ and $[C_4C_1im][NTf_2]$ (Sun et al., 2012; Dietz, 2006):

$$M^{n+}_{aq} + nNO^{-}_{3\ aq} + L_{o} + n[ZX]_o \rightleftarrows [ML][X_n]_o + nNO^{-}_{3\ aq} + nZ^{+}_{aq} \qquad (4.6)$$

where Z^+ and X^- denote the cation and anion of an RTIL, respectively. A partitioning mechanism (4.6) in which metal ion transfer is accompanied by the loss of an RTIL cation to the aqueous phase indicates a serious problem for potential applications of ILs in liquid–liquid extraction. First of all, this mechanism questions the standing of RTILs as "green" extractants. Indeed, in solving the problem of air pollution caused by conventional solvents, RTILs can give rise to an increase in aqueous phase pollution by their cations. At the same time, this process leads to RTIL decomposition, shortens the number of possible extraction cycles, and therefore makes the process more expensive.

One solution to this problem is possible if ion exchange is somehow suppressed in favor of traditional extraction. Recently, three approaches to diminish the ion exchange mechanism's contribution have been described (Sun et al., 2012; Dietz, 2006): (1) an increase of alkyl chain length in the 1-alkyl-3-methylimidazoluim cation of the RTIL; (2) fluorination of this alkyl chain, and (3) use of a sacrificial cation exchanger.

For the series of $[C_nC_1im][NTf_2]$ it was demonstrated that the partitioning mode of Sr^{2+} into the RTIL in the presence of DC18C6 shifts from Scheme (4.6) to (4.5) as alkyl *R* changes from pentyl to decyl (e.g., as the hydrophobicity of the RTIL cation increases; Dietz et al., 2003). Indeed, an increase of the alkyl chain length makes the transfer of $[C_nC_1im]^+$ into the aqueous phase more difficult and therefore shifts the balance of relative contribution of the ion exchange mechanism, and of a neutral complex extraction, in favor of the latter. However, total extraction efficiency in this case decreases significantly. A similar result was obtained by fluorination of the alkyl chain C_n (Dietz, 2006). Therefore, both approaches (1) and (2) seem to be dead ends. The alternative approach (3) does not influence the ion exchange mechanism, but simply aims to reduce losses of imidazolium cations at the expense of "sacrificial" cations specifically added to the RTIL. This cation partitions into the aqueous phase instead of the RTIL cation. For this purpose sodium tetraphenylborate was found to reduce $[C_nC_1im]^+$ losses by 24% as Cs^+ was extracted into $[C_nC_1im][NTf_2]$ by "BOB-CalixC6" without any decrease of D_{Cs} values (Luo et al., 2004). This approach looks significantly better, but does not seem a universal and satisfactory solution. A novel approach in this field was proposed by Janssen et al. (2014). This research group added $LiNTf_2$ salt to the strontium and potassium chlorides aqueous phase in equilibrium with DC18C6 in $[C_8C_1im][NTf_2]$. It was demonstrated that $Sr(NTf_2)^+$ is preferably extracted in this case: the ratio of potassium to strontium in the IL phase is shifted from 4:1 (in the absence of lithium bistriflimide) to 3:4. Hence, further research is needed.

The use of anionic complexing agents seems to be more advantageous. In this way the general extraction scheme may eliminate the ion exchange mechanism if a polydentate chelating agent H_nL is properly selected in order to form a zero-charged species $[ML]^\circ$:

$$M^{n+}_{aq} + nNO^-_{3\ aq} + H_nL_{o/aq} \rightleftarrows [ML]_o + nNO^-_{3\ aq} + nH^+_{aq} \quad (4.7)$$

or for a ligand HL of a lower denticity and basicity:

$$M^{n+}_{aq} + nNO^-_{3\ aq} + nHL_{o/aq} \rightleftarrows [ML_n]_o + nNO^-_{3\ aq} + nH^+_{aq} \quad (4.8)$$

where the index "o/aq" indicates the presence of L in both the organic and aqueous phases. For example, for Am^{3+} and Ln^{3+} ethylenediaminetetraacetic acid (EDTA) is undesirable, but the dialkyl ester of diethylenetriaminepentaacetic acid (DTPA, H_5dtpa) – H_3R_2dtpa – is expected to form the same zero-charged species: $[Ln(R_2dtpa)]^\circ$, where R indicates an alkyl chain. The recent data of Shkrob et al. (2014) reveal that an appropriate choice of R in R_2dtpa and of a RTIL ion composition (*N*,*N*,*N*-trimethylglycine bistriflimide) can provide good lanthanide complex integrity in ILs.

It was also demonstrated that for such *d*-metal ions as Ni^{2+}, Cu^{2+}, and Pb^{2+}, extraction by RTILs in the presence of β-diketonates (HL) proceeds in the same way as in chloroform, nitrobenzene, or toluene – as a neutral hydrated complex $[ML_2(H_2O)_n]^o$, where $n = 1, 2$ (Hirayama et al., 2005). A similar observation was reported for nonchelating inorganic ligands: CN^-, OCN^-, SCN^-, and halide ions (Visser et al., 2001b).

At the same time, improper selection of an anionic ligand for a particular cation (Visser et al., 2001b) can lead to an anionic complex formation, which has the same disadvantages concering the ion exchange mechanism as with neutral ligands. In fact, it was demonstrated (Jensen et al., 2003) that Eu^{3+} with 2-thenoyltrifluoroacetone (Htta) forms $[Eu(tta)_4]^-$ in addition to the neutral species $[Eu(tta)_3]^o$, and an undesirable anion exchange mechanism of $[Eu(tta)_4]^-$ for $[NTf_2]^-$ prevails at the water/$[C_4C_1im][NTf_2]$ interface. This example indicates the importance of stability constant–based speciation of complexes formed by a ligand with a target cation in both the aqueous and RTIL phases.

Most of the extraction results listed earlier could be rather successfully interpreted on the grounds of complex formation equilibrium constants and corresponding chemical speciations. Unfortunately, published log K values for RTILs are still very rare (Table 4.3).

However, for Cs^+ such data do exist and some comments are possible. Formation of crown ether complexes promotes cesium extraction (Vendilo et al., 2009a) into hydrophobic RTILs from water if complex stability in the RTIL is higher than that in water (Table 4.8). Indeed, for $[C_4C_1im][NTf_2]$, $[C_6C_1im][NTf_2]$, $[(C_2)^2C_6C_1im][NTf_2]$, $[C_4C_1im][PF_6]$, and $[N_{8881}][Sal]$,

Table 4.8 Stability constants of cesium complex formation with 18C6 and DB18C6 in RTIL and extraction efficacy from water at 25°C*

		18C6		DB18C6	
Solvent	**log D_{Cs}**	**log K_1**	**log D_{Cs}^{18C6}**	**log K_1**	**log D_{Cs}^{DB18C6}**
$[C_6C_1im][NTf_2]$	−1.24	4.4	0.82		
$[(C_2)^2C_6C_1im][NTf_2]$	−0.81	3.4	0.56		
$[C_4C_1im][NTf_2]$	−0.67	3.4	1.56	3.47	2.28
$[C_4C_1im][PF_6]$	−0.59	2.4	−0.20		
$[N_{8881}][Sal]$	0.69	1.43	0.83	3.1	1.91
$[N_{6666}][DHSS]$	1.21	0.76	0.25	2.7	1.56
1,2-Dichloroethane	−7.7**	7.98†	0.8‡		

* Extraction data for RTIL are taken from Vendilo et al. (2009a), and from Popov et al. (2011) for 25°C, pH water = 7, $[Cs]_o = 0.0015$ mol L^{-1}, $[DB18C6]_o = [18C6]_o = 0.15$ mol L^{-1}.

** Levitskaia et al. (2007), picrate salt.

† Kikuchi and Sakamoto (2000).

‡ Abramov (2000), picrate salt.

$\log K_1^{\mathrm{RTIL}} > \log K_1^{\mathrm{water}}$ and the cesium content in the RTIL increases due to 18C6 administration. For $[N_{6666}]$[DHSS], $\log K_1^{\mathrm{RTIL}} < \log K_1^{\mathrm{water}}$ and the crown ether decreases the cesium content in the RTIL. For $[C_4C_1im]$ $[NTf_2]$, the $\log D_{Cs}^{18C6}$ values exceed the corresponding values for 1,2-DCE and other volatile organic diluents (Abramov, 2000; Levitskaia et al., 2007), while the hazardous properties of the latter are much higher.

It should be noted that complex stability in RTILs increases as the RTIL ability to extract cesium without crown ether (D_{Cs}) decreases (Figure 4.3). At the same time, for a number of crown ethers (L) dissolved in $[C_4C_1im]$ $[NTf_2]$, a definate linear correlation between $\log K_1$ and $\log D_{Cs}^{L}$ is observed (Vendilo et al., 2015), indicating a sufficient increase of $\log D_{Cs}^{L}$

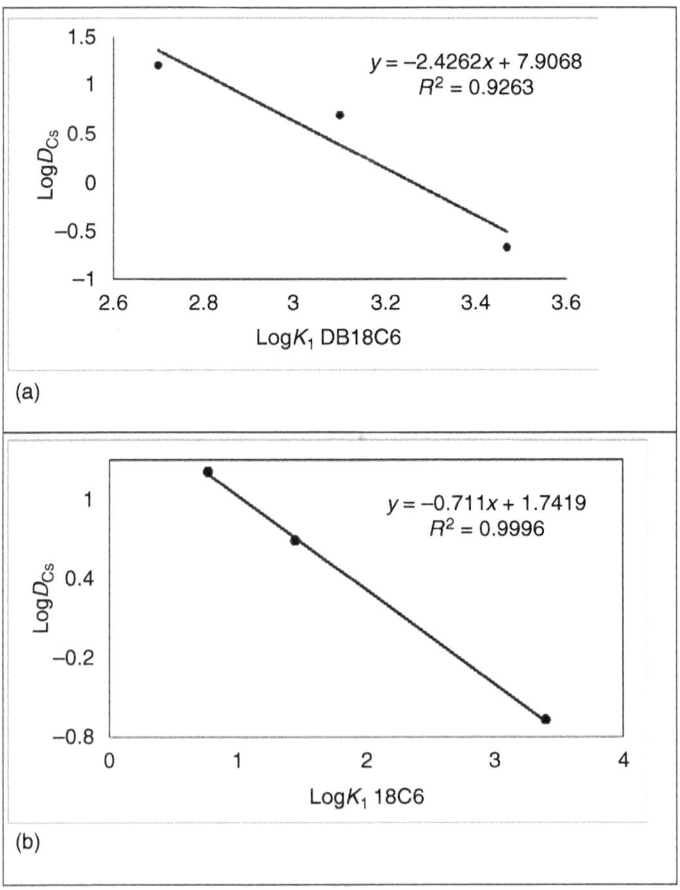

Figure 4.3 ***Plots of log D_{Cs} Versus log K_1 for Cesium Complex with DB18C6 (a) and 18C6 (b) in $[N_{8881}]$[Sal], $[N_{6666}]$[DHSS], and $[C_4C_1im][NTf_2]$ at 25°C (Popov et al., 2011).***

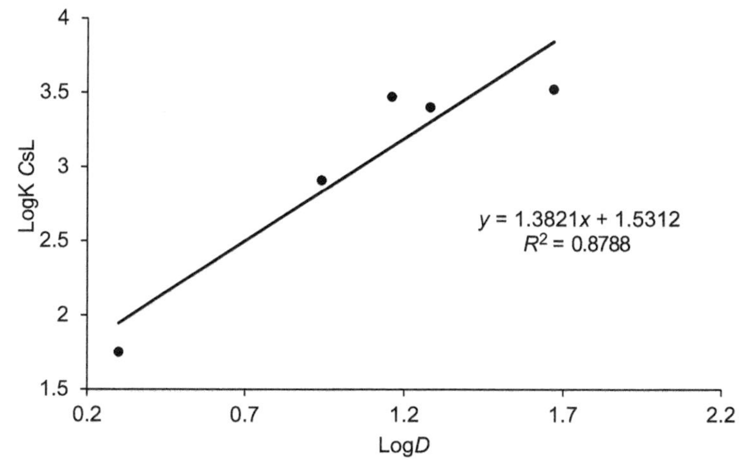

Figure 4.4 ***Correlation Between log D_{Cs}^{L} and log K_{CsL}^{RTIL} Values for L: 15C5, 18C6, DB18C6, DB21C7, DB24C8, and $[C_4C_1im][NTf_2]$ at 22 –25°C (Vendilo et al., 2015).***

as log K_1 increases (Figure 4.4). Moreover, an even better correlation was found for cesium if one-crown ether (18C6 of DB18C6) is used for several RTILs ($[N_{8881}][Sal]$, $[N_{6666}][DHSS]$, and $[C_4C_1im][NTf_2]$; Figure 4.5).

This linearity (Figures 4.4 and 4.5) has a specific background. Such a correlation indicates a dominating role of complex formation in an RTIL phase in the crown ether–assisted extraction process. For cesium transfer from a neutral aqueous solution into an RTIL phase an ion exchange mechanism prevails (4.6). The correlation, demonstrated in Figure 4.4 for $[C_4C_1im][NTf_2]$), is the simplest one. Neglecting the small contribution of $[Cs(18C6)_2]^+$ and $[Cs(DB18C6)_2]^+$ species in $[C_4C_1im][NTf_2]$ to total thermodynamic stability, one can express the distribution coefficient of cesium between aqueous and $[C_4C_1im][NTf_2]$ phases in the presence of crown ethers D_{Cs}^{L} by a rather simple equation:

$$D_{Cs}^{L} = ([Cs]^{RTIL} + [CsL]^{RTIL})/([Cs]^{w} + [CsL]^{w}) \tag{4.9}$$

where the superscripts "RTIL" and "w" denote the organic phase and aqueous phase, respectively. As for total concentration of L$[L]^{o} >> [Cs]^{o}$ (total cesium concentration), the stability constants of CsL complexes in an RTIL phase are high enough (log $K_{CsL}^{RTIL} \geq 2$), $[L]^{RTIL} >> [Cs]^{RTIL}$, then $[Cs]^{RTIL} << [CsL]^{RTIL}$ and equilibrium concentrations $[Cs]^{RTIL}$ could be neglected. At the same time, for the aqueous phase the log K_{CsL}^{w} are

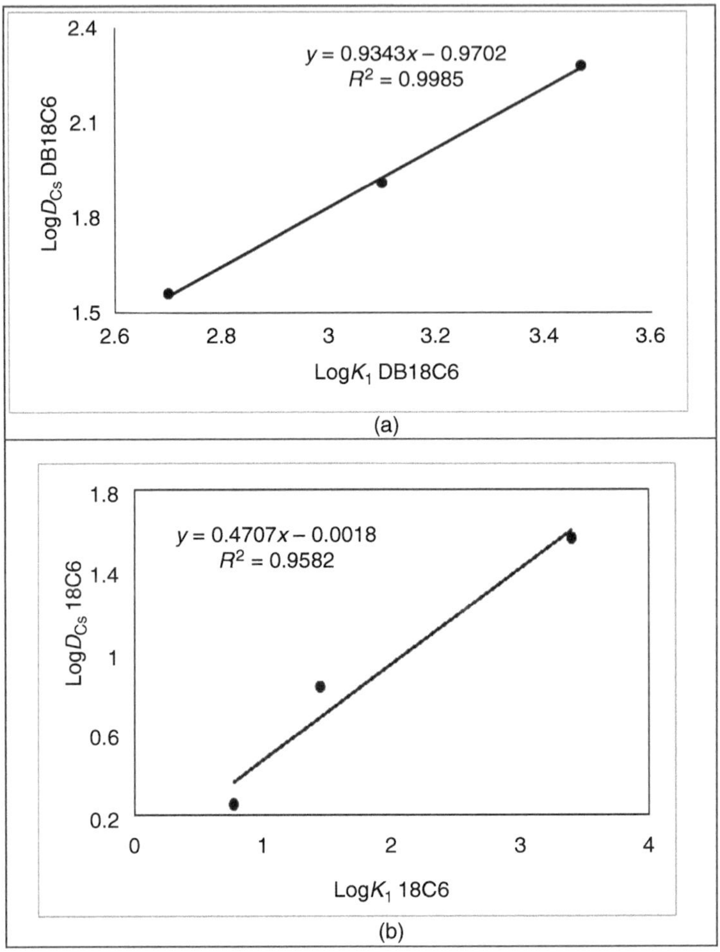

Figure 4.5 ***Plots of log D_{Cs}^{DB18C6} Versus log K_1 for DB18C6 (a) and log D_{Cs}^{18C6} Versus log K_1 for 18C6 (b) in $[N_{8881}]$[Sal], $[N_{6666}]$[DHSS], and $[C_4C_1im][NTf_2]$ at 25°C (Popov et al., 2011).***

comparatively small, and $[CsL]^w << [Cs]^w$. Thus, $[CsL]^w$ values can be neglected. Therefore, (4.9) can be transmitted into:

$$D_{Cs}^{L} = [CsL]^{RTIL}/[Cs]^{w} \tag{4.10}$$

or

$$D_{Cs}^{L} = (K_{CsL}^{RTIL}[Cs]^{RTIL}[L]^{RTIL})/[Cs]^{w} \tag{4.11}$$

Taking into account that $[Cs]^{RTIL}/[Cs]^{w} = D_{Cs}$ the following expression is valid:

$$D_{Cs}^{\ L} = K_{CsL}^{\ RTIL} D_{Cs} [L]^{RTIL} \tag{4.12}$$

or

$$\log D_{Cs}^{\ L} = \log K_{CsL}^{\ RTIL} + \log D_{Cs} + \log([L]^{RTIL}) \tag{4.13}$$

As long as we have a single RTIL, log D_{Cs} is a constant for all the L series. At the same time, $[L]° >> [Cs]°$, and most of the crowns used are poorly soluble in water, then an equilibrium concentration $[L]^{RTIL}$ should not change significantly for different crowns and would be close to $[L]°$.

Therefore, a dependence of log $D_{Cs}^{\ L}$ on log $K_{CsL}^{\ RTIL}$ should be linear with slope 1. Some deviation from this slope (Figure 4.4) is probably due to some change of crown solubility in the aqueous phase relative to each other. This relative solubility should also correlate with CsL complex stability: the fewer water-soluble crowns that form the greater number of stable complexes CsL in an RTIL phase. This statement is valid only for hydrophobic crowns. For example, 12C4 is water soluble; therefore, it reveals no extraction enhancement although the corresponding $[Cs(12C4)]^+$ complex is rather stable in the $[C_4C_1im][NTf_2]$ phase (Vendilo et al., 2015).

Moreover, if (1) log D_{Cs} is a linear function of log K_1^{RTIL}, and (2) log D_L values are nearly the same for the number of RTILs under consideration (the case of DB18C6 in $[N_{8881}][Sal]$, $[N_{6666}][DHSS]$, and $[C_4C_1im][NTf_2]$) or correlate linearly with log K_1 (the case of 18C6 in $[N_{8881}][Sal]$, $[N_{6666}]$ [DHSS], and $[C_4C_1im][NTf_2]$), then log $D_{Cs}^{\ L}$ becomes a linear function of log K_1 (Figure 4.5).

At the same time, there is generally no simple relationship between log K_1 and crown ether–assisted extraction efficiency ($D_{Cs}^{\ 18C6}$) if we consider a broad spectrum of RTILs (Table 4.8). This happens due to a superposition of at least two phenomena: the relative stability of complexes is superimposed on the relative solubility of the crown ether and its complexes in RTIL/water systems. However, when ligand extractions (D_L) and D_{Cs} are taken into account then a perfect linear relationship between [log $D_{Cs}^{\ 18C6}$ − log D_{Cs} − log D_L] and log $\beta_{1,2}$ is observed (Vendilo et al., 2010a).

This clearly indicates the important contribution of complex stability to hydrophobic RTILs in extraction processes. Indeed, for $[C_4C_1im][NTf_2]$

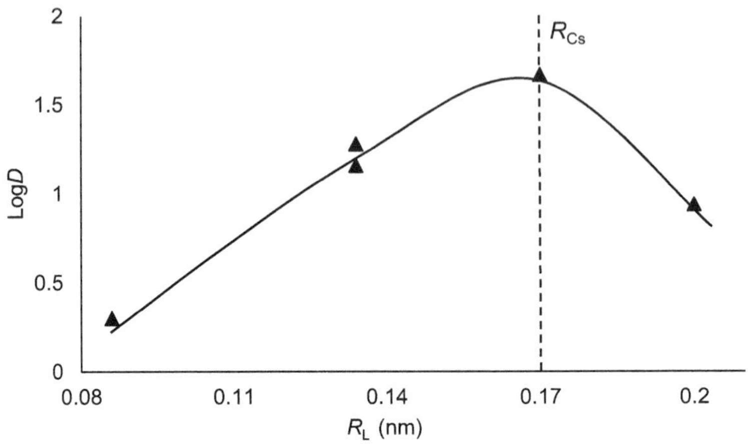

Figure 4.6 ***Log*** D_{Cs}^{L} ***(log D) Dependence on the Crown Cavity Size*** R_L ***for [C_4C_1im][NTf_2] and L: 15C5, 18C6, DB18C6, DB21C7, DB24C8 (Vendilo et al., 2015).***

there is an excellent fit of D_{Cs}^{L} values and the cavity size R_L of a crown ether (Figure 4.6), which arises from a corresponding log K_1 dependence on R_L. In fact, log K_1 values for an aqueous phase do not demonstrate such a trend as [C_4C_1im][NTf_2] does.

At the same time, this trend is strongly modulated by the relative solubility of the chelating agent, its complex with the target cation, and the metal nitrate in water and in the RTIL phase. Our group has found that the more hydrophobic DB18C6 reveals better extraction of Cs^+ into [C_4C_1im][NTf_2], [N_{8881}][Sal], and [N_{6666}][DHSS] than 18C6, while the stability constants of complex formation are approximately the same (Djigailo, 2010).

Among the other issues that challenge researchers in the field of RTIL extraction is the problem of extracted cation regeneration from the IL and therefore RTIL regeneration. The low vapor pressure becomes an evident disadvantage because RTILs cannot be removed via evaporation like conventional solvents. However, it was demonstrated for lanthanides that stripping from [C_4C_1im][PF_6], in the presence of *n*-octyl(phenyl)-*N,N*-diisobutylcarbamoylmethylphosphane oxide (CMPO), can be accomplished by using aqueous solutions of guanidine carbonate with either DTPA or EDTA (Levitskaia et al., 2007). For the Am(III) complex with CMPO in [C_4C_1im][NTf_2] more than 95% of americium was stripped by using 1.0 M oxalic acid (Nakashima et al., 2003). Recovery of metal ions from ILs could also be achieved by simply lowering the pH (Murding and Tang, 2010 and references therein), or by back extraction by supercritical CO_2 (Murding and Tang, 2010). It should be noted that extraction can be combined with electrochemical recovery, as was demonstrated for UO_2^{2+} and Sm^{3+} (Murding and Tang, 2010), or with electrodialysis (Vendilo et al., 2010d; see later). At the same time, it is reasonable to note that for some RTILs distillation is possible, although at incredibly low pressures (Earle et al., 2006).

It is also important to note that RTILs can generally withstand α- and γ-radiation better than conventional molecular solvents (Sun et al., 2012; Murding and Tang, 2010; Allen et al., 2002), although in some cases irradiation of RTILs and of chelating agents can diminish radioactive cation partitioning (Chao et al., 2010). Allen et al. (2002) reported that 1,3-dialkylimidazolium nitrate and chloride reveal the same α- and γ-radiation stability as benzene, and are much more stable than mixtures of tributyl phosphate and kerosene under similar conditions. It is noted that $[C_4C_1im][NTf_2]$ is stable at high-irradiation doses, while imidazolium and pyridinium RTILs associated with $[NTf_2]$ anions reveal much smaller irradiation yields than *n*-alkanes (Sun et al., 2012).

Recently, the broad application of RTILs in extraction processes has been limited by their relatively high cost, as yet unclear toxicity, and unclear long-term stability. The number of publications on RTIL toxicity has rapidly increased recently (Calza et al., 2015; Mulkiewicz and Stepnowski, 2015; Liu et al., 2015; Docherty and Kulpa, 2005; Jastorff et al., 2003) but a lot of work in this field is still to be done. The same situation applies to RTIL solubility in water, which is critical for their recyclability and lifetime. Solubility data are still available for a small number of ILs (Sun et al., 2012). At the same time, the cost of some recently developed RTILs (e.g., $[N_{8881}][Sal]$) is about an order of magnitude lower than others such as $[C_4C_1im][NTf_2]$ and can be diminished if production increases as a result of market demand.

SEPARATIONS OF INORGANIC CATIONS FROM AQUEOUS SOLUTIONS BY RTIL-BASED LIQUID–LIQUID EXTRACTION, LIQUID MEMBRANES, AND SUPPORTED LIQUID MEMBRANES

The separation of inorganic cations can be performed by liquid–liquid extraction, the use of RTIL-based liquid membranes, and supported membranes.

Liquid/Liquid Extraction

Compared with molecular solvents, metal–ion partitioning in RTIL-based extraction systems exhibits different and enhanced selectivity, although higher complexity.

The higher selectivity of dibenzo-18-crown-6 (DB18C6) to K^+ over Na^+ in *N*-octadecylisoquinolinium tetrakis[3,5-bis(trifluoromethyl)phenyl] borate compared with that in molecular solvents suggests that RTILs provide a unique solvation environment for the complexation of crown ethers with ions (Nishi et al., 2006). Luo et al. (2006b) demonstrated that the extraction selectivity for Sr^{2+} over Na^+, K^+, and Cs^+ can be significantly improved through the variation of RTIL hydrophobic anions with constant either $[C_4C_1im]^+$ or $[C_2C_1im]^+$ cations. In particular, using DC18C6 in $[C_nC_1im]$[BETI] one can change the D_{Sr}/D_{Cs} ratio from 7.95 ($[C_2C_1im]$[BETI]) to 0.16 ($[C_8C_1im]$[BETI]). A similar flip-flop occurs for D_{Sr}/D_K as cation hydrophobicity increases from $[C_2C_1im]$[BETI] to $[C_8C_1im]$[BETI] or an anion is changed from $[C_4C_1im][PF_6]$ to $[C_4C_1im]$[BETI]. These results characterize RTILs as "designer solvents." It is possible to design an RTIL composition with specific chemical and physical properties through the appropriate combination of cations and anions.

It has been observed that the partitioning of lanthanides in $[C_4C_1im]$ $[PF_6]$, in the presence of CMPO, was higher and had better selectivity than that in dodecane (Levitskaia et al., 2007). Reversed extraction selectivity in RTILs relative to isooctane, for a series of lanthanides in the presence of *N,N,N′,N′*-tetra(*n*-octyl)diglycolamide (Murding and Tang, 2010) and in the presence of tri-*n*-octylphosphine oxide relative to *n*-dodecane (Yang et al., 2013) has also been reported. The unusual behavior of reduced extraction efficiency of Eu^{3+} with *N,N,N′,N′*-tetraoctyl diglycolamide (TODGA) in $[C_6C_1im][NTf_2]$ compared to that of Am^{3+} has been found (Sengupta et al., 2013).

The separation of Sc^{3+} from a mixture of Sc^{3+}, La^{3+}, Y^{3+}, and Yb^{3+} in an aqueous phase by Cyanex 925 (bis(2,4,4-trimethylpentyl)octylphosphine oxide) into $[C_8C_1im][PF_6]$ has been reported (Sun et al., 2007).

Vander Hoogerstraete et al. (2015) recently reported a single-step separation of a mixture of three metal ions Sn^{2+}, Y^{3+}, and Sc^{3+} by a triphasic RTIL–water/RTIL–liquid solvent extraction system whereas at least two steps are required to separate three metals in the case of two liquid–phase solvent extraction. In the triphasic system, the lower organic phase is comprised of the IL betainium or choline bis(trifluoromethylsulfonyl)imide, whereas the upper organic phase is comprised of the IL trihexyl(tetradecyl) phosphonium bis(trifluoromethylsulfonyl)imide.

Betainium or choline bis(trifluoromethylsulfonyl)imide ILs with DTPA–betainium and DTPA–choline conjugates were sucsessfully used for lanthanide and actinide ion separation (Shkrob et al., 2014). Large separation factors for Eu/Am partitioning between two phases are observed, approaching 120–150 for DTPA–betainum and 250–270 for DTPA–choline.

Along with noncomplexing RTILs, numerous and effective TSIL applications have been reported (Li et al., 2016; Sun et al., 2012). Selective recovery of Zn(II) and Fe(III) from sulfate media by liquid–liquid extraction has been achieved using Cyphos 109. The separation factor (SZn/Fe = 110) indicated the possibility of selectively recovering 98% of Zn(II) from Fe(III) when they are in the same solution at pH = 9.2, which can have a wide range of applications in industries such as electroplating, where these metals are present in low concentrations and can be recovered. At the same time, in sulfuric acid medium (pH = 1.8) 96% of Fe(III) can be extracted but only 4.8% of zinc (Perales et al., 2014). Some task-specific RTILs provide almost quantitative uranyl ion extraction (Srncik et al., 2009).

However, the problem of RTIL phase loss and aqueous phase contamination as metal ion is extracted still remains an obstacle to their broad application. Another disadvantage is associated with $[PF_6]$-based RTIL decomposition at the acidic aqueous phase (Sun et al., 2012).

Liquid Membrane Separation

The use of RTIL-based liquid membranes under a voltage gradient for cation separations is less common than extraction and supported membrane applications (Vendilo et al., 2010b; Chen, 2007). Our group has tested Cs^+ transfer from its nitrate salt in aqueous solution from the anode chamber to a cathode chamber filled with water under a voltage gradient through a $[C_4C_1im][NTf_2]$ layer with 18C6 (Vendilo et al., 2010b) (Figures 4.7–4.9). In the absence of crown ether ca. 50% of Cs^+ is accumulated in the RTIL phase within 4 h (Figure 4.7). The process seems to stop when 50% of the cations still remain in the initial solution. At the same time, no increase of cesium concentration at the cathode was observed. This clearly indicates the formation of either neutral or anionic species of cesium with RTIL anions (e.g., $[CsNTf_2]^0$, $[Cs(NTf_2)_2]^-$), which is in good agreement with NMR data (Popov et al., 2011). The neutral or negatively charged species do not strip into the cathode compartment and stay partly in the RTIL phase in

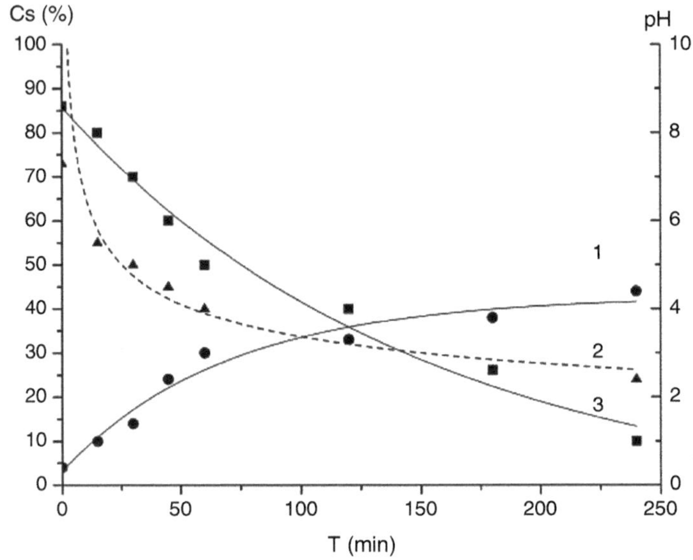

Figure 4.7 ***Cs^+ Distribution in the Electrokinetic Cell with a $[C_4C_1im][NTf_2]$-Based Liquid Membrane: (1) Membrane; (2) Feeding Solution in Anode Compartment; and (3) pH in Anode Compartment (Vendilo et al., 2010b).***

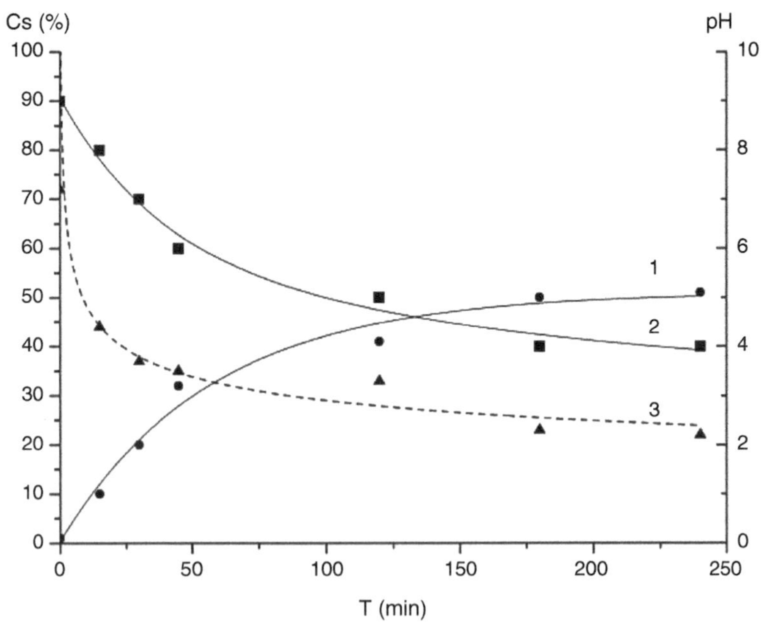

Figure 4.8 ***Cs^+ Distribution in the Electrokinetic Cell with a $[C_4C_1im][NTf_2]$-Based Liquid Membrane: (1) Membrane with 0.15 mol L^{-1} 18C6; (2) pH in Anode Compartment; and (3) Cs^+ Content in a Feeding Solution in Anode Compartment (Vendilo et al., 2010b).***

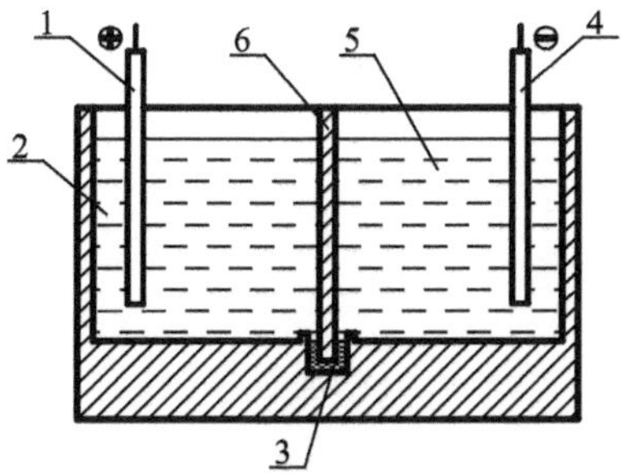

Figure 4.9 ***The Electrokinetic Cell: (1) Platinated Titanum Anode; (2) Anode Chamber with Feeding Solution; (3) Liquid RTIL-Based Bridge; (4) Platinated Titanum Cathode; (5) Cathode Chamber with Deionized Water (Permeate); (6) Screen.***

equilibrium with the anode compartment when the voltage gradient becomes compensated by the concentration gradient.

When 18C6 was added to the IL phase, 90% of cesium, placed initially in the anode chamber, moved into the RTIL phase (44%) and the cathode compartment (46%) within the same period of time (Figure 4.8). This is also consistent with stability constants and the NMR-based chemical speciation model, which assumes the formation of cationic species in excess of 18C6 over $[Cs(18C6)_2]^+$ (Figure 4.2), Therefore, there are no obstacles for a cationic complex to strip into the cathode compartment.

Application of chemical speciation makes the situation clearer (Figure 4.10). Indeed, for the stability levels of 18C6 complexes with Cs^+ in $[C_4C_1im][NTf_2]$ a 1000-fold excess of 18C6, over Cs^+, is needed to provide absolute $[Cs(18C6)_2]^+$ species domination (93%).

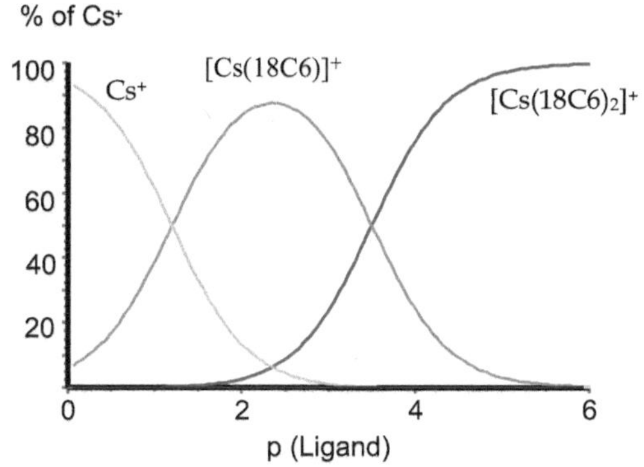

Figure 4.10 ***Chemical Speciation of 0.001 mol L^{-1} Cs^+ in $[C_4C_1im][NTf_2]$ as a Function of 18C6 Concentration (p[Ligand]), Obtained by Species Software (IUPAC, 2007).***

When only a 100-fold excess takes place, as happens after 4 h of applied voltage, then almost 44% of cesium exist in either the $[CsNTf_2]^{\circ}$ or $[Cs\{NTf_2\}_2]^-$ forms, which are not capable of moving to the cathode compartment and process intensity declines.

It should be noted that the pH of the feeding solution changes dramatically from 7 to 2 (Figures 4.7 and 4.8). This should affect the distribution coefficients (*D* values). For RTILs, D_{Cs}^{18C6} normally decreases as acidity approaches pH 2 (Vendilo et al., 2009a). Experiments on the selectivity of Cu^{2+}/Cs^+ separations have been performed in the same electrokinetic cell (Figure 4.9). The model aqueous solution of equimolar copper and cesium nitrates (0.001 mol L^{-1}) was treated under a 1 V cm^{-1} gradient for 4 h. It is worth to note that the separation process can be significantly enhanced if different chelating agents are added into a RTIL phase and into the anode compartment. We ran some experiments where a solution of EDTA was added (pH ~5) to a mixture of Cu^{2+} and Cs^+ in the anode chamber, while 18C6 was dissolved in the RTIL phase. Indeed, the speciation scheme (Figure 4.11) clearly demonstrates that under these experimental conditions copper forms negatively charged species

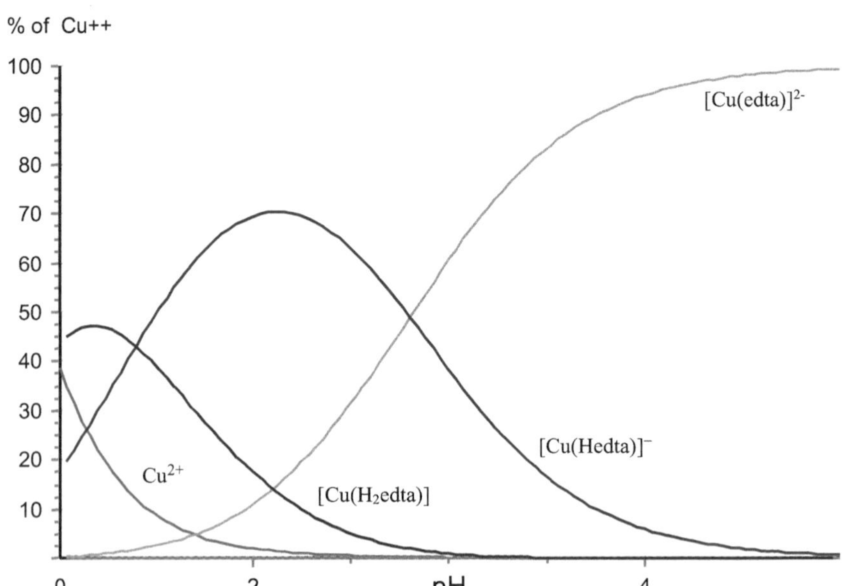

Figure 4.11 ***Chemical Speciation of 0.005 mol L^{-1} Cu^{2+} in Aqueous Solution as a Function of pH Obtained by Species Software (IUPAC, 2007).***

Table 4.9 Results of electrokinetic separation of Cu^{2+} and Cs^+ ions using $[C_4C_1im][NTf_2]$ as a liquid membrane at 25°C*

	Cation content (%)				
	Anode (feeding) chamber		Membrane,	Cathode (receiving) chamber	
Cations	Initial	Final	final	Initial	Final
Cu^{2+}	100	74.7	15.1	0	10.2
Cs^+	100	39.9	54.0	0	6.1
With Na_2H_2edta in anode chamber					
Cu^{2+}	100	98.3	0	0	1.7
Cs^+	100	58.4	39.3	0	2.5

* Data obtained by V. Chistov and J. Dikareva.

$[Cu(edta)]^{2-}$, while Cs^+ ions do not interact with EDTA. Thus, Cs^+ strips into the membrane and then into the cathode compartment, while copper is retained in the anode compartment.

The results of some "blank" experiments without 18C6 in the RTIL bridge are presented in Table 4.9. It is demonstrated that Cs^+ is better at entering the RTIL phase than Cu^{2+} from an aqueous solution, but both cations are retarded in the bridge. Following speciation-based predictions, EDTA effectively holds all copper in the feeding solution, while Cs^+ migrates into the RTIL phase. However, cesium transfer is also diminished due to competition from Na^+ ions, administrated along with EDTA (Na_2H_2edta) in the anode chamber. Some of the negative impact of pH decrease on copper EDTA complexes (at low pH $[Cu(edta)]^{2-}$ is transformed into less stable $[Cu(Hedta)]^-$) does not seem to affect copper retardation in the feeding solution.

Surprisingly, administration of 18C6 into the RTIL phase does not affect cation transfer from a feeding solution into the RTIL phase (Table 4.10). At the same time, a combination of 18C6/RTIL and EDTA in the anode chamber gives the best separation results, although the accumulation of cesium mostly takes place in the RTIL bridge. Thus, additional work is still required to understand the mechanisms of cationic transport.

Supported Membrane Separation

Supported liquid membranes are attracting attention for separation and preconcentration of target cations (Ruhela et al., 2015; Pospiech and Kujawski, 2015; Sun et al., 2012).

Table 4.10 Results of electrokinetic separation of Cu^{2+} and Cs^+ ions using $[C_4C_1im][NTf_2]$ and 18C6 as a liquid membrane at 25°C*

	Cation content (%)				
	Anode (feeding) chamber		Membrane,	Cathode (receiving) chamber	
Cations	Initial	Final	final	Initial	Final
Cu^{2+}	100	75.4	15.1	0	7.8
Cs^+	100	43.5	54.0	0	5.7
With Na_2H_2edta in anode chamber					
Cu^{2+}	100	97.6	0	0	2.5
Cs^+	100	28.7	63.5	0	7.8

* Data obtained by V. Chistov and J. Dikareva.

The main problem in such a liquid membrane process is RTIL decomposition due to its slow dissolution in water in both chambers. A possible solution could be associated with fixation of the RTIL on to some solid porous frame. Various forms of RTIL immobilization on solid phases have recently been reported (Ruhela et al., 2015; Xing et al., 2015; Gega and Otrembska, 2014; Martinis et al., 2014; Gallardo et al., 2008; Vincent et al., 2008; Campos et al., 2008; Guibal et al., 2008). These research results provide numerous analytical applications of RTILs (Hawkins et al., 2015; Martinis et al., 2014).

Supported membrane separation applications have recently been summarized in a comprehensive review (Pospiech and Kujawski, 2015) and therefore are not considered here in detail. Our preliminary experiments with nanoporous Al_2O_3 filled with $[C_4C_1im][NTf_2]$ indicate this path to be a promising one (Table 4.11).

Table 4.11 Results of electrokinetic separation of Cu^{2+} and Cs^+ ions using Al_2O_3-supported membrane impregnated by $[C_4C_1im][NTf_2]$ with 18C6 at 25°C*

	Initial cation content in anode (feeding) chamber (mol L^{-1})		K_{sep} (cathode chamber)
Membrane	Cs^+	Cu^{2+}	
Al_2O_3	0.050	0.050	2.4
Al_2O_3 + $[C_4C_1im][NTf_2]$ + 18C6	0.050	0.050	4.2

* Data obtained by V. Chistov and J. Dikareva.

However, a recent attempt to determine whether this increase in liquid–liquid extraction efficiency will lead to improved strontium sorbents was not successful. Several extraction chromatography (EXC) resins and sol–gel glasses incorporating di-*tert*-butylcyclohexano-18-crown-6 (DtBuCH18C6) in $[C_{10}C_1im][NTf_2]$ unexpectedly revealed the performance of these materials was not uniformly better than that of a conventional EXC resin (Hawkins et al., 2015).

The facts listed above indicate that ILs are very promising solvents for liquid–liquid extraction of a wide variety of metal ions. This opens up a good opportunity for the replacement ofVOCs currently used in the separation of metal–ions from aqueous solutions by RTILs. In many cases these processes are better understandable when the thermodynamic stability of complexes is taken into consideration and corresponding stability constant-based chemical speciations are applied (Figures 4.10, 4.11). At the same time, a lot of physicochemical data on the properties of complexes formed in this medium have yet to be obtained before practical application of RTILs in this field becomes a reality.

ACKNOWLEDGMENT

The authors acknowledge support from the Russian Foundation for Basic Research (Project #13-08-00246).

REFERENCES

Abebe, A., Admassie, S., Villar-Garcia, I.J., Chebude, Y., 2013. 4-Bipyridinium ionic liquids exhibiting excellent solubility for metal salts: potential solvents for electrodeposition. Inorg. Chem. Commun. 29, 210–212.

Abramov, A.A., 2000. Extraction of cations with crown ethers. Vestn. Mosk. Univ. 41 (2), 3–15.

Ali, Sk.M., Joshi, J.M., Singha Deb, A.K., Boda, A., Shenoy, K.T., Ghosh, S.K., 2014. Dual mode of extraction for Cs^+ and Na^+ ions with dicyclohexano-18-crown-6 and bis(2-propyloxy)calix[4]crown-6 in ionic liquids: density functional theoretical investigation. Roy. Soc. Chem. Adv. 4, 22911–22925.

Allen, D., Baston, G., Bradley, A.E., Gorman, T., Haile, A., Hamblett, I., Hatter, J.E., Healey, M.J.F., Hodgson, B., Lewin, R., Lovell, K.V., Newton, B., Pitner, W.R., Rooney, D.W., Sanders, D., Seddon, K.R., Sims, E., 2002. An investigation of the radiochemical stability of ionic liquids. Green Chem. 4, 152–158.

Ansari, S.A., Liu, L., Rao, L., 2015a. Binary lanthanide (III)/nitrate and ternary lanthanide (III)/nitrate/chloride complexes in an ionic liquid containing water: optical absorption and luminescence studies. Dalton Trans. 44 (6), 2907–2914.

Ansari, S.A., Mohapatra, P.K., Verboom, W., Zhang, Z., Dau, P.D., Gibson, J.K., Rao, L., 2015b. Binding of pyrazine-functionalized calyx[4]arene ligands with lanthanides in a ionic liquid: thermodynamics and coordination modes. Dalton Trans. 44 (14), 6416–6422.

Arnaud-Neu, F., Delgado, R., Chaves, S., 2003. Critical evaluation of stability constants and thermodynamic functions of metal complexes of crown ethers. Pure Appl. Chem. 75, 71–102.
Babai, A., Murding, A.-V., 2006. Homoleptic alkaline earth metal bis(trifluoromethanesulphonyl) imide complex compounds obtained from an ionic liquid. Inorg. Chem. 45, 3249–3255.
Banić, N., Vraneš, M., Abramović, B., Csanádi, J., Gadžurić, S., 2014. Equilibrium of contact and solvent-separated ion pairs in mixtures of protic ionic liquids and molecular solvents controlled by polarity. Dalton Trans. 43 (41), 15515–15525.
Bayley, P.M., Lane, G.H., Lyons, L.J., MacFarlane, D.R., Forsyth, M., 2010. Undoing lithium ion association in ionic liquids through the complexation by oligoethers. J. Phys. Chem. C. 114, 20569–20576.
Billard, I., Gaillard, C., 2009. Actinide and lanthanide speciation in imidazolium-based ionic liquids. Radiochim. Acta 97, 355–359.
Billard, I., Moutiers, G., Labet, A., El Azzi, A., Gaillard, C., Mariet, C., Lutzenkirchen, K., 2003. Stability of divalent europium in an ionic liquid: spectroscopic investigations in 1-methyl-butylimidazolium hexafluorophosphate. Inorg. Chem. 42, 1726–1733.
Billard, I., Ouadi, A., Gaillard, C., 2013. Is a universal model to describe liquid–liquid extraction of cations by use of ionic liquids in reach? Dalton Trans. 42, 6203–6212.
Calza, P., Vione, D., Fabbri, D., Aigotti, R., Medana, C., 2015. Imidazolium-based ionic liquids in water: assessment of photocatalytic and photochemical transformation. Environ. Sci. Technol. 49, 10951–10958.
Campos, K., Domingo, R., Vincent, T., Ruiz, M., Sastre, A.M., Guibal, E., 2008. Bismuth recovery from acidic solutions using Cyphos IL-101 immobilized in a composite biopolymer matrix. Water Res. 42, 4019–4031.
Cecchetto, L., Salomon, M., Scrosati, B., Crose, F., 2012. Study of a Li–air battery having an electrolyte solution formed by a mixture of an ether-based aprotic solvent and an ionic liquid. J. Power Sources 213, 233–238.
Chao, X., Liyong, Y., Xinghai, S., Maolin, Z., 2010. Efficient removal of caesium ions from aqueous solutions using a calix crown ether in ionic liquids: mechanism and radiation effect. J. Chem. Soc. Dalton Trans. 39, 3897–3902.
Chen, P.-Y., 2007. The assessment of removing strontium and cesium cations from aqueous solutions based on the combined methods of ionic liquid extraction and electrodeposition. Electrochim. Acta 52, 5484–5492.
Chun, S., Dzyuba, S.V., Bartsch, R.A., 2001. Influence of structural variation in room-temperature ionic liquids on the selectivity and efficiency of competitive alkali metal salt extraction by a crown ether. Anal. Chem. 73, 3737–3741.
Cocalia, V.A., Gutowsky, K.E., Rogers, R., 2006a. The coordination chemistry of actinides in ionic liquids: a review of experiment and simulation. Coord. Chem. Rev. 250, 755–764.
Cocalia, V.A., Holbrey, J.D., Gutowski, K.E., Bridges, N.J., Rogers, R.D., 2006b. Separations of metal ions using ionic liquids: the challenges of multiple mechanisms. Tsinghua Sci. Technol. 11, 188–193.
Critical surveys of stability constants of metal complexes – guidelines for prospective authors, 1975. Coord. Chem. Rev. 17, p. 358.
Dai, S., Ju, Y.H., Barnes, C.E., 1999. Solvent extraction of strontium nitrate by a crown ether using room-temperature ionic liquids. J. Chem. Soc. Dalton Trans., 1201–1202.
de Boer, J.A.A., Reinhoudt, D.N., Harkema, S., van Hummel, G.J., de Jong, F., 1982. Thermodynamic constants of complexes of crown ethers and uncharged molecules and X-ray structure of the 18-crown-6 $(CH_3NO_2)_2$. J. Am. Chem. Soc. 104, 4073–4076.
Dietz, M.L., 2006. Ionic liquids as extraction solvents: where do we stand? Sep. Sci. Technol. 41, 2047–2063.
Dietz, M.L., Dzielawa, J.A., Lazak, I., Young, B.A., Jensen, M.P., 2003. Influence of solvent structure variations on the mechanism of facilitated ion transfer into room-temperature ionic liquids. Green Chem. 5, 682–685.

Djigailo, D., 2010. Extraction of cations from aqueous solutions into RTILs and task specific RTILs. PhD Thesis, Lomonosov State University, Moscow (in Russian).

Docherty, K.M., Kulpa, C.F., 2005. Toxicity and antimicrobial activity of imidazolium and pyridinium ionic liquids. Green Chem. 7, 185–189.

Earle, M.J., Esperança, J.M.S.S., Gilea, M.A., Lopes, J.N.C., Rebelo, L.P.N., Magee, J.W., Seddon, K.R., Widegren, J.A., 2006. The distillation and volatility of ionic liquids. Nature 439, 831–834.

Ellerman, J., Bauer, W., Schuetz, M., Heinemann, F.W., Moll, M., 1998. Chemie polyfunctioneller Moleküle 130 Mitt. [1]: Spaltprodukte, Kristallstrukturen und Festkoerper-NMR-Spektern. Monatsh. Chem. 129, 547–566.

Eyring, E.M., Cobranchi, D.P., Garland, B.A., Gerhard, A., Highley, A.M., Huang, Y.H., Konya, G., Petrucci, S., Van Eldik, R., 1993. Lithium ion – crown ether complexes in a molten salt. Pure Appl. Chem. 65, 451–454.

Gallardo, V., Navarro, R., Saucedo, I., Ávila, M., Guibal, E., 2008. Zinc (II) extraction from hydrochloric acid solutions using amberlite XAD-7 impregnated with Cyphos IL 101 (tetradecyl (trihexyl)phosphonium chloride). Sep. Sci. Technol. 43, 2434–2459.

Garrell, R.L., Smyth, J.C., Fronczek, F.R., Gandour, R.D., 1988. Crystal structure of the 2:1 acetonitrile complex of 18-crown-6. J. Inclusion Phenom. 6, 73–78.

Gega, J., Otrembska, P., 2014. Separation of Ni(II) and Cd(II) ions with supported liquid membranes (SLM) using D2EHPA as carrier. Sep. Sci. Technol. 49, 1746–1760.

Georg, S., Billard, I., Quadi, A., Gaillard, C., Petitjean, L., Picquet, M., Solov'ev, V., 2010. Determination of successive complexation constants in an ionic liquid: complexation of UO_2^{2+} with NO_3^- in C_4-mimTf_2N studied by UV–Vis spectroscopy. J. Phys. Chem. B 114, 4276–4282.

Gjikaj, M., Adam, A., 2006. Complexation of alkali triflates by crown ethers: synthesis and crystal structure of Na (12-crown-4)$_2$[SO_3CF_3], Na (15-crown-5)[SO_3CF_3], [Rb (18-crown-6)] [SO_3CF_3] and [Cs (18-crown-6)] [SO_3CF_3]. Z. Anorg. Allg. Chem. 632, 2475–2480.

Guibal, E., Gavilan, K.C., Bunio, P., Vincent, T., Trochimczuk, A., 2008. CYPHOS IL 101 (tetradecyl (trihexyl)phosphonium chloride) immobilized in biopolymer capsules for Hg (II) recovery from HCl solutions. Sep. Sci. Technol. 43, 2406–2433.

Gutmann, V., Vichera, E., 1966. Coordination reactions in non aqueous solutions – the role of the donor strength. Inorg. Nucl. Chem. Lett. 2, 257–260.

Harjani, J.R., Friscic, T., MacGillivray, L.R., Leonard, R., Singer, R.D., 2008. Removal of metal ions from aqueous solutions using chelating task-specific ionic liquids. Dalton Trans., 4595–4601.

Hawkins, C.A., Momen, M.A., Garvey, S.L., Kestell, J., Kaminski, M.D., Dietz, M.L., 2015. Evaluation of solid-supported room-temperature ionic liquids containing crown ethers as media for metal ion separation and preconcentration. Talanta 135, 115–123.

Hirayama, N., Deguchi, M., Kawasumi, H., Honjo, T., 2005. Use of 1-alkyl-3-methylimidazolium hexafluorophosphate room temperature ionic liquids as chelate extraction solvent with 4,4,4-trifluoro-1-(2-thienyl)-1,3-butanedione. Talanta 65, 255–260.

IUPAC, 2007. Stability Constants Database. Academic Software, Timble, UK.

Janssen, C.H., Sanchez, A., Kobrak, M.N., 2014. Selective extraction of metal ions from aqueous phase to ionic liquids: a novel thermodynamic approach to separations. Chem. Phys. Chem. 15, 3536–3543.

Jastorff, B., Störmann, R., Ranke, J., Mölter, K., Stock, F., Oberheitmann, B., Hoffmann, W., Nüchter, M., Ondruschka, B., Filser, J., 2003. How hazardous are ionic liquids? Structure–activity relationships and biological testing as important elements for sustainability evaluation. Green Chem. 5, 136–142.

Jensen, M.P., Dzielawa, J.A., Ricket, P., Dietz, M.L., 2002. EXAFS investigations of the mechanism of facilitated ion transfer into a room-temperature ionic liquid. J. Am. Chem. Soc. 124, 10664–10665.

Jensen, M.P., Neuefind, J., Beitz, J.V., Skanthakumar, S., Soderholm, L., 2003. Mechanisms of metal ion transfer into room-temperature ionic liquids: the role of anion exchange. J. Am. Chem. Soc. 125, 15466–15473.

Jones, P.G., Hiemisch, O., Blaschette, A., 1994. Bildung und Kristallstruktur des Komplexes [(18-Krone-6) $(CH_2Cl_2)_2$]. Z. Naturforsch. B. 49, 852–854.

Keypour, H., Zebarjadian, M.H., Rezaeivala, M., Afkhami, A., 2014. Competitive ^{7}Li NMR study of the stoichiometry, stability and thermodynamic data for the complexation of Li^+, Mn^{2+}, Zn^{2+} and Cd^{2+} ions with two asymmetrical branched pentadentate (N5) amines containing pyridine moiety in ionic liquid–acetonitrile mixtures. J. Mol. Struct. 1075, 525–533.

Kikuchi, Y., Sakamoto, Y., 2000. Complex formation of alkali metal ions with 18-crown-6 and its derivatives in 1,2-dichloroethane. Anal. Chim. Acta 403, 325–332.

Köddermann, T., Wertz, C., Heintz, A., Ludwig, R., 2006. The association of water in ionic ligands: a reliable measure of polarity. Angew. Chem. 45, 3597–3702.

Langmaier, J., Samec, Z., 2009. Voltametry of ion transfer across a polarized room-temperature ionic liquid membrane facilitated by valinomycin: theoretical aspects application. Anal. Chem. 81, 6382–6389.

Lertlapwasin, R., Bhawawet, N., Imyim, A., Fuangswasdi, S., 2010. Ionic liquid extraction of heavy metal ions by 2-aminothiophenol in 1-butyl-3-methylimidazolium hexafluorophosphate and their association constants. Sep. Purif. Technol. 72, 70–76.

Levitskaia, T.G., Maya, L., van Berkel, G.J., Moyer, B.A., 2007. Anion partitioning and ion-pairing behavior of anions in the extraction of cesium salts by 4,5″-bis (*tert*-octylbenzo) dibenzo-24-crown-8 in 1,2-dichloroethane. Inorg. Chem. 46, 261–272.

Lewandowski, A., Swiderska-Mocek, A., 2009. Ionic liquids as electrolytes for Li–ion batteries – an overview of electrochemical studies. J. Power Sources 194, 601–609.

Lewandowski, A., Osinska, M., Steniak, I., 2005. Stability of Ag^+ complexes with Cryptand 222 in ionic liquids. J. Inclus. Phenom. Macrocyclic Chem. 52, 237–240.

Li, Z., Xia, S., Wang, J., Bian, C., Tong, J., 2016. Determination of trace mercury in water based on *N*-octylpyridinium ionic liquids preconcentration and stripping voltammetry. J. Hazard. Mater. 301, 206–213.

Li, H., Wang, B., Liu, S., 2015. Synthesis of pyridine-based task-specific ionic liquid with alkyl phosphate cation and extraction performance for uranyl ion. Ionics 21, 2551–2556.

Li, N., Fang, G., Zhang, J., Zhao, L., Wang, S., 2010. A novel hydrophobic task specific ionic liquid for the extraction of Cd (II) from water and food samples as applied to AAS determination. Anal. Sci. 26, 455–459.

Liu, L., Tian, G., Rao, L., 2013. Effect of solvation? Complexation of neodymium (III) with nitrate in an ionic liquid ($BumimTf_2N$) in comparison with water. Solvent Extr. Ion Exc. 31, 384–400.

Liu, H., Zhang, X., Chen, C., Du, S., Dong, Y., 2015. Effects of imidazolium chloride ionic liquid and their toxicity to *Scenedesmus obliquus*. Ecotoxicol. Environ. Safety 122, 83–90.

Luo, H., Dai, S., Bonnesen, P.V., 2004. Solvent extraction of Sr^{2+} and Cs^+ based on room-temperature ionic liquids containing monoaza-substituted crown ethers. Anal. Chem. 76, 2773–2779.

Luo, H., Dai, S., Bonnesen, P.V., Buchanan, A.C., 2006a. Separation of fission products based on ionic liquids: task-specific ionic liquids containing an aza-crown ether fragment. J. Alloys Compounds 418, 195–199.

Luo, H., Dai, S., Bonnesen, P.V., Haverlock, T.J., Moyer, B.A., Buchanan, A.C., 2006b. A striking effect of ionic-liquid anions in the extraction of Sr^{2+} and Cs^+ by dicyclohexano-18-crown-6. Solvent Extr. Ion Exc. 24, 19–31.

Majlesi, K., Rezaienejad, S., 2013. Investigation on the complexation of dioxovanadium (V) with D-(−)-quinic acid in water + 1-butyl-3-methylimidazolium tetrafluoroborate and water + methanol mixed solvents using the KAT model. J. Sol. Chem. 42, 1305–1319.

Martinis, E.M., Berton, P., Wuilloud, R.G., 2014. Ionic liquid-based microextraction techniques for trace-element analysis. Trends Anal. Chem. 60, 54–70.

Mei, E., Popov, A.I., Dye, J.L., 1977. Complexation of the cesium cation by macrocyclic polyethers in various solvents. A cesium-133 Nuclear Magnetic Resonance study of the thermodynamics and kinetics of exchange. J. Phys. Chem. 81, 1677–1681.

Murding, A.-V., Tang, S., 2010. Ionic liquids for lanthanide and actinide chemistry. Eur. J. Inorg. Chem., 2569–2581.

Murding, A.-V., Babai, A., Arenz, S., Giemoth, R., 2005. The "noncoordinating" anion Tf_2N^- coordinated to Yb^{3+}: a structurally characterized Tf_2N^- complex from the ionic liquid [mppyr][Tf_2N]. Angew. Chem. Int. Edit. 44, 5485–5488.

Mulkiewicz, E., Stepnowski, P., 2015. Recent developments in ionic liquid toxicity assessme. Int. J. Condensed Matter, Adv. Mater. Superconductivity Res. 13, 391–428.

Nakashima, K., Kubota, F., Maruyama, T., Goto, M., 2003. Ionic liquids as a novel solvent for lanthanide extraction. Anal. Sci. 19, 1097–1098.

Nishi, N., Murakami, H., Imakura, S., Kakiuchi, T., 2006. Facilitated transfer of alkali–metal cations by dibenzo-18-crown-6 across the electrochemically polarized interface between an aqueous solution and a hydrophobic room-temperature ionic liquid. Anal. Chem. 78, 5805–5812.

Nockemann, P., Thijs, B., Lunstroot, K., Van Hecke, K., Van Meervelt, L., Binnemans, K., 2008. Polynuclear metal complexes obtained from the task-specific ionic liquid betainium bistriflimide. Cryst. Growth Des. 8, 1353–1363.

Nockemann, P., Thijs, B., Lunstroot, K., Parac-Vogt, T.N., Görller-Walrand, C., Binnemans, K., Van Hecke, K., Van Meervelt, L., Nikitenko, S., Daniels, J., Hennig, C., Van Deun, R., 2009. Speciation of rare-earth metal complexes in ionic liquids: a multiple-technique approach. Chem. Eur. J. 15, 1449–1461.

Ohtsu, K., Ozutsumi, K., 2003. Thermodynamics of solvation of 18-crown-6 and its alkali–metal complexes in various solvents. J. Inclusion Phenom. Macrocyclic Chem. 45, 217–224.

Ozutsumi, K., Natsuhara, M., Ohtaki, H., 1989. An X-ray diffraction study on the structure of 18-crown-6 ether complexes with alkali metal ions in aqueous solutions. Bull. Chem. Soc. Jpn. 62, 2807–2818.

Perales, L.H., Cruz, L.E.H., Garcia, F.L., Garcia, B.M., Rodriguez, M.A., 2014. Selective recovery of Zn(II) and Fe(III) by liquid–liquid extraction using cyphos IL 109 as extractant. Adv. Mater. Res. 976, 217–221.

Pereiro, A.B., Araujo, J.M.M., Oliveira, F.S., Esperanca, J.M.S.S., Canongia, L.J.N., Marrucho, I.M., Rebelo, L.P.N., 2012. Solubility of inorganic salts in pure ionic liquids. J. Chem. Thermodyn. 55, 29–36.

Popov, K.I., Wanner, H., 2005. Stability constants data sources: critical evaluation and application for environmental speciation. In: Nowack, B., Van Briesen, J.M. (Eds.), Biogeochemistry of Chelating AgentsACS Symp. Ser., Washington DC, pp. 50–75, vol. 910.

Popov, K.I., Rönkkömäki, H., Hannu-Kuure, M., Kuokkanen, T., Lajunen, M., Vendilo, A., Glazkova, I.V., Lajunen, L.H.J., 2007a. Stability constant of the sodium complex with dibenzo-18-crown-6 in mixed water–ionic liquid solvent. Russ. J. Coord. Chem. 33, 393–395.

Popov, K., Rönkkömäki, H., Hannu-Kuure, M., Kuokkanen, T., Lajunen, M., Vendilo, A., Oksmann, P., Lajunen, L.H.J., 2007b. Stability of crown-ether complexes with alkali–metal ions in ionic liquid–water mixed solvents. J. Inclusion Phenom. Macrocyclic Chem. 59, 377–381.

Popov, K., Vendilo, A., Pletnev, I., Lajunen, M., Rönkkömäki, H., Lajunen, L.H.J., 2011. Formation of complexes in RTIL and ion separations. In: Kokorin, A. (Ed.), Ionic Liquids: Theory, Properties. New Approaches. InTech, Rijeka, Croatia, pp. 483–510.

Popov, K.I., Vendilo, A.G., Chistov, V., Rönkkömäki, H., Lajunen, M., Privalov, V., Dikareva, J., 2013. Stability constants of lithium complexes with 15-crown-5 and 18-crown-6 in a hydrophobic ionic liquid 1-butyl-3-methylimidazolium bis[(trifluoromethyl)sulphonyl] imide. Polyhedron 63, 50–54.

Pospiech, B., Kujawski, W., 2015. Ionic liquids as selective extractants and ion carriers of heavy metal ions from aqueous solutions utilized in extraction and membrane separation. Rev. Chem. Eng. 31, 179–191.

Reichardt, C., 2005. Polarity of ionic liquids determined empirically by means of solvatochromic pyridinium *N*-phenolate betaine dyes. Green Chem. 7, 339–351.

Richter, K., Lorbeer, C., Murding, A.-V., 2015. A novel approach to prepare optically active ion doped luminescent materials via electron beam evaporation into ionic liquids. Chem. Commun. 51, 114–117.

Richter, K., Campbell, P.S., Baecker, T., Schimitzek, A., Yaprak, D., Murding, A.-V., 2013. Ionic liquids for the synthesis of metal nanoparticles. Phys. Status Solidi B: Basic Solid State Physics 250, 1152–1164.

Rogers, R.D., Richards, P.F., Voss, E.J., 1988. Neutral solvent/crown ether interactions, 4. Crystallization and low temperature (−150°C) structural characterization of 18-crown-6•2 (CH_3CN). J. Incl. Phenom. 6, 65–71.

Rosol, Z.P., German, J., Gross, S.M., 2009. Solubility, ionic conductivity and viscosity of lithium salts in room temperature ionic liquids. Green Chem. 11, 1453–1457.

Ruhela, R., Iyer, N., Yadav, M., Singh, A.K., Hubli, R.C., Chakravartty, J.K., 2015. *N,N,N*-trimethyl benzyl ammonium bis-(2-ethylhexyl)-phosphonate grafted polymer – a solid supported ionic liquid for the separation of uranium from aqueous processing streams. Green Chem. 17, 827–830.

Seddon, K.R., Stark, A., Torres, M.-J., 2000. Influence of chloride, water and organic solvents on the physical properties of ionic liquids. Pure Appl. Chem. 72, 2275–2287.

Sengupta, A., Murali, M.S., Mohapatra, P.K., 2013a. A comparative study of the complexation of Am(III) and Eu(III) with TODGA in room temperature ionic liquid. J. Radioanal. Nucl. Chem. 298, 405–412.

Sengupta, A., Mohapatra, P.K., Iqbal, M., Huskens, J., Verboom, W., 2013b. A diglycolamide-functionalized task specific ionic liquid (TSIL) for actinide extraction: solvent extraction, thermodynamics and radiolytic stability studies. Sep. Purif. Technol. 118, 264–270.

Samsipur, M., Irandoust, M., 2012. ^7Li-NMR study of the stoichiometry, stability and exchange kinetics of Li^+ ion with 12-crown-4, 15-crown-5 and cryptands C222, C221 and C211 in 50% ionic liquid–acetonitrile mixtures. Polyhedron 31, 395–401.

Shimojo, K., Goto, M., 2004. Solvent extraction and stripping of silver ions in room-temperature ionic liquids containing calixarenes. Anal. Chem. 76, 5039–5044.

Shkrob, I., Marin, T.W., Jensen, M.P., 2014. Ionic liquid based separations of trivalent lanthanide and actinide ions. Ind. Eng. Chem. Res. 53, 3641–3653.

Song, X., Li, H., Li, J., Liu, J., Li, Z., Liu, S., Li, C., Wu, P., 2005. The study on the coordination equilibrium of Cu^{2+}-ethylenediamine in ionic liquid. Polym. Prepr. 46, 842.

Srncik, M., Kogelnig, D., Stojanovic, A., Körner, W., Krachler, R., Wallner, G., 2009. Uranium extraction from aqueous solution by ionic liquids. Appl. Radiat. Isotopes 67, 2146–2149.

Steinrueck, H.-P., Wasserscheid, P., 2015. Ionic liquids in catalysis. Catal. Lett. 145, 380–397.

Sun, X., Wu, D., Chen, J., Li, D., 2007. Separation of scandium(III) from lanthanides(III) with room temperature ionic liquid based extraction containing Cyanex 925. J. Chem. Technol. Biotechnol. 82, 267.

Sun, X., Luo, H., Dai, S., 2012. Ionic liquids-based extraction: a promising strategy for the advanced nuclear fuel cycle. Chem. Rev. 112, 21000–22128.

Takao, K., Bell, T.J., Ikeda, Y., 2013. Actinide chemistry in ionic liquids. Inorg. Chem. 52, 3459–3472.

Tang, S., Babai, A., Murding, A.-V., 2008. Europium-basierte ionische Flussigkeiten als lumineszierende weiche Materialien. Angew. Chem. 120, 7743–7746.

Umecky, T., Takamuku, T., Kanzaki, R., Takagi, M., Kawai, E., Matsumoto, T., Funazukuri, T., 2014. Role of water in complexation of 1,4,7,10,13,16-hexaoxacyclooctadecane (18-crown-6) with Li^+ and K^+ in hydrophobic 1-ethyl-3-methylimidazolium bis(trifluoromethanesulfonyl)amide ionic liquid. J. Incl. Phenom. Macrocyclic Chem. 80, 401–407.

Vander Hoogerstraete,T., Blockx,J., De Coster, H., Binnemans, K., 2015. Selective single-step separation of a mixture of three metal ions by a triphasic ionic-liquid-water- ionic-liquid solvent extraction system. Chem.-Eur. J. 21, 11757–11766.
Vendilo, A.G., Rönkkömäki, H., Hannu-Kuure, M., Lajunen, M., Asikkala, J., Petrov, A.A., Krasovsky,V.G., Chernikova, E.A., Oksman, P., Lajunen, L.H.J., Popov, K.I., 2008. Stability constants of cesium complexes with 18-crown-6 in ionic liquids. (Koord. Khim. 34, 645-650). Russ. J. Coord. Chem. 34, 635–640.
Vendilo, A.G., Djigailo, D.I., Smirnova, S.V., Torocheshnikova, I.I., Popov, K.I., Krasovsky, V.G., Pletnev, I.V., 2009a. 18-Crown-6 and dibenzo-18-crown-6 assisted extraction of cesium from water into room temperature ionic liquids and its correlation with stability constants for cesium complexes. Molecules 14, 5001–5016.
Vendilo, A.G., Rönkkömäki, H., Hannu-Kuure, M., Lajunen, M., Asikkala, J., Chernikova, E.A., Lajunen, L.H.J., Tuomi, T., Popov, K.I., 2009b. First data on thermodynamics of complex formation in ionic liquids: cesium complexes with 18-crown-6. Mendeleev Commun. 19, 196–197.
Vendilo, A.G., Djigailo, D.I., Rönkkömäki, H., Lajunen, M., Chernikova, E.A., Lajunen, L.H.J., Pletnev, I.V., Popov, K.I., 2010a. A correlation of caesium-18-crown-6 complex formation constants with the extraction capability for hydrophobic ionic liquids. Mendeleev Commun. 20, 122–124.
Vendilo, A.G., Popov, K.I., Artem'eva, A.A., Kovaleva, N.E., 2010b. Hydrophobic ionic liquid and crown-ether based liquid membrane. Kriticheskie Tekhnologii. Membrany (Crit. Technol. Membr.), Iss. 1, 27–29 (in Russian).
Vendilo, A.G., Rönkkömäki, H., Hannu-Kuure, M., Lajunen, M., Asikkala, J., Krasovsky, V.G., Chernikova, E.A., Oksmann, P., Lajunen, L.H.J., Popov, K.I., 2010c. Thermodynamics of cesium complexes formation with 18-crown-6 in ionic liquids. J. Incl. Phenom. Macrocyclic Chem. 66, 223–230.
Vendilo, A.G., Rönkkömäki, H., Kotova, N.N., Hannu-Kuure, M., Kovaleva, N.E., Lajunen, M., 2010d. Complexes of strontium with dichlormethylenbis (phosphonic acid triisopropyl ester. Russ. J. Inorg. Chem. 55, 1652–1655.
Vendilo, A.G., Djigailo, D.I., Pletnev, I.V., Popov, K.I., Torocheshnikova, I.I., 2011. Thermodynamics of cesium complex formation with dibenzo-18-crown-6 in hydrophobic ionic liquids. Russ. J. Inorg. Chem. 56, 1010–1016.
Vendilo, A.G., Chistov, V.I., Privalov, V.I., 2012. Thermodynamics of cesium complexes formation with dibenzo-21-crown-7 and dibenzo-24-crown-8 in 1-butyl-3-methylimidazolium bis[(trifluoromethyl)sulphonyl]imide. Russ. J. Inorg. Chem. 56, 1010–1016.
Vendilo, A., Popov, K., Lajunen, M., Chistov, V., Djigailo, D., Ronkkomaki, H., Privalov, V., Pletnev, I., 2014. A cesium-133 nuclear magnetic resonance study of the cesium cation complexation by macrocyclic polyethers in hydrophobic RITLs. Polyhedron 81, 341–348.
Vendilo, A.G., Chistov, V.I., Dikareva, J.M., Popov, K.I., 2015. Crown ethers assisted cesium extraction from aqueous solutions into a hydrophobic room temperature ionic liquid 1-butyl-3-methylimid-azolium bis[(trifluoromethyl)sulfonyl]imide. Macroheterocycles 8, 181–184.
Vincent, T., Parodi, A., Guibal, E., 2008. Pt recovery using Cyphos IL-101 immobilized in biopolymer capsules. Sep. Purif. Technol. 62, 470–479.
Visser, A.E., Rogers, R.D., 2003. Room-temperature ionic liquids: new solvents for F-element separations and associated solution chemistry. J. Solid State Chem. 171, 109–113.
Visser, A.E., Swatloski, R.P., Reichert, W.M., Griffin, S.T., Rogers, R.D., 2000. Traditional extractants in nontraditional solvents: groups 1 and 2 extraction by crown ethers in room-temperature ionic liquids. Ind. Eng. Chem. Res. 39, 3596–3604.
Visser, A.E., Swatloski, R.P., Griffin, S.T., Hartman, D.H., Rogers, R.D., 2001a. Liquid–liquid extraction of metal ions in room-temperature ionic liquids. Sep. Sci. Technol. 36, 785–804.

Visser, A.E., Swatloski, R.P., Reichert, W.M., Mayton, R., Sheff, S., Wierzbicki, A., Davis, J.H., Rogers, R.D., 2001b. Task-specific ionic liquids for the extraction of metal ions from aqueous solutions. Chem. Commun., 135–136.

Visser, A.E., Swatloski, R.P., Reichert, W.M., Mayton, R., Sheff, S., Wierzbicki, A., Davis, J.H., Rogers, R.D., 2002. Task-specific ionic liquids incorporating novel cations for the coordination and extraction of Hg^{2+} and Cd^{2+}: synthesis, characterization, and extraction studies. Environ. Sci. Technol. 36, 2523–2529.

Wimberg, J., Stiemke, F.M., Schnetzer, H., Schubert, T.J.S., 2014. Ionic liquids and nanoparticles – a successful symbiosis of highly innovative materials. Chimica Oggi 32, 27–30.

Xing, C., Wang, Y., Zhang, C., Li, L., Li, Y., Li, J., 2015. Immobilization of ionic liquids onto the poly(vinylidene fluoride) by electron beam irradiation. Ind. Eng. Chem. Res. 54, 9351–9359.

Yakshin, V.V., Tsarenko, N.A., Koshcheev, A.M., Tananaev, I.G., Myasoedov, B.F., 2012. Extraction of metal ions from aqueous solutions of mineral acids with crown ether–ionic liquid systems. Radiochemistry 54, 54–58.

Yakshin, V.V., Vilkova, O.M., Laskorin, B.N., 1992. Extractive separation of radionuclides of cesium and strontium with crown ethers. Dokl. Akad. Nauk SSSR 325, 967–969.

Yakshin, V.V., Vilkova, O.M., Tsarenko, N.A., Tsivadze, A.Yu., Demin, C.V., Zhilov, V.I., 2006. Control of the extractive power of crown ethers by alkyl substitution. Radiokhimiya 48, 153–157, Radiochemistry (New York, NY, United States), 48, 170-174.

Yan, N., Fei, Z., Scipelliti, R., Laurenczy, G., Kou, Y., Dyson, P.J., 2010. Crystallization of inorganic salts containing 18-crown-6 from ionic liquids. Inorg. Chim. Acta 363, 504–508.

Yang, F., Kubota, F., Kamiya, N., Goto, M., 2013. A comparative study of ionic liquids and a conventional organic solvent on the extraction of rare-earth ions with TOPO. Solvent Extr. Res. Develop. 20, 225–232.

Yim, T., Kwon, M.-S., Mun, J., Lee, K.T., 2015. Room temperature ionic liquid-based electrolytes as an alternative to carbonate-based electrolytes. Israel J. Chem. 55, 586–598.

Zhao, Y., Bostrom, T., 2015. Application of ionic liquids in solar cells and batteries: a review. Curr. Organic Chem. 19, 556–566.

CHAPTER 5

Methods of IL Recovery and Destruction

Olga Kuzmina
Department of Chemistry, Imperial College London, UK

RECOVERY VERSUS REGENERATION AND RECYCLING VERSUS REUSE

Before discussing the possible ways to recover and reuse ionic liquids (ILs) from the waste streams, it is important to understand the meaning of all the relevant terms, as many of them are used incorrectly.

As defined by the IUPAC Gold Book (IUPAC, 2012):

- *Recovery* is a term used in analytical and preparative chemistry to denote the fraction of the total quantity of a substance recoverable following a chemical procedure.
- *Solvent regeneration* in extraction processes is a treatment of the solvent for recycling, for example, by removal of degradation products or non-strippable solutes. This term is synonymous with solvent purification.
- *Recycling* is defined (by the Dictionary of Environment and Conservation (Park, 2007)) as a process to change materials (waste) into new products to prevent the waste of potentially useful materials, reduce the consumption of fresh raw materials, reduce energy usage, reduce air pollution (from incineration) and water pollution (from landfilling) by reducing the need for "conventional" waste disposal, and lower greenhouse gas emissions as compared to plastic production.

Thus, ionic liquids are used in the reaction, and then they appear in the waste stream. Ionic liquids can be recovered or regenerated from wastes and reused again. All of these processes together are the recycling of the ionic liquids.

The purification stage is needed to avoid the deterioration of ILs, while regeneration is needed if a compound cannot be reused directly. During the processing stages and continued reuse some impurities can accumulate and the concentration of IL in the final extract can be significantly lower than expected. Most impurities found in ILs are traces of unreacted starting materials. However, after the processing of the IL the waste stream can contain

Application, Purification, and Recovery of Ionic Liquids
http://dx.doi.org/10.1016/B978-0-444-63713-0.00005-5

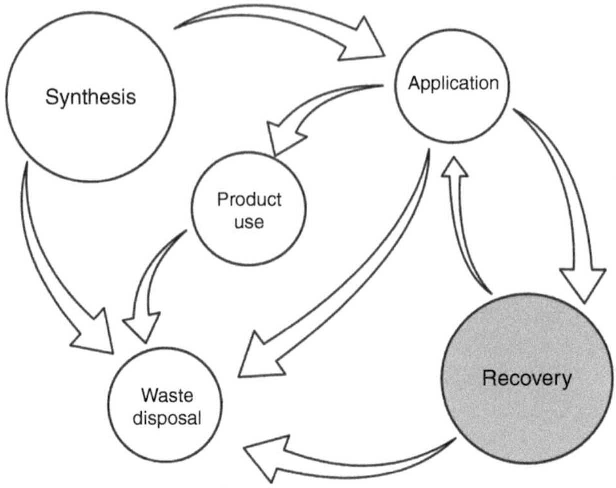

Figure 5.1 ***A Simplified Lifecycle of ILs.***

not only IL with its initial impurities, but also traces of the new reaction products, other chemicals, and their impurities. It is known that even small amounts of impurities influence the IL properties significantly, and the quality of ILs is an important consideration for many applications (Seddon et al., 2000). Without recycling or removal ionic liquids can become a persistent hazard in the aquatic environment because their low volatility prevents them from being released into the atmosphere (Haerens et al., 2010).

Generally, the lifecycle of ionic liquids can be presented as follows (Figure 5.1): synthesis of ILs is accompanied by production of waste and by-products, then ILs are often transported to be used in other chemical reactions. Then ILs can appear in waste streams (e.g., a water waste after the dissolution process, where ILs were used as a solvent or reaction media and water as an antisolvent) or as a part of a final product (e.g., polysaccharide gels containing ILs). Secondary waste streams can be produced at the stage of IL recovery. The appropriate waste treatment can significantly decrease production costs, which define the overall price of the final product and the benefits for the producer. During product usage, the IL ends up as a waste with very little possibility of reuse.

The best way to decrease the final cost of waste disposal is to reduce the amount of waste produced. A specific example of application of ILs in closed production cycle proposed for biomass pretreatment by Dibble et al. (2011) is shown in (fig. 5.2). The ionic liquid reclamation step is one of the most difficult and least explored aspects of this process. Possible ways to reduce the

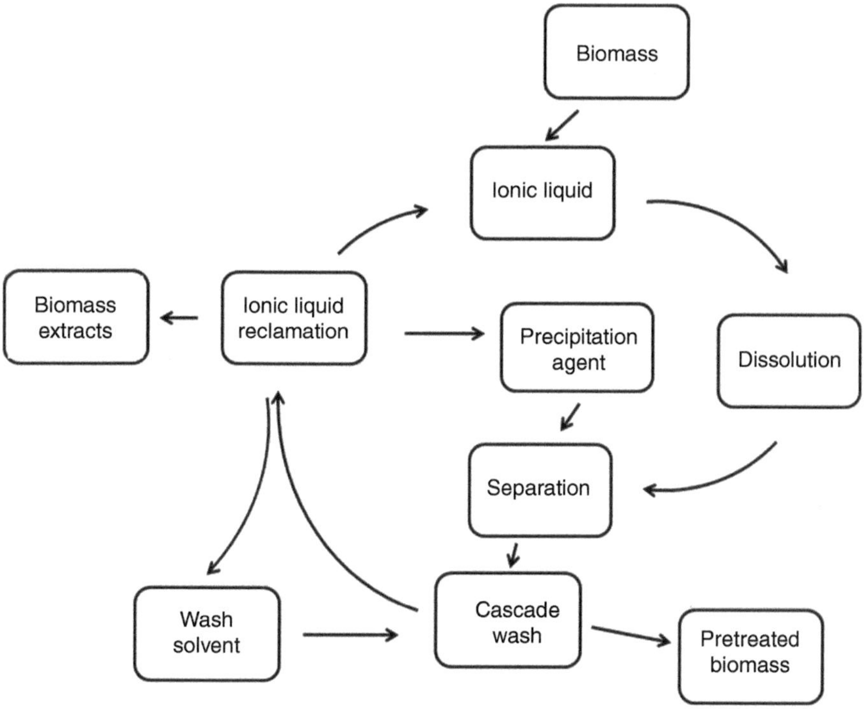

Figure 5.2 ***Simplified Diagram of an Ionic Liquid/Biomass Pretreatment Process.*** *Adapted from Dibble et al. (2011).*

amount of industrial waste streams containing ILs are suggested by Siedlecka et al. (2011) (Figure 5.3). Generally, removal of the ILs from waste streams, their regeneration, and recovery need to be implemented wherever and whenever possible.

The recovery processes are limited to the concentration of contaminants. The exceeding of a reasonable limit of impurities prevents IL use and turns it to IL waste. From this point of view the recovery processes for IL wastes are of special interest.

The main impurities that can be found in the postreaction mixture containing ILs include but are not limited to the aqueous solution (or water), traces of unreacted starting materials (mainly cellulose derivatives and organics), the residues after galvanic treatment, and catalysts. They all have different natures and properties and require unique methods for IL separation.

Despite it being clearly stated in the literature earlier that ILs are recyclable, their purification, especially for those ILs soluble in water, has proved to be more difficult than expected. The first method that comes to mind

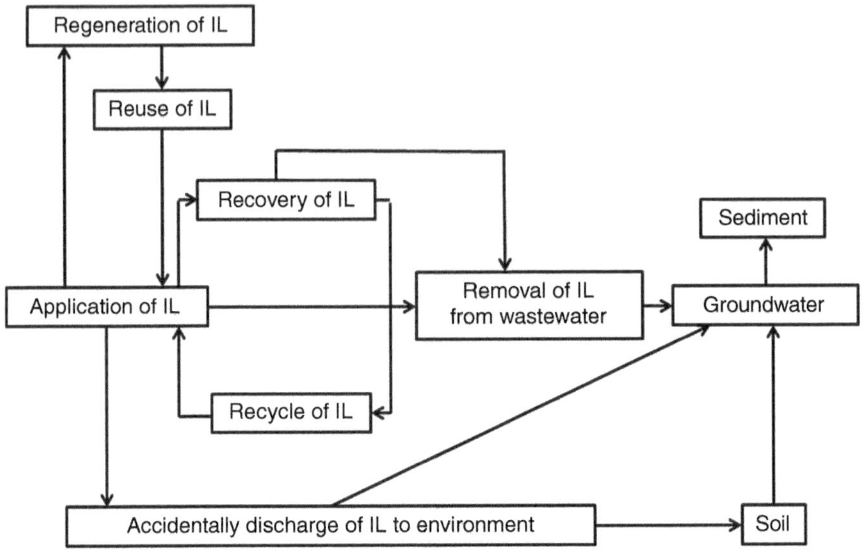

Figure 5.3 ***Reduction of Wastes Containing Ionic Liquids in Industrial Applications, and Potential Sources of ILs Released to the Environment.*** *Adapted from Siedlecka et al. (2011).*

when thinking about separation of ILs from water is evaporation. However, this method requires high temperatures and therefore is very energy-consuming. Moreover, the IL itself can be degraded if it is subjected to high temperatures for a prolonged period of time. Despite most ILs having high decomposition temperatures, based on TGA scans, in real environments their decomposition temperature can be significantly lower. Evaporation may provoke some other undesirable consequences, such as evaporation of the conjugate acid or base species from protic ILs, therefore leaving hydroxide or proton impurities in the solution (Haerens et al., 2010).

The lowest energy consumption separation methods are barrier separation methods such as membrane separation and nanofiltration. Nanofiltration can be applied to IL separation from other solvents because ILs consists only of ions. Charged and neutral compounds or mono- and divalent ions can be separated by nanofiltration membranes. The charge and size of the ions or molecules and the choice of membrane influence the retention (Mulder et al., 2005). This method is especially promising for separation of nonvolatile compounds or, at low concentration, ILs from waste streams when distillation methods are not applicable. The principle of nanofiltration is that the membranes are permeable to the ILs but not to the other products in the waste stream. Another barrier method is pervaporation, which is more energy-consuming in comparison to nanofiltration, but is often used as a less energy-demanding alternative to vacuum or extractive distillation. During pervaporation the feed component is interacting with the membrane material combined with the gradient in a chemical potential (Mulder, 1996). Some of the main difficulties of nanofiltration and pervaporation are the viscosity of ILs and the requirement of large membrane areas.

This overview proves that only through the combination of currently existing separation methods can we guarantee the efficient recovery of ILs. The development of each of the described methods and investigation of new ones is necessary in order to increase the sustainability of ILs processing.

Recently, scientists stopped calling ILs "green solvents," because their "greenness" is doubtful. Negligible vapor pressure is a reason why ILs are regarded as environmentally nonhazardous chemicals. However, IL processing in synthesis, catalysis, separation/extraction reactions, and the recycling of ILs usually involve other solvents (e.g., VOCs); this diminishes the "green" aspect of their usage due to the cross-contamination arising from the use of organic solvents. Moreover, some ILs are defined as toxic and have a detrimental effect on aquatic organisms and will tend to persist in local environments, on land/sea, and so on (Li et al., 1998; Pernak et al., 2001; Ranke et al., 2004; Docherty and Kulpa, 2005). Methods such as evaporation of volatiles under vacuum, extractions with VOC solvents, supercritical CO_2 ($scCO_2$) extraction, and distillation/stripping of the solute from the thermally stable ILs have been known for some time (Kanel et al., 2003) and have been applied to ILs.

Wu et al. (2009) raised the question of how to understand the recyclability of ILs. Methods for separation and recycling of ILs were discussed in a minireview combined with their own experience on separation of ILs from their "working" environment (Wu et al., 2009). The authors stated that the choice of the appropriate separation method should be based on the specific properties of each system. Also, they questioned the need to obtain absolutely pure ILs, which makes sense only for academic researchers, who do not want even negligible amounts of impurities to have an effect on their results. Usually grades of purity are greater than 95% "for synthesis," greater than 99% "high purity," and greater than 99.9% "ultrapure." However, some applications do not need such purity, and additional purification steps would significantly increase the cost of ILs when produced on a commercial scale (Abu-Eishah, 2010). To summarize all these, broad and deep research investigating the physical and chemical properties of ILs and their behavior in solution and waste streams is needed in order to create an efficient method for the separation of ILs to support recyclability.

All of the separation and removal methods are described in this chapter in detail. Temperature-dependent methods, such as distillation and crystallization, are marked as suitable for separation of hydrophilic and hydrophobic ILs. However, these energy-consuming techniques should not be used for the separation of hydrophobic ILs, where methods, such as liquid-liquid and membrane separation, are available.

Methods for the recovery of ILs using absorption will be discussed in the part dedicated to the removal of ILs from their waste streams. The recovery of the IL after absorption is poorly studied, thus it is more relevant to describe the achievement of this technology in the last section.

RECOVERY METHODS SUITABLE FOR ILs

Distillation

The lack of significant vapor pressure makes the purification of ionic liquids by distillation challenging. The counterpoint to this is that theoretically any volatile impurity could be separated from an ionic liquid by distillation (Swatloski et al., 2002; Welton, 1999). The simplest method to separate an IL from its water solution is distillation (Swatloski et al., 2002). In this method water, due its lower boiling point, is distilled at high temperature or pressure. By this method, the distilled pure water and an IL mixture with all the contaminants are obtained. The disadvantages of this method are evident: (1) the main objective was to obtain pure IL, but not pure water; and (2) vacuum distillation is an energy-consuming method. Also, the application of this method is limited because of its unsustainable use for water waste streams containing low concentrations of IL and for the separation of mixtures of thermally unstable ILs, which have a tendency to undergo hydrolytic decomposition (Wu et al., 2009).

In my own research (Kuzmina, 2012), alkylimidazolium ILs were used as a solvent for cellulose dissolution and then washed from the cellulose solution by water. Then the IL/water solution was separated by water distillation using a rotary evaporator fitted with a water bath and the IL was reused. Such a recycling procedure could be repeated up to three to four times without a deterioration of the IL's dissolving ability. However, in this study the mechanical properties of cellulose films produced using recycled IL, especially the strength of cellulose films, were slightly worse than the mechanical properties of cellulose films produced with fresh IL.

There is also information that 3-methyl-1-octylimidazolium tetrafluoroborate IL used for one-pot aminomethylation of electron-rich aromatic

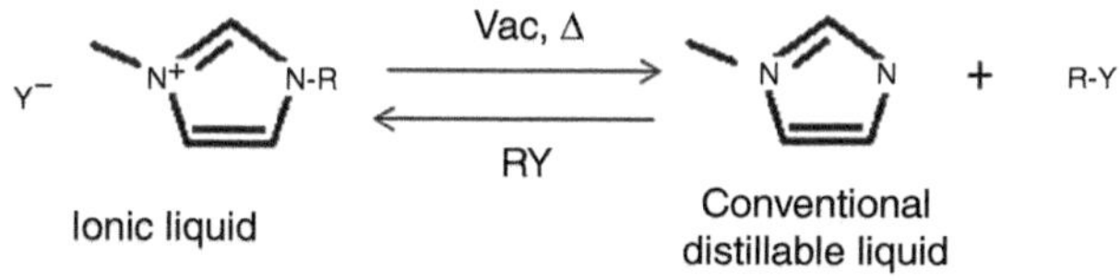

Figure 5.4 ***ILs Switching off by Deprotonation and Back-Alkylation of the Anion.***

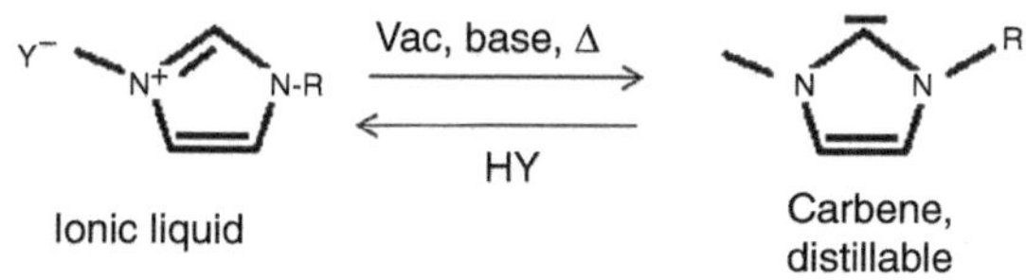

Figure 5.5 ***Formation of Distillable Carbenes.***

compounds was reported to be recycled four times by removing the volatile solvent under reduced pressure using a rotary evaporator (Arzephoni et al., 2013).

If we refer to "BASF FAQs on How to Recycle ILs," (BASF, 2013) the information relates to the distillation of the amine or imidazole produced after deprotonation. This method is suitable for ILs with protonated and alkylated cations (Figure 5.4). Moreover, for ILs with alkylated cations apart from the back-alkylation of the anion (Figure 5.4) the formation of distillable carbenes has also been reported (Earle and Seddon, 2001; Jeapes et al., 2001) (Figure 5.5).

BASF recommends that imidazolium cations should be deprotonated by bases to form neutral carbenes. These stable carbenes can be distilled and the IL can be recycled by the further reaction of the carbene with an acid. Thus, this controlled decomposition reaction can be used to purify and recycle ILs. The IL is thermally disrupted and the neutral imidazole and alkylating agent are distilled, collected and re-reacted by this method.

Despite common distillation of low molecular weight fractions, there are also some impressive studies of IL distillation. Maase for BASF's WO 05/019183 (Maase, 2005; Maase and Massonne, 2005; Maase et al., 2005) and consequently Earle et al. (2006) distilled pure ILs without degradation. Maase used in a first step the pressure set to a value below the ambient pressure and in a subsequent step, the ionic liquid was heated to a temperature ranging between 60°C and 350°C. The method is suggested to be used in particular for purifying ionic liquids. Meanwhile, Earle et al. proved that some selected families of commonly used aprotic ionic liquids can be distilled at 200–300°C and low pressure, with concomitant recovery of significant amounts of pure substance. Moreover, Earle et al. proved the possibility of distillation of ILs mixtures. Most of the used ILs were distilled without significant decomposition by volatilization of these ionic liquids as not altered ions, either alone or aggregated. Chen et al. (2013) presented a new class of low-cost amide-derived distillable aprotic ILs synthesized by amide *O*-alkylation with alkyl triflate. As the synthesis described (Chen et al., 2013) is reversible, the regenerated volatile precursors can be easily vacuum distilled at about 200°C and also revert to ILs at room temperature, with a maximum recovery rate of up to 99 wt% and a purity of 97%. This is the first case demonstrating the feasibility of the easy recycling of aprotic ILs by the distillation of the regenerated molecular precursors at elevated temperatures without the need for further processing.

Dal and Lancaster (2005) demonstrated a method for recovery and reuse of 1-butyl-1-methylpyridinium bis(trifluoromethylsulfonyl)imide ($[C_4C_1py][NTf_2]$). The authors dissolved the postreaction mixture into dichloromethane and added some water. Then any unreacted nitric acid and acetic acid, generated by the reaction, is extracted into the water phase. After removing the dichloromethane, steam distillation was used to remove the organics from the IL. Compounds extracted from the water and the ionic liquid were recovered by extraction with dichloromethane and then dried under vacuum (Figure 5.6).

It was possible to recover up to 95.9% of the ionic liquid after the first run and up to 83.83% after the fifth run by this method. However, the main losses of ionic liquid via this method were mechanical (including withdrawal of a sample for NMR from each run). The role of dichloromethane (which diminishes the environmental friendliness of this process) used in

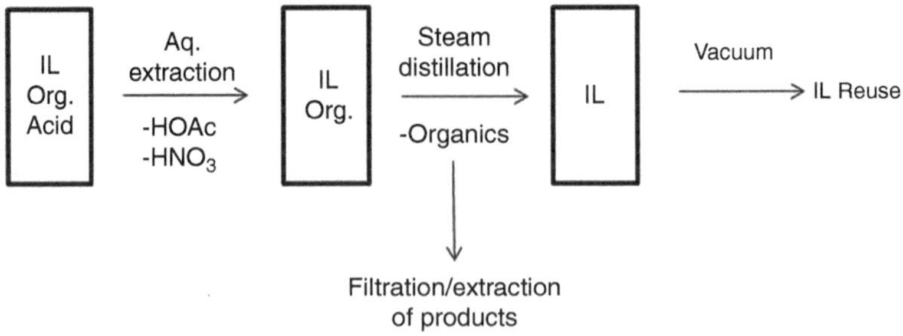

Figure 5.6 ***Schematic Representation of the Ionic Liquid Recycling Process.*** *Adapted from Dal and Lancaster (2005).*

this experiment was to enable product separation and IL recovery. The authors do not doubt the possibility of performing steam distillation on the entire reaction mixture in an industrial setting. Dal and Lancaster stated that there is a possibility to separate distilled organic and aqueous acid phases by simple decantation. Also, decanting can be used to remove water from hydrophobic ILs and subsequently recycling of the IL would not require the use of dichloromethane.

Recently the class of distillable ILs has been complemented by Parviainen et al. (2013). They aimed to design distillable IL that were efficient for biomass treatment and processing. The dissolution efficiency and processability of these ILs allowed researchers to develop a new method for cellulose fibers spinning, namely Ioncell-F (Michud et al., 2015; Parviainen et al., 2015; Sixta et al., 2015). These ILs contain cations with higher acidity relative to ILs with imidazolium cations, and dissociate at low temperatures to give neutral volatile species. These cations included, but are not limited to: 1,1,3,3-tetramethylguanidine (TMG), 1,2-dimethyl-1,4,5,6-tetrahydropyrimidine (DMP), 1,5-diazabicyclo[4.3.0]non-5-ene (DBN), and 1,8-di-azabicyclo[5.4.0]undec-7-ene (DBU). The authors described the recyclability of one of the most efficient ILs for cellulose dissolution, [DBNH][CO_2Et] (Parviainen et al., 2013) by heating it under reduced pressure to 170°C while cooling the collecting flask to −40°C. Despite the promising claim of an average ≥95% recovery of [DBNH][OAc] during 7 cycles of cellulose hydrolysis (Parviainen et al., 2015) and ≥99% recovery of ≥99% pure [TMGH][CO_2Et] King et al. (2011), the efficiency of recyclability of these ILs are still subject to further study.

Molecular Distillation

Molecular or short-path distillation (MD) is a mild distillation method suitable for the separation and purification of thermally unstable liquids as well as liquids with relatively high viscosity, low vapor pressure, and higher molecular weight, without the threat of thermal decomposition. MD is mostly used for the recovery of compounds with high boiling points. This method was used for the regeneration of a phosphonium IL and its solution in dodecane as a solvent loaded with butanoic acid (BA) by Blahušiak et al. (2011). The authors used trihexyl(tetradecyl)phosphonium bis 2,4,4-(trimethylpentyl) phosphinate (commercial name Cyphos IL-104; Cytec, Canada), a hydrophobic IL exhibiting good thermal stability below a temperature of about 250°C. The main advantage of this process compared with the classic stripping of acids with alkali solutions is that the product is separated in the form of the acid, not its salt, hence the consumption of chemicals is reduced and the regeneration process becomes simpler. The yield of BA via MD was 88% (major portion of the distillate condenses at the temperature of cooling water) and almost 90% of dodecane present in the solvent was also evaporated. The difference of this method from others described here is that in all other methods pure ILs are recovered from the waste stream, even if not completely. By this method, pure impurities are recovered from ILs, and the residual IL still contains some of those and probably other impurities. Work on further improvement in the yield of the stripping must be done in order to make this method widely applicable for IL recovery.

Extraction With Supercritical CO_2

Hydrophobic and hydrophilic room-temperature ionic liquids can be separated from aqueous solutions with relatively low-pressure gaseous carbon dioxide (Scurto et al., 2003). The application of supercritical carbon dioxide ($scCO_2$) in solute extraction from ILs has been known since 2001. The advantage of this technique is that it combines two potentially environmentally benign solvents to perform the separation. CO_2 is inexpensive, nonflammable, nontoxic, and environmentally friendly (Blanchard and Brennecke, 2001). This "green solvent" is a possible replacement for traditionally hazardous and environmentally harmful organic solvents. Because the studied 1-butyl-3-methylimidazolium hexafluorophosphate ($[C_4C_1im][PF_6]$) does not dissolve appreciably in CO_2, the supercritical phase is essentially an organic solute and CO_2. The liquid phase is a mixture of solute and IL. An important thermodynamic property is the distribution coefficient, *K*, the ratio of the solute mole fractions in the supercritical (*y*) and IL phases (*x*), respectively (Eq. (5.1)):

$$K = \frac{y}{x} \tag{5.1}$$

It was found that high volatility and low polarity will favor the solubility of a solute in CO_2, increasing "*y*." In contrast, high polarity and aromaticity will favor solute solubility in the IL-rich liquid phase, increasing "*x*." It was found that different organic solutes can be separated from IL with more than 95% efficiency. The main advantage of this process is the lack of cross-contamination between the IL and $scCO_2$.

Another method implied $scCO_2$ is used to extract IL 1-ethyl-3-metylimidazolium tetrafluoroborate ($[C_2C_1im][BF_4]$) itself from its mixtures

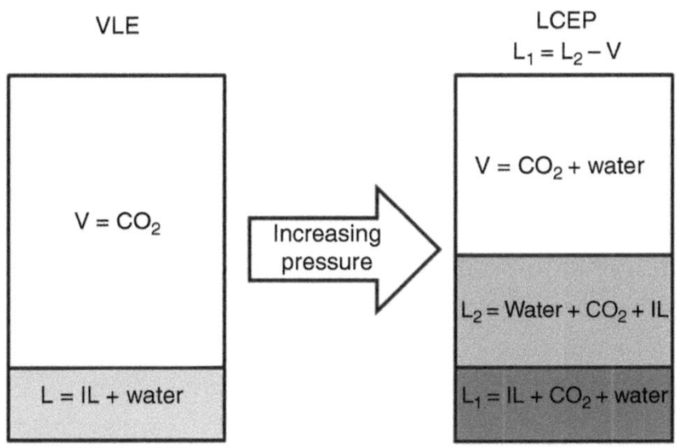

Figure 5.7 ***Phase Behavior of IL/Water/CO_2 Mixture at Increasing Pressure at Near-Ambient Temperatures.*** *Adapted from Scurto et al. (2002).*

(Scurto et al., 2002). The liquid/liquid/vapor equilibrium in the IL/methanol/CO_2 system depends mainly on IL concentration. Following these results Zhang et al. (2004), while studying a similar starting 1-butyl-3-methylimidazolium tetrafluoroborate $[C_4C_1im][BF_4]$/water/CO_2 system, found that increasing the CO_2 pressure while decreasing the temperature enhanced the separation of IL and water.

Meanwhile, Scurto et al. (2003) were using supercritical CO_2 ($scCO_2$) to separate this system. It was found that the IL/H_2O/CO_2 mixture forms a three-phase solution at increasing pressure at near-ambient temperatures. The upper phase consisted of water and CO_2, the middle phase included traces of IL, and the lower phase was rich in IL; however, it did contain traces of water and CO_2 (Figure 5.7).

It was not possible to achieve an IL-rich phase with more than 75.63 wt% of IL as this phase still contained water and CO_2. This could be explained by the low intersolubility of CO_2 and water even at high pressure (20 MPa) (Zhang et al., 2004). Therefore, at that time the potential of use of $scCO_2$ was still poorly explored and there are no technically promising methods created on the basis of these investigations. Wu et al. (2009) in their work disagreed with the Scurto group that the reason for such separation is a "pressure phenomenon" and changes in dielectric constant upon addition of CO_2. Wu et al. assumed that the formation of carbonate in solution is of key importance, as this leads to the formation of carbonic acid and its dissociation products, which is followed by a salting-out effect as CO_2 reacts with water.

Salting-Out

The salting-out technique was demonstrated by Gutowski et al. (2003). They used the water-structuring salt potassium phosphate K_3PO_4 to produce a biphasic system in the $[C_4C_1im]Cl$/water solution. The formation of an upper IL-rich phase and lower K_3PO_4-rich phase was attributed to the increasing dielectric constant of the aqueous phase. This change was caused by the addition of kosmotropic salts, which stimulated low dielectric IL cations to move to the upper phase concurrent with the transfer of Cl^-. One of the main disadvantages of this method, which prevents its use on a large scale, is the production of secondary waste streams containing the inorganic ions (K^+ and PO_4^{3-}). Therefore, Wu's group in 2007 suggested replacing the inorganic structuring salts with covalent compounds – kosmotropic saccharides (sucrose, glucose, xylose, and fructose) – which can be more easily removed. Their study (Bridges et al., 2007) indicated that separation of an IL-rich upper phase and saccharides-rich lower phase could be enhanced by decreasing the temperature. Unfortunately however, the recycling efficiency of this method is still not reported to be high.

In recent years, IL/salt aqueous biphase or two-phase systems (ABS or ATPSs, respectively) have been introduced as a "new" method for the recovery of 1-allyl-3-methylimidazolium and 1-butyl-3-methylimidazolium tetrafluoroborates with recovery efficiencies of up to 96.8% and 98.8%, respectively (Deng et al., 2009; Li et al. 2010). Xiong et al. (2012) found that the ILs in aqueous solutions can be significantly enriched if CO_2 and amines are introduced into aqueous solutions of ILs to produce ATPSs consisting of an ammonium salt-rich upper phase and an IL-rich lower phase (at room temperature and atmospheric pressure). Heating the solution will then remove CO_2 and bubbling with Ar or N_2 regenerates the amines in the salt-rich phase. It was found that after four cycles the recovery efficiency of $[C_4C_1im][BF_4]$ decreased from 82.4% to 74.2%. This drop could be related to the amine regeneration efficiency, which needs to be improved. The recovery efficiency of ILs is found to increase with decreasing temperature of the system and increasing time because of the enhancement of the salting-out strength.

Remarkable results on back-extraction of IL using salting-out phase separation were obtained by Cláudio et al. (2014). The researchers extended their previous study on gallic acid extraction with ATPSs containing ILs (Cláudio et al., 2012) to the investigation of the back-extraction of gallic acid and the recyclability of the ILs in a combined approach. The authors investigated 11 imidazolium-based ILs using two cycles, each involving a first step with Na_2SO_4+ IL ATPS to extract gallic acid from an aqueous solution to the IL-rich phase, followed by a second step, using Na_2CO_3+ IL aqueous systems to perform the back-extraction of the biomolecule and to "clean" the IL-rich phase for its subsequent reutilization. The process was described for aqueous solutions with two charged species (ionic liquid + inorganic salt). Thus, the back-extraction procedure is defined by the possibility of ion exchange between the salts and the ionic liquid. The described two-step cycle was repeated only twice with IL recovery efficiencies between 94% and 95%. The authors also claim that the recyclability of ILs was established without losses in the extractive performance of the corresponding ATPS. However, more prolonged experiments are needed to confirm this conclusion.

A similar separation process based on ionic liquids that has been recently developed is ionic liquid-based three-phase partitioning (ILTPP). ILTPP is a promising technique to recover proteins by means of their accumulation at the liquid–liquid interface between the aqueous ionic liquid-rich phase and the aqueous salt-rich phase. ILTPP combines the advantages of three-phase partitioning (TPP) and ATPS: it does not use volatile organic solvents and does not require any back-extraction steps as the target protein is accumulated at the interface. Alvarez-Guerra et al. (2014a) modeled the thermodynamic equilibrium of the ILTPP systems with 1-butyl-3-methylimidazolium trifluoromethanesulfonate that led to successful protein (lactoferrin, LF) recovery and concluded that the fraction of ionic liquid that cannot be reused/recycled in each cycle varies from almost 5%, at low protein concentrations in the feed stream, to values around 0.8% at high protein mass fractions. The same group continued this study (Alvarez-Guerra et al., 2014b) and investigated the reduction of the ionic liquid losses in ILTPP by increasing the salt content and by water evaporation. Figure 5.8 shows schematic diagrams of the ILTPP process when: (1) extra salt is added to the system, or (2) evaporation is used to remove water from the salt-rich phase. It was found that the effect of additional salt is limited at high protein concentrations in the feed stream, and that higher salt content may lead to higher ionic liquid losses under certain operating conditions. However, with vacuum evaporation of water the recovery and reuse of all of the ionic liquid and the salt was possible. Thus, ILTPP is a promising alternative to recover ILs from water/protein mixtures. The authors claimed that by varying and optimizing technological parameters they could potentially reuse more than 99% of the IL involved in the process.

Barrier Method or Membrane Separation

Figure 5.9 shows the principle of membrane separation. The sample is in contact with one side of the membrane – the feed (or donor) side. The

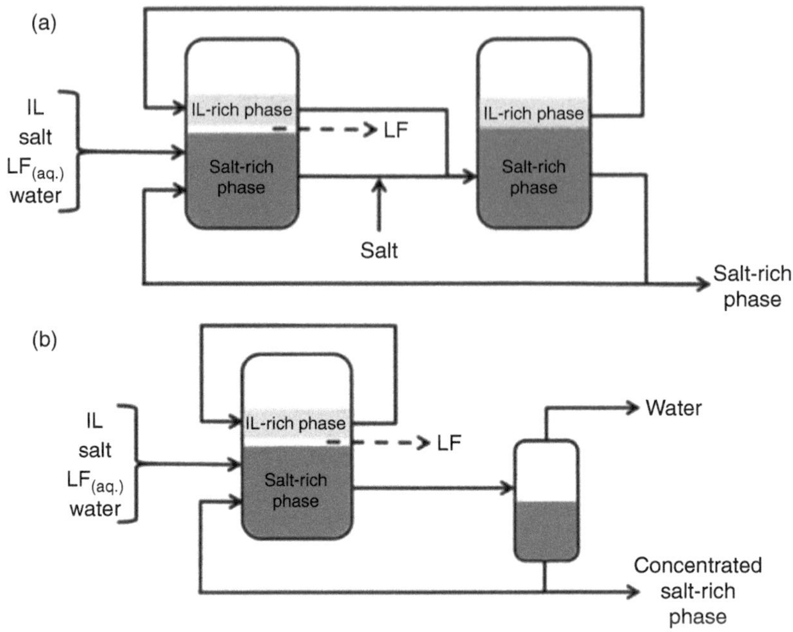

Figure 5.8 ***Schematic Diagrams of the ILTPP Process With Two Pathways for Increasing IL Recovery.*** (a) Extra salt is added to the system and (b) evaporation is used.

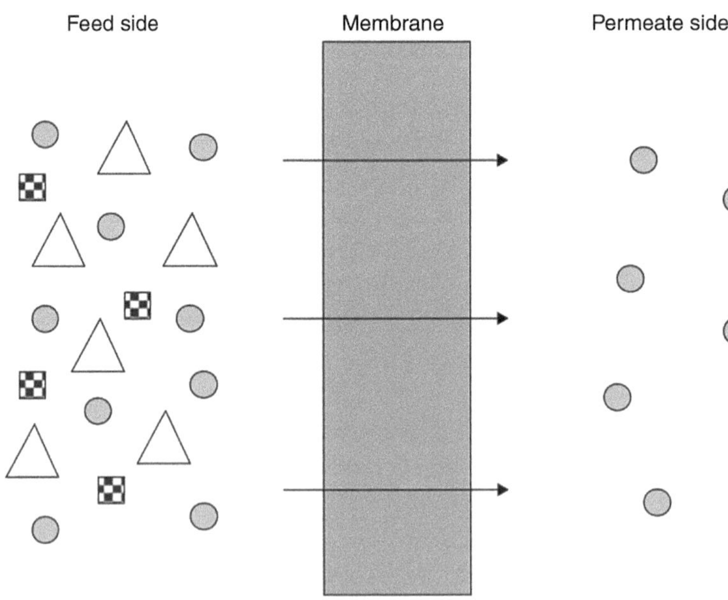

Figure 5.9 ***Concept of Membrane Separation; the Circles are the Analytes (Slack et al., 2003).***

membrane acts as a selectively permeable barrier and allows the analyte of interest to pass through to the other side – permeate side (Slack et al., 2003).

As noticed by Wu et al. (2009), the common drawback of the already mentioned methods is their amenability to industrial implementation. On the other hand, membrane technology is already widely applied to wastewater treatment in many different industrial processes. The main advantage of membrane technology is its high efficiency. The only equivalent separation method to achieve as high a recovery is the high-cost distillation method. This method is essential for the separation of ILs from nonvolatile compounds that are not suitable for distillation.

More than a decade ago nonporous membranes were applied to the recovery of solutes with different physico-chemical properties from the room-temperature ionic liquid 1-butyl-3-methylimidazolium hexafluorophosphate ($[C_4C_1im][PF_6]$) by Schäfer et al. (2001). This membrane had selective layers consisting of hydrophilic or hydrophobic polymers. Wasserscheid et al. (2003) patented the membrane separation of volatile substances from solutions with ionic liquids with membranes used for selectively separating of hard and nonvolatile substances. The selectivity is achieved by the difference in the size of the components or their charge or their different solubility in the membrane materials. Since this time membranes started to be widely studied to purify and concentrate ionic liquids.

Pervaporation is a membrane separation method based on the selective transition of a solute from a liquid feed phase into a dense, nonporous membrane through which it diffuses according to its chemical potential gradient. This method is based on condensing the solutes permeated at an established vacuum. Pervaporation consumes less energy than distillation and it is very efficient as it is applied only to the permeating solute because no bulk solvent permeates the membrane. The rate of water and some other organic solutes' removal from ILs is very impressive, up to 99.2%. One of the main drawbacks is the timescale of the process. This was, as described by the authors, an "uneconomically long" filtration because of the "small membrane area used in the laboratory." However, the real difficulty in implementation of this method (in an industrial setting), is the high viscosities of ILs, which would limit the rate of separation. Kröckel and Kragl (2003) used commercial Desal DVA 00 membranes to separate water-miscible ILs from dyes and lactose. They found it necessary to dilute the ILs with water to decrease their viscosities.

Gan et al. (2006) noted that the viscosities of ILs were typically two to three orders of magnitude higher than water and investigated the potential

Table 5.1 Characteristics of nuclepore MF membranes (Gan et al., 2006)

Nominal pore size (μm)	0.1
Porosity (%)	<15
Thickness (μm)	8.0 ± 1.2
Water flux at 0.50 bar and 20°C (kg m^{-3} h^{-1})	1825
Methanol flux at 0.50 bar and 20°C (kg m^{-3} h^{-1})	2718
Ethanol flux at 0.50 bar and 20°C (kg m^{-3} h^{-1})	1236

application of a microfiltration process to two ionic liquids 1-decyl-3-methylimidazolium bis(trifluoromethylsulfonyl)imide ([$C_{10}C_1$im][NTf_2]) and methyltrioctylammonium bis(trifluoromethylsulfonyl)imide ([N_{8881}][NTf_2]). The use of the same anion, [NTf_2]$^-$, attached to different cations allows a direct comparison of the influence of the cation on the fluid properties and microfiltration characteristics. Due to the high viscosity of the pure ILs, they demonstrated a very low permeation rate through a Nuclepore polycarbonate microfiltration membrane (Whatman). These membranes possess an extremely narrow size distribution and smooth flat surface.

Table 5.1 summarizes some membrane properties and the permeation rate through the Nuclepore membrane of some pure solvents (Gan et al., 2006). Gan et al. (2006) reported that the absolute permeation rate of the IL was exceedingly low in comparison to the permeation rates of water, methanol, and ethanol. The fact that pure ILs have low permeation rates due to their high viscosities questions the economic viability of microfiltration processes for undiluted ILs. The viscosity of ionic liquids can be dramatically reduced by mixing them with a small amount of common diluting agents such as methanol or ethanol. It is necessary to know the miscibility of any diluting agents used with other solvents in order to predict the extractability in extraction and membrane separation processes. Mixed solutions easily permeated through the microfiltration membranes. Gan et al. (2006) found that the amount of added diluting agent significantly influenced the permeation rate of the IL solution. The most effective filtration rate was found at a 20% (v/v) ratio of diluting agents to ionic liquid, the flux increase amounted to around 10–20 times for the two ionic liquids investigated, respectively. After permeation methanol and ethanol can be easily evaporated from the nonvolatile ILs. However, the use of distillation for postseparation IL/alcohol mixtures makes this process environmentally harmful and diminishes the purpose of using an IL in the first place.

There is still room for nanomembrane utilization in the separation and purification of ILs. Han et al. (2005) started their work on the implementation of organic solvent nanofiltration (OSN) membranes STARMEM™ for

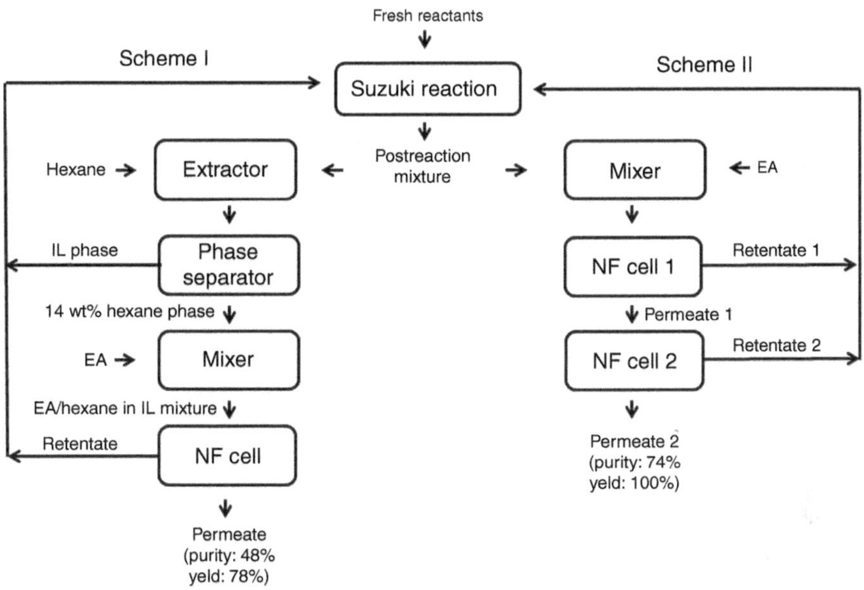

Figure 5.10 ***Configurations for the Product Purification of the Ionic Liquid-Mediated Suzuki Reaction.*** NF, nanofiltration; EA, ethyl acetate *(Adapted from Han et al., 2005).*

recycling ILs and purification of the products from the IL-mediated Suzuki cross-coupling reaction. The authors showed two schemes (Figure 5.10) that could be applied. In Figure 5.10, Scheme I, hexane was used to extract the products from the postreaction mixture. The IL-rich phase, containing the catalyst, was recycled for the next reaction with the addition of fresh reactants. It should be pointed out that multiple-stage filtration can be used for further purifying the product in the final permeated solution (e.g., the purity will reach 81% for two-stage filtration). In Figure 5.10, Scheme II, ethyl acetate is used to dilute the postreaction mixture directly and OSN is used to purify the product and recycle ILs.

As can be seen from Figure 5.10, recycling of IL by Scheme I gives significantly less pure ILs than the extraction method shown in Scheme II. Also, Scheme I has some disadvantages, when compared with Scheme II, such as mandatory removal of ethyl acetate from the retained solutions and the use of only hexane (or other apolar solvents) to create phase separation. This can be avoided by using nanofiltration as suggested in Scheme II. Han's results show that recovery of ILs using nanofiltration allows the recycling of the IL up to six times with the same efficiency (up to 95 wt%); however, the purity of the IL is impacted. This definitely affects the possible industrial implementation of this method. Currently, there are intensive studies on improving the membrane technology for the separation of ILs. Based on the hypothesis that ILs should show a similar behavior to surfactants in membrane filtration, Fernández et al. (2008) analyzed and proved the potential of aggregation control to improve the performance of downstream membrane separations, applied to the system 3-octyl-1-methylimidazolium chloride ($[C_8C_1im]Cl$), water, and iron(III) chloride ($FeCl_3$). Wu et al. (2009) mentioned the importance that the membranes could have charge and the concentration polarization effect have on IL filtration.

Haerens et al. (2010) investigated the use of pressure-driven membrane processes, nanofiltration, reverse osmosis, and pervaporation, as possible methods for recycling ILs from water. Reverse osmosis membranes show slightly higher retention than for the nanofiltration membrane. Nanofiltration and reverse osmosis were found to be limited by the osmotic pressure of the water/ionic liquid solution. Due to that fact, purification of ILs from the concentrated solutions by these methods is highly problematic. The highest reported concentration of ILs obtained after membrane filtration is 30 vol% (Haerens et al., 2010).

The mentioned membrane separation methods can be used as a preconcentration process for other separation techniques that require a concentrated ionic liquid solution. With the existing membranes only small amounts of water can be removed. Many authors are calling for more specific membranes and combinations of different separation methods to allow recycling of ILs with high efficiencies and purities (Haerens et al., 2010).

Melt Crystallization

Melt crystallization is a technique suited for purification of organic chemicals as well as ionic liquids. The principle of the technique is to cool down a melt in a controlled way in order to crystallize a fraction of the melt. Choudhury et al. (2005) mentioned this method for the first time regarding ILs. They were cooling ILs down to $-150°C$ at $\sim 6°C/min$ and then heating it up with 10°C steps. It was found for some ILs that they failed to crystallize on cooling, but did crystallize upon heating (approx. $-65 \pm 30°C$) before their phase transition. In their later studies, Choudhury et al. (2006) demonstrated that 1-ethyl-3-methylimidazolium trifluoromethanesulfonate and bis(trifluoromethylsulfonyl)imide ($[C_2C_1im][OTf]$) and $[C_2C_1im][NTf_2]$) form single crystals during *in situ* crystallization. König and Wasserscheid (2006) purified 1-ethyl-3-methylimidazolium chloride ($[C_2C_1im]Cl$) and 1-ethyl-3-methylimidazolium bromide ($[C_2C_1im]Br$) by melt crystallization

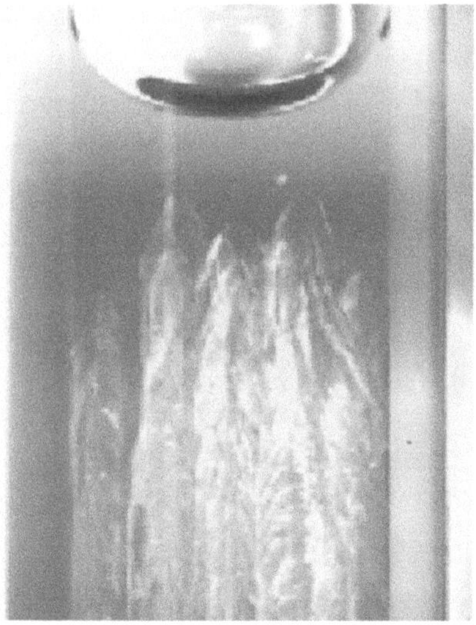

Figure 5.11 ***Phase Separation of [C_2C_1im]Cl at Static Crystallization (König and Wasserscheid, 2006).***

up to "ultrapure" grade with a purity of >99.9%. König and Wasserscheid reported successful IL purification via crystallization using the following:

1. Zone melting,
2. Static and dynamic layer crystallization,
3. Dry sweating, and
4. Suspension crystallization.

Figure 5.11 represents the phase separation during static crystallization of [C_2C_1im]Cl, where the residue has an intensive brown color and is a liquid at room temperature, indicating the collection of impurities in the liquid phase.

The layer melt crystallization technique has been successfully applied to the binary system of [C_2C_1im]Cl–$AlCl_3$ and to real ternary electrolytes [C_2C_1im][NTf_2]–$AlCl_3$ (Solà Cervera and König, 2010) using a layer melt crystallizer (Figure 5.12), and the recycling flow chart was proposed as well (Figure 5.13). Through crystallization of the compound [C_2C_1im][$AlCl_4$], the two main components in the spent electrolyte [C_2C_1im]Cl and $AlCl_3$ are recovered and the impurities reduced to very low concentrations. The authors reported recovery yields of at least 65%, though this does mean that different amounts of $AlCl_3$ and [C_2C_1im][NTf_2] have to be added to the crystallized product in order to generate a reusable electrolyte with the same composition as the starting fresh electrolyte.

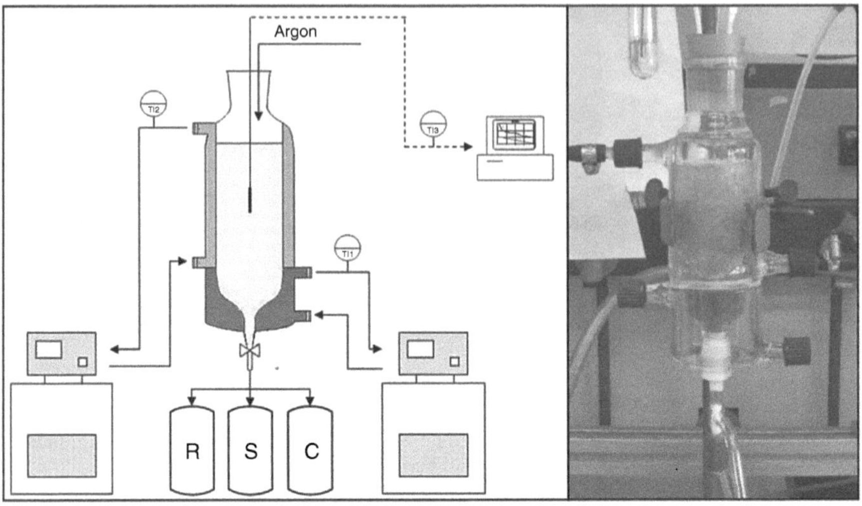

Figure 5.12 ***Sketch and View of the Layer Melt Crystallizer (Solà Cervera and König, 2010): R - residue, S - sweating fraction and C - product fraction from crystal layer are collected at different stages.***

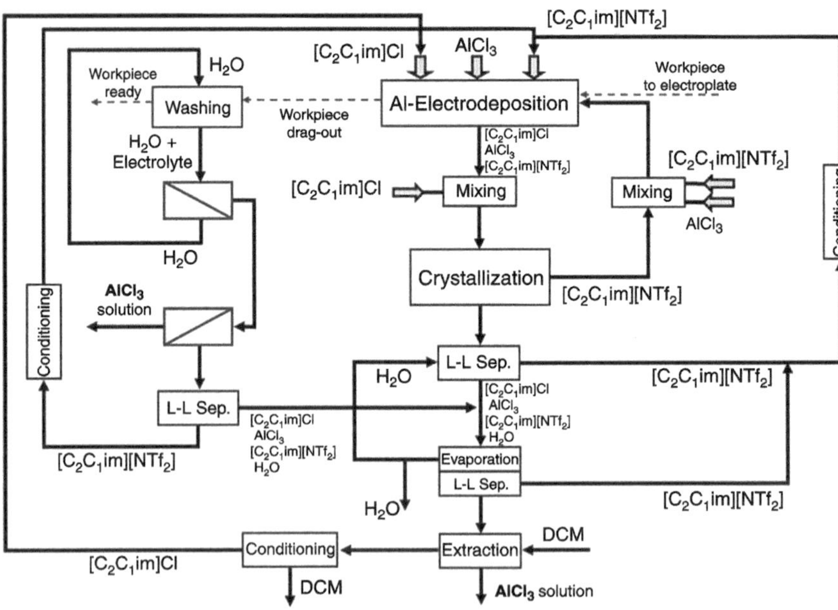

Figure 5.13 ***Complete Recycling Concept for the [C_2C_1im][NTf_2]-$AlCl_3$-Based Electrolytes in Aluminum Electrodeposition (Solà Cervera and König, 2010).***

Liquid–Liquid Extraction

For those systems containing hydrophobic ILs, simple liquid–liquid (L/L) extraction, based on the relative solubility of separated substances in two different immiscible liquids, usually water and an organic solvent, is one of the easies separation methods. The recyclability of these ILs is a result of the lack of their solubility in some common organic solvents. This allows impurities to be extracted by using this organic solvent or water as the extraction solvent. However, the use of other organic solvents, except water, diminishes the "greenness" of ILs application. ILs have a strong ability to interact with organic molecules through various mechanisms. According to published information, ILs are used mainly as extractants in liquid-liquid separation and it has been applied to extract different compounds into IL phase (Billard et al., 2011; Ren et al., 2011). A liquid–liquid extraction is described for the novel ILs choline-indole-3-acetate (chol-IAA) and choline-indole-3-butyrate (chol-IBA), which are able to dissolve DNA, and the ILs are recycled with 90–95% yield up to three times (Mukesh et al., 2013). DNA is proposed to be isolated or regenerated from the ionic liquid by the addition of ice-cold isopropyl alcohol (IPA). The IL itself could be regenerated upon the addition of ethyl acetate as depicted in Figure 5.14. The upper ethyl-acetate-contained layer became transparent upon standing for 3–4 h. DNA up to 3.3 wt% could again be solubilized in the recycled ILs.

Another example is 1-(4-sulfobutyl)-3-methylimidazolium ionic liquid recovery by liquid-liquid separation with cyclohexane was demonstrated by Titze-Frech et al. (2013). These Brønsted acidic ionic liquid systems were

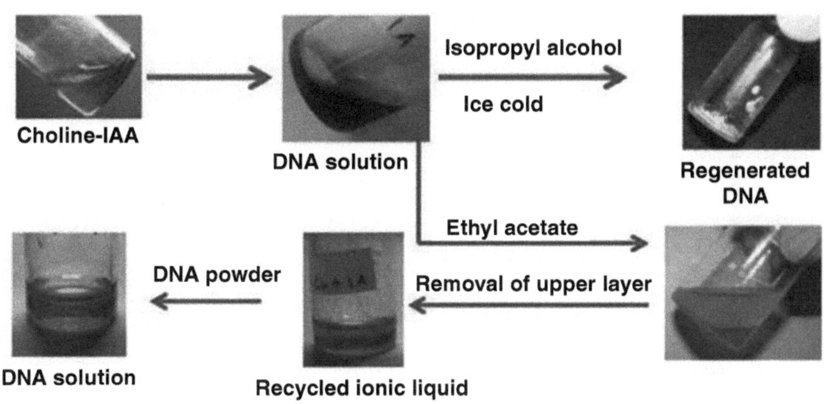

Figure 5.14 ***Dissolution of DNA in Chol-IAA and the Recycling of the IL for Subsequent DNA Dissolution.*** *Adapted from Mukesh et al. (2013).*

used in phenol alkylation. After extraction of the reaction mixture, the remaining ionic liquid was dried under vacuum, and fresh reactants were added. The authors reported six recycle runs were performed without loss of IL activity or selectivity.

Electrodialysis

It is known that ILs possess very interesting electrochemical properties; they are electrochemically stable and have wide electrochemical (potential) windows ranging from 2 V to 6 V (Galiński et al., 2006). The electrochemical (potential) window is the range over which the ionic liquid is electrochemically inert (Kroon et al., 2006). However, in the presence of water or other diluents, the potential window of ILs can be decreased dramatically (Wasserscheid and Welton, 2002). Wang et al. (2012) reported their investigation of the performance of an electrodialysis (ED) process used to concentrate aqueous IL solutions. They used anion-exchange membranes and cation-exchange membranes with titanium-plated ruthenium cathode and anode electrodes. Wang et al. (2012) enriched $[C_4C_1im]Cl$ solutions using the simple scheme presented in Figure 5.15. The recovery ratio reached 85.2% and the energy consumption was reported to be composed of the following parts: high energy consumption by the membrane stack energy consumption and circulation system energy consumption, plus comparably negligible power consumption of the DC power supply unit and the power consumption at the electrodes.

Later, Wang's group (Bai et al., 2013) focused their research efforts on the recovery of the 1-butyl-3-methylimidazolium chloride ($[C_4C_1im]Cl$)

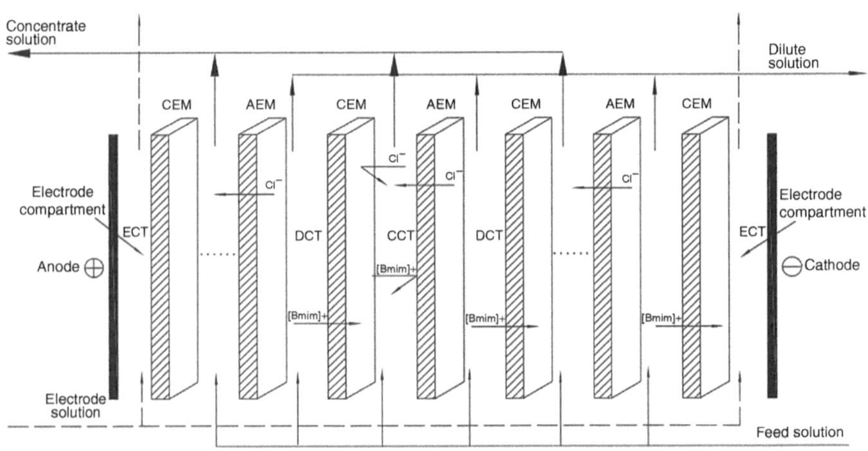

Figure 5.15 ***Schematic Diagram Showing the ED Method for Concentrating Aqueous IL Solutions.***

IL by ED with regard to possible fouling of ion-exchange membranes by ionic components, which will lead to an increase of membrane resistance and hence energy consumption. They used UV-spectroscopy to define the concentration of $[C_4C_1im]^+$, which has a maximum absorption wavelength at 211.4 nm, due to the imidazolium ring structure. Fourier transform infrared spectrometry (FT-IR) together with scanning electron microscopy and nitrogen content analysis were performed in order to characterize the fouling of cation-exchange membrane. They found that in the ED process the removal and desalination ratios increase with increased initial IL concentration and applied voltage. Membrane fouling was also found after the desalination of the IL. The deposits on the surface or into the membrane were assumed to be caused by $[C_4C_1im]^+$. This method was also successfully applied by Trinh et al. (2013) to separate the same ionic liquid, $[C_4C_1im]Cl$, from an aqueous solution containing hydrolysate of softwood biomass pretreated with IL. The influence of the operating parameters and IL concentration on the process performance was investigated in the ED experiments. The influence of the electrical potential on IL removal is of special interest (Figure 5.16). The study confirmed that ED could be used as one of the

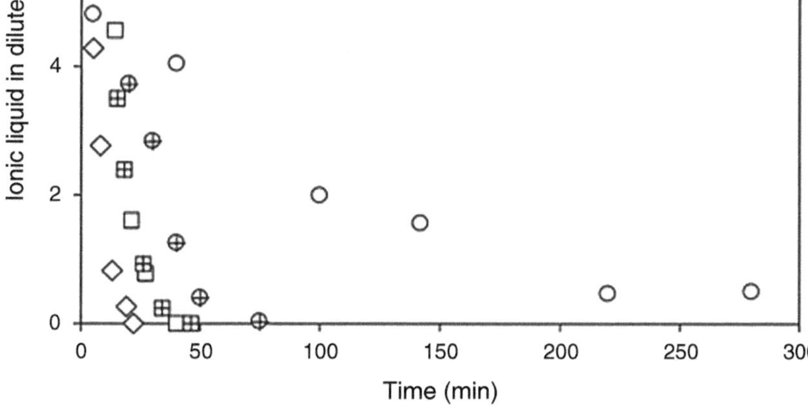

Figure 5.16 ***The Influence of Electrical Potential on the Removal of Ionic Liquid in the ED of the Synthetic Solutions.*** Initial dilute: synthetic solution with 6.90 g/L of $[C_4C_1im]Cl$; initial concentrate: distilled water (Trinh et al., 2013).

Figure 5.17 ***Images Demonstrating the Response of [C_4C_1im][$FeCl_4$]/Water to a Magnet (Hayashi and Hamaguchi, 2004).***

applicable methods to recover ionic liquids, in hydrolysate, for reutilization in the pretreatment of biomass.

Application of Force Fields

Two excellent examples of the recovery of ILs from their water or salt solutions are gravity settling (Birdwell et al., 2006) and the application of a magnetic field for ILs with an iron-containing anion (Hayashi and Hamaguchi, 2004; Lee et al., 2007). However, it should be noted that Birdwell et al. applied gravity settling combined with subsequent L/L separation. Generally, they just used the following rule to make L/L separation easier: "Efficient separation of dispersions containing ionic liquid solvents is essential for utilization of these compounds in liquid–liquid extraction applications to maximize both solute transfer efficiency and solvent recovery" (Birdwell et al., 2006).

Hayashi and Hamaguchi (2004) described the magnetic properties of 1-butyl-3-methylimidazolium tetrachloroferrate ([C_4C_1im][$FeCl_4$]) and showed the possibility to separate this water-immiscible IL from water by using a magnetic field (Figure 5.17). However, Hayashi and Hamaguchi did not explain this mechanism and a few years later Lee et al. (2007), presented their study on the recovery of [C_4C_1im][$FeCl_4$] from its mixture with water. They confirmed the successful separation of IL from water using an applied magnetic field. These studies open an opportunity to investigate the possibility to separate other nonmagnetic ILs by incorporating magnetic elements.

REMOVAL OF ILs

The methods discussed earlier are designed to separate IL waste streams and recover ILs for their future reuse. However, in a situation where ILs are not reusable, or recoverable anymore, it is important to remove them from the waste stream. Removing ILs is the important final technological stage for every process. The methods for final removal of IL from waste streams can be divided into physico-chemical adsorption and degradation (including biodegradation) (Siedlecka et al., 2011).

Physico-Chemical Removal or Absorption

Physico-chemical removal is based on the absorption of the IL from waste streams, mainly wastewater. Absorption is widely applied to remove a number of different organic impurities from water. The most common adsorbents used to adsorb ILs from their aqueous solutions are the following (Anthony et al., 2001; Vijayaraghavan et al., 2009):

- Activated carbon (AC),
- Ion-exchange resin,
- Activated sludge, and
- Fermentation waste.

The first studies on the application of AC (Anthony et al., 2001) showed that this method is relatively inefficient and does not provide the recovery of the IL. The researchers confirmed that the cation and the anion are removed from the solution. However, the removal of IL from water (using activated carbon) takes about 50 times longer that the time to remove other organic solvents, for example, toluene, used as a reference in their study. The stated reason for such differences is the better efficiency of AC at removing small nonpolar compounds, such as toluene, rather than polar or ionic species, such as ILs. Further studies (Palomar et al., 2009) have shown that factors strongly influencing the adsorption capacities were the size and the hydrophobic nature of the cation and anion of the IL, and the surface chemistry of AC. In this case the authors used acetone to regenerate the exhausted AC and to recover the ILs. Lemus et al. (2012) expanded the study of absorption using AC for 27 ILs. Lemus et al. found that microporous/narrow mesoporous activated carbons (with high pore content of diameters lower than 8 nm) had the highest adsorption capacities with a maximum uptake of up to 1 g of IL per gram of adsorbent. It was also found that the adsorption process can be favored by proper modification of the adsorbent chemical surface. Therefore, ACs with a high content of polar groups at their surface removed hydrophilic ILs from aqueous waste streams more effectively because of the promoted hydrogen bonding, adsorbate–adsorbent, interactions. Conversely, hydrophobic ILs are more effectively removed by thermally treated ACs with low polarity. The authors also reported the evidence of restricted absorption of ILs with large molecular volumes. Successful results were obtained for regenerating the exhausted adsorbent and recovering the IL by acetone extraction upon adsorption–desorption cycles with two commercial ACs. The used acetone was separated by atmospheric distillation, and the IL was recovered without structural modification.

At the same time, Farooq et al. (2012) were investigating the adsorption isotherms and kinetics of three ionic liquids (1-methyl-3-octylimidazolium chloride, 1-butyl-3-methylimidazolium chloride, and *N*-octylpyridinium bromide) on several types of activated carbon: a microporous granular, a microporous–mesoporous AC fabric, and an AC prepared from Artichoke leaves using phosphoric acid activation. The authors studied the kinetics of adsorption and found that plateaus were obtained after 3 h for Artichokes AC, 4 h for fabric AC, and more than 24 h for microporous AC. Thus, the authors attributed the efficiency of IL adsorption to the activated carbon type, pH, and the IL chain length and cation type. The adsorption uptakes were correlated to the AC porous structures, and their study into pH effects showed that the calculated thermodynamic parameters, from temperature-dependence of adsorption isotherms, confirmed more spontaneous adsorption of the longer chain cations and pyridinium-like cations compared with the methylimidazolium ILs. Moreover, the adsorption at basic pH (pH = 9) was entropically favored.

Some other studies on adsorption of ILs by different soil organics are summarized in Table 5.2 and showed that the hydrophobic chains of the IL cation contribute significantly to the sorption capacity hence indicating the role of van der Waals interactions between the nonpolar groups of the cation and the organic matter of soils and sediments.

Regarding all the mentioned adsorption methods for the removal of ILs these methods are promising not only for removal but also for recovery of ILs. According to Wells (2003), the recovery depends on the sorption and desorption efficiency:

$$\text{Recovery} = \text{Sorption efficiency} \times \text{desorption efficiency}$$

If sorption is 50% efficient but desorption is 100% efficient, the recovery measured is 50% and it is impossible to know whether sorption or desorption was inefficient or if reduced recovery was produced by a combination of both. Therefore, either optimizing sorption while controlling desorption, or vice versa – using an iterative approach – is required.

Won et al. (2013) have shown the results of sorption of 1-ethyl-3-methylimidazolium cations with bacterial biosorbents from aqueous media.

Table 5.2 The state of art in absorption methods of IL by different organic solids

IL	Adsorbent	Findings	References
1-Butyl, 3-methylimidazolium chloride	Bacterial and mineral surfaces	$[C_4C_1im]Cl$ has no affinity to adsorb to alumina or silica mineral surface sites or to the Gram-positive bacteria *Bacillus subtilis*. The clay was the only sorbent that displayed significant ability to adsorb $[C_4C_1im]Cl$. Solution pH, solute: sorbent ratio, and ionic strength did not have any effect on the lack of adsorption on the other solids within experimental conditions used.	Gorman-Lewis and Fein (2004)
A range of alkyl imidazolium ILs	Soil and sediments	The sorption coefficients of ILs in soils vary between 6.0 mL g^{-1} and 226 mL g^{-1} and depend primary on their structures. They are strongly sorbed on the sediment from the southern Baltic Sea (up to 2467 mL g^{-1}).	Stepnowski (2005)
Alkylimidazolium and alkylpyridinium ILs	Natural soils (clayey agricultural, fluvial meadow, forest and fluvial agricultural soil)	The sorption coefficient of ionic liquids was the highest for fluvial and clayey agricultural soils, and the lowest for forest and fluvial meadow soils. The sorption strengths of ionic liquids are likely to be related primarily to the cationic exchange capacity, which is related directly to the fine clay particle fraction of the soil. The fate of ionic liquid anions was not studied in this paper.	Stepnowski et al. (2007)

(Continued)

Table 5.2 The state of art in absorption methods of IL by different organic solids *(cont.)*

IL	Adsorbent	Findings	References
Alkylimidazolium chlorides	Soils	Chloride anion, which builds in the studied ionic liquids, is not sorbed in any types of used soils. Ionic liquid anions do not influence the sorption and were not taken under consideration in the later investigation and conclusions. Soil with more organic carbon sorbs ionic liquid cations better than a soil with little or no organic carbon.	Studzińska et al. (2008)
Alkylimidazolium chlorides and bromides	Aquatic sediments	None of the ILs were measurably sorbed to sand (error bars overlapped 0), while at an initial IL concentration of 0.5 mM, Pond Large sediment sorbed 75–90% of the added IL, Pond Small sorbed 75–92%, and Lake sediment sorbed 40–73%.	Beaulieu et al. (2008)
1-Butyl-3-methylimidazolium chloride	ACs (SPS-100; SPC-200), an ion-exchange resin (IER), a fermentation waste, and a dried activated sludge	Removal of $[C_4C_1im]Cl$ was only possible using IER and the two forms of activated carbon. The removal efficiency of different sorbents can be summarized in the following order: IER > SPS-200 > SPC-100 > *Corynebacterium glutamicum* > activated sludge. Up to 98% of IL was removed from sample in 15 min by AC and 180 were needed for IER to exceed this rate.	Vijayaraghavan et al. (2009)

These results were not very promising – only 5% of ILs was removed from the solution, possibly due to the sorbent uptake of 72.6 mg/g. The great desorption efficiency, >91.3%, and dependence of sorption on pH does open further possibilities to improve this method of IL separation.

Biodegradation

As long as ILs are considered as environmentally benign solvents, the most appropriate way to remove them from waste streams is by using living organisms. Not all ILs can be considered as biodegradable; however, degradation of chemicals by microorganisms has great potential due to broad, naturally occurring, microbial metabolic variety allowing for the transformation, degradation or even accumulation of an enormous range of compounds (Markiewicz et al., 2009).

First of all, the term "biodegradable compound" should be defined. Compounds that evolve more than 60% of the total CO_2 measured by the modified Sturm test (OECD 301B) are considered to be biodegradable (OECD, 1981). In this test, the chemicals are added to an aerobic aqueous medium and inoculated with wastewater microorganisms. Then the emission of CO_2 is measured for a defined period and reported as a percentage of the theoretical maximum value (Gathergood et al., 2004). When the depletion of dissolved molecular oxygen is measured for a defined period of time and reported as a percentage of the theoretical maximum, this method is called the "closed bottle test" (OECD 301D) (OECD, 1993). Gathergood et al. (2004) found that the one of the most widely used ionic liquids in academia and industry is based on the 1-butyl-3-methylimidazolium corecation, that is, $[C_4C_1im][BF_4]$. The "closed bottle test" (OECD 301D) reveals that $[C_4C_1im][BF_4]$ is not biodegradable while $[C_4C_1im]Br$ has negligible biodegradation (1%). Generally, for the 3-methyl-1-(alkyloxycarbonylmethyl) imidazolium bromide series, the greatest biodegradation (>30% after 28 days) was observed when alkyl chain is either butyl, pentyl, or octyl. Gathergood et al. (2004) agreed with Scott and Jones (2000) that the reason for this phenomenon is that biodegradation is initiated by enzymatic cleavage leading to the separation of the imidazolium fragment and the corresponding primary alcohol that can readily be metabolized further via the pathway of fatty acid β-oxidation.

Further studies on the influence of the anion on biodegradation of ILs showed that a higher percentage of biodegradation is observed for octyl sulfate anions, as linear alkyl sulfates are well known to be readily biodegradable. Ionic liquids containing an ester moiety in the side chain of the imidazolium cation and the octylsulfate anion (1-(propoxycarbonylmethyl))-3-methylimidazolium octylsulfate) had the highest extent of biodegradation (49% after 28 days). However, this value is still too low and the time for biodegradation is too long to be implemented on an industrial scale.

Soon after, Gathergood et al. (2006) finally found a "readily degradable" IL. They studied three groups of ILs according to molecular structure and biodegradability:

1. Conventional ionic liquids based on a 1-butyl-3-methylimidazolium core ($[C_4C_1im]Br$ and $[C_4C_1im][BF_4]$), which showed a negligible extent of biodegradation (0–1%).
2. Ionic liquids with an ester group in the side chain, which reached biodegradation extents of 23–33%.
3. Ionic liquids with an ester bond in the side chain and octyl sulfate as counter ion, which presented the highest biodegradation values (49–56%).

1-Pentoxycarbonyl-3-methylimidazolium octylsulfate IL, containing an ester group in the side chain and octylsulfate as counter anion with a minor effect of the alkyl side chain length and the number of methyl groups, was recognized as a "readily degradable" with biodegradation 60–67% after 10–28 days. Similar trends were also found by Docherty et al. (2007). They used different tests to determine biodegradability and found that biodegradation rates increase with the alkyl chain length. However, this was attributed to the enhanced selection of the microbial community used. They found another IL, 1-octyl-3-methylpyridinium bromide, to be readily biodegradable; it degraded up to 96% after 25 days. Stasiewicz et al. (2008) also confirmed that biodegradability improves with alkyl chain lengthening, but only in the range from 4 to 11 carbon atoms. They also noticed that the biodegradation rate of most of these compounds does tend to increase with time. Markiewicz et al. (2009) related biodegradability results to adsorption levels, sludge concentration, and IL concentration in water waste streams. Contrary to Stasiewicz result, a dramatic decrease of IL concentration during only the first few days was observed. During the rest of the treatment time the levels remained constant. It was also found that 0.2 mM of 1-methyl-3-octylimidazolium chloride is a critical concentration for primary degradation. Increased concentrations lead to sorption onto the sewage sludge flocks and inhibit cell activity.

A later study (Esquivel-Viveros et al., 2009) showed the possibility of degradation of up to 80% of 1-butyl-3-methylimidazolium hexafluorophosphate ($[C_4C_1im][PF_6]$), after 28 days at 30°C from wastewater using a membrane-aerated biofilm reactor (Pyrex tube-bioreactor) as well as Petri

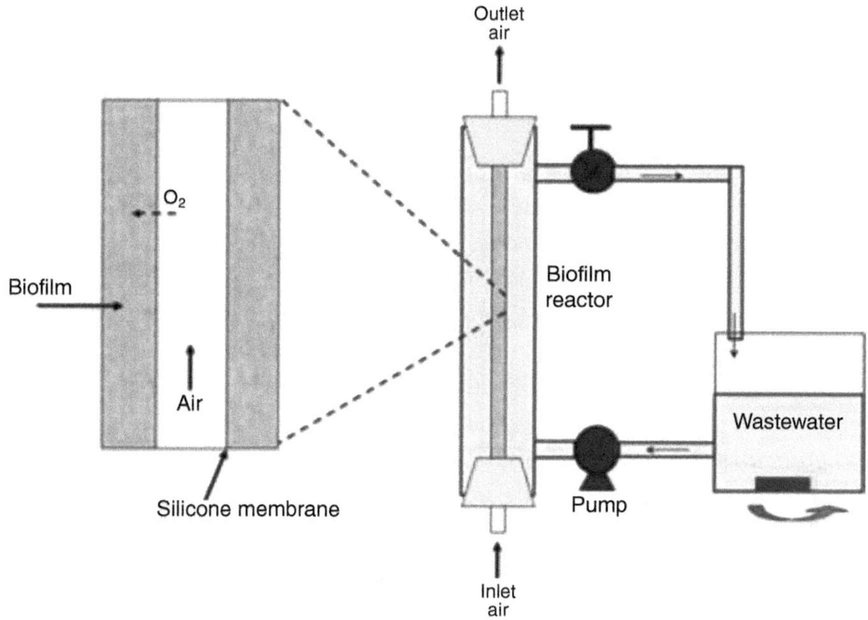

Figure 5.18 ***Membrane-Aerated Biofilm Reactor for Biodegradation Studies of Synthetic Ionic Liquid Contaminated Wastewater (Esquivel-Viveros et al., 2009).***

dishes and Erlenmeyer flasks. The Pyrex tube-bioreactor (Figure 5.18) has a 200 mL capacity, and is 300 mm long with a 30 mm outside diameter and 2 mm thickness. An oxygen-permeable silicon rubber membrane was used and the silicone rubber membrane tube was immersed in the glass bioreactor. One side of the membrane was connected to the air and oxygen supply valves and the other side was partially blocked to force the gases to flow through the membrane wall. The biofilm was developed for 15 days at 30°C. The reactor was operated with a recycle rate of 12.6 mL/min and an overall hydraulic retention time of approx. 11 min, and the ascending liquid velocity parallel to the membrane was 0.5 mph. Some other studies (Harjani et al., 2008) investigated a few more ILs with high (up to 89%) biodegradability. This study confirmed once again the dependence of biodegradation on the characteristics of the cation.

To summarize the achievements of this method it is worth mentioning that despite relatively low efficiency and long-term processing, biodegradation is a promising method for the degradation of some ILs; however, the toxicity to microorganisms has the potential to limit biodegradation rates and should always be taken into account.

Chemical, Photochemical, and Electrochemical Degradation

Most ILs are thermally, chemically, and electrochemically stable. However, they could be decomposed during an industrial process. Chemical, photochemical, and electrochemical oxidation by several agents, such as UV radiation, Fenton's reagent, or ozone, have been used to successfully allow removal not only of ILs but also of their degradation products (if degradation has occurred) (Pernak and Branicka, 2004; Stepnowski and Zaleska, 2005; Li et al., 2007; Czerwicka et al., 2009).

Pernak and Branicka (2004) investigated ozonation of some pyridinium salts and they found that 82 out of the 98 pyridinium salts with alkoxymethyl and alkylthiomethyl substituents showed excellent ozonation effects while the remaining 16 showed a lack of ozonation effects. Degradation by ozonation was very rapid during the first 15 min when the degree of destruction reached 84%. Within 30 min pyridinium salts completely decayed in the presence of ozone. The pyridinium cation became oxidized and produced organic acids. Thus, the decomposition resulted in a complex mixture, containing no pyridine rings and the degradation products of these compounds showed no toxic effects and were less harmful than the untreated compounds.

Following these impressive results Stepnowski and Zaleska (2005) studied three common advanced oxidation processes (UV, UV/H_2O_2, and UV/TiO_2) for the degradation of imidazolium ionic liquids in aqueous solutions. The H_2O_2/UV system was found to be the most effective for the degradation of all of the compounds. Czerwicka et al. (2009) studied ionic liquid breakdown products produced by this oxidation process. They found that the main products formed (amines, imines, aldehydes, and ketones) can undergo further chemical or enzymatic reactions. Some Fenton and Fenton-like reactions where the hydroxyl radical, with a very strong oxidative potential (E_o = +2.8 V), degrades organic impurities were also implemented to IL separation (Siedlecka and Stepnowski, 2009; Siedlecka et al., 2009). It was demonstrated that the degradation is structurally dependent. For the H_2O_2/UV system an increase of H_2O_2 concentration in Fenton reactions (H_2O_2/Fe^{2+}) accelerates the IL degradation rate. These reactions are especially effective for imidazolium-based ILs. In a Fenton system with 1 mmol/L of Fe(III) and 100 mmol/L of H_2O_2, the levels of degradation were 97%, 92%, 88%, and 68%, respectively, for [C_4C_1im]Cl, [C_4C_1pyr]Cl, [C_6C_1im]Cl, and [C_8C_1im]Cl after a 90 min degradation reaction. However, the degradation rates even this high are still too low to implement this on an industrial scale. Further studies should be undertaken to find an oxidation process or combination of processes to increase its effectiveness.

An ultrasonic oxidative process gave more promising results with degradation of 93% of the 1,3-dialkylimidazolium cation with the corresponding Cl^-, Br^-, $[BF_4]^-$, and $[PF_6]^-$ counter-anions at a concentration of 2.5 mM at 50°C within 12 h, while at 72 h the conversions approached 99% (Li et al., 2007). Li et al. (2007) proposed a three-step degradation mechanism for 1,3-dialkylimidazolium ILs using the hydrogen peroxide/acetic acid oxidation system with the aid of ultrasonic treatment (Schemes 5.1-5.3):

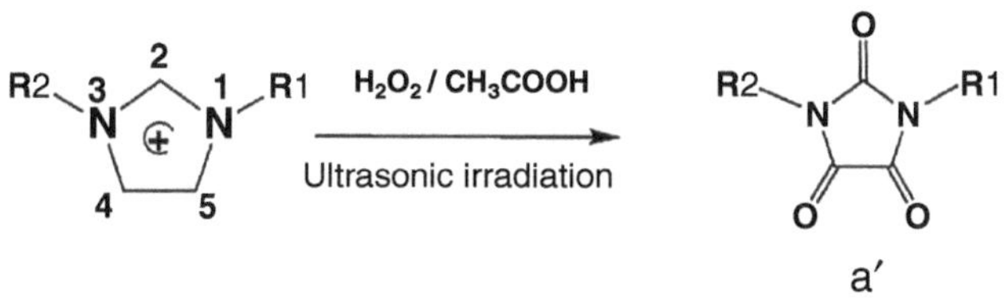

Scheme 5.1 1,3-Dialkylimidazolium ring backbone hydrogen atoms are oxidized generating transient or intermediate 1,3- dialkylimidazolidine-2,4,5-trione (**a′**).

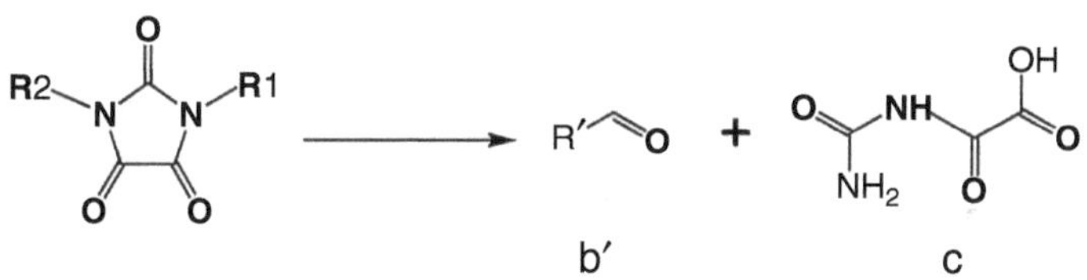

Scheme 5.2 The cleavage of *N*-alkyl side chains together with the destruction of either the N1—C5 or N3—C4 bond in **a′** simultaneously forming secondary transient/intermediate products as alkyl aldehyde **b′** and ring open degradation products **c**.

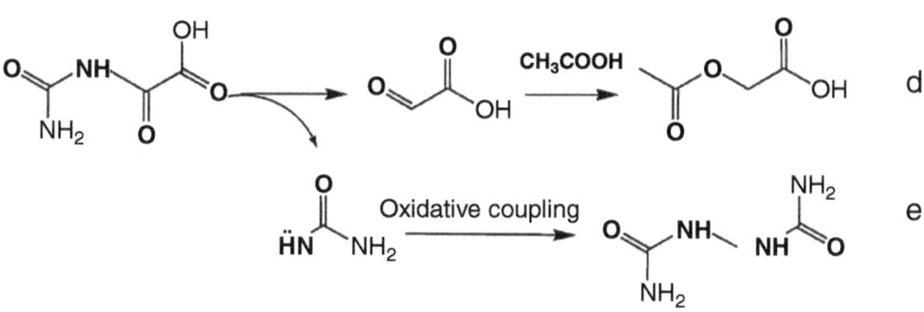

Scheme 5.3 Further cleavage of the remaining C—N bond from the intermediate **c** generates some reactive species that are capable of reacting with other active chemicals or undergoing self-coupling to form degradation products such as acetoxyacetic acid (**d**) and biurea (**e**).

Compounds **d** and **e** are common nontoxic organic species and biurea can be found within the plastics industry. The authors suggest that increasing the amount of ILs investigated in this study would take into account the possibility of industrial implementation of this method.

Another successful complete decomposition method for the removal of ionic liquids from aqueous solution was introduced by Itakura et al. (2008). They demonstrated a complete decomposition method for ionic liquids (the organic cation and inorganic anion) in aqueous media by combining a hydrothermal mineralization (HM), with $Ca(OH)_2$ mineralizer, with a photocatalytic decomposition (PD) approach. Hydrothermal mineralization provoked decomposition of the cation of up to 87.8% and subsequent PD decomposed the organic cation part up to 99.8% after 24 h. One of the disadvantages of this approach is the requirement of relatively high temperatures, 473 K for HM, which will result in high energy consumption and consequently high cost if implemented industrially.

Apparently, from the electrodeposition approach for the recovery of ILs, electrochemical treatment could be used to decompose ionic liquids when voltage differences larger than their electrochemical/potential window (2–6 V) are applied (Kroon et al., 2006). It is believed that at these limits, the cathodic reaction is the reduction of the cation and the anodic reaction is an oxidation of the anion. Voltage differences larger than the electrochemical potential window (8 V) were applied to $[C_4C_1py][NTf_2]$ at room temperature for 3 h and the decomposed products were extracted with toluene. Using quantum chemical calculations it was found found that when an electron was added to the $[C_4C_1py]^+$ cation, the resulting radical is not stable and will decompose into a neutral molecule and a smaller radical. The most likely explanation is that the 1,1-butylmethylpyrrolidinium radical will decompose into methylpyrrolidine and a butyl radical (Figure 5.19). The authors suggested three decomposition mechanisms with different reaction products and confirmed the reaction rates for each mechanism. $[C_4C_1py][NTf_2]$ decomposes into methylpyrrolidine,

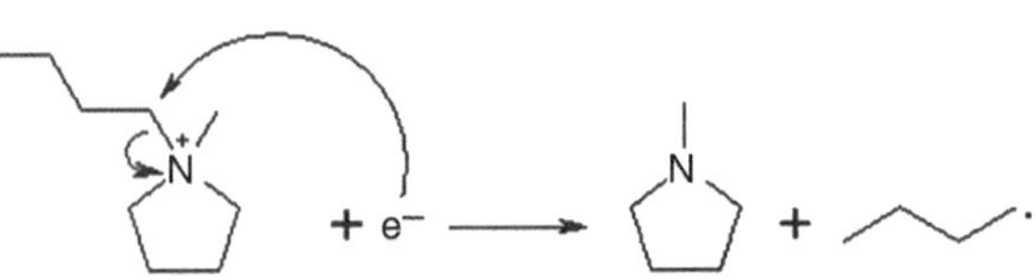

Figure 5.19 ***Formation of Methylpyrrolidine and the Butyl Radical at Electrodeposition (Kroon et al., 2006).***

octanes, octenes, 2-butanol, dibutylmethylamine, and butylpyrrolidine. During the electrochemical breakdown of $[C_4C_1im][BF_4]$, 1-butyl-3-methylimidazolium radicals are formed; they react with each other in a radical–radical coupling reaction and in a disproportionation reaction. Thus, electrochemical deposition seems to be a predictable and efficient method of degradation.

Almost complete degradation of 1-butyl-3-methylimidazolium chloride was reported by Stolte et al. (2008). The authors used a bipolar boron-doped diamond-coated electrode for electrochemical oxidation of imidazolium ILs. Four hours of electrolysis completely decomposed the positively charged imidazolium core, which is responsible for the nonbiodegradability of imidazolium-based ILs (Stolte et al., 2008). There is unpublished information that another "nonactive" anode – PbO_2 – also successfully degrades imidazolium- and pyridinium-based ILs (Siedlecka et al., 2011).

Degradation by Ultrasound and Zero-Valent Iron Activated Carbon

Zero-valent iron activated carbon (ZVI/AC) microelectrolysis was developed for wastewater treatment during the 1970s. In this method numerous microscopic galvanic couples were naturally formed between ZVI and carbon, and then Fe(II), hydroxyl and nascent hydrogen, and nonionized, monatomic hydrogen that exists transiently after the reduction of hydrogen ions, were released from the galvanic corrosion reaction, which were highly active in the decomposition of organic pollutants. Zhou et al. (2013a) verified for the first time the usefulness of ZVI/AC microelectrolysis assisted by ultrasonic irradiation for the degradation of ionic liquid residues in water. 1-Butyl-3-methylimidazolium chloride ($[C_4C_1im]Cl$) was chosen as a model ionic liquid to study the degradation efficiency in such a system. It was found that under the optimized conditions, more than 95% of $[C_4C_1im]Cl$ was degraded in 3 h and the degree and the rate of degradation may be increased by adding NaCl, hence improving the electrical conductivity of the system. The imidazolium ring of $[C_4C_1im]Cl$ was first oxidized to 1-butyl-3-methyl-2,4,5-trioxoimidazolidine and then the ring was opened to form 1-butyl-3-methylurea, which was further decomposed to form *N*-butylformamide and small molecular compounds (Zhou et al., 2013b). At the end of 2013, Zhou's group presented a study of eight kinds of ILs. They included 1-alkyl-3-methylimidazolium bromide ($[C_nC_1im]Br$, n = 2, 4, 6, 8, 10), 1-butyl-1-methylpyrrolidinium bromide ($[C_4C_1pyr]Br$), 1-butyl-1-methylpiperidiniumbromide ($[C_4C_1pip]Br$), and *N*-butyl-*N*-methylmorpholinium bromide ($[C_4C_1mor]Br$) (Zhou et al., 2013b). This complete broad study demonstrated that the chosen ILs could be efficiently degraded by the coupling of ultrasound and ZVI/AC microelectrolysis methods (separately these methods have a much lower efficiency). Zhou et al. characterized the different mechanisms of ILs degradation and its degradation products, most of which are environmentally friendly.

CONCLUSIONS

The separation of ILs after use from their waste streams and consequent recovery of the ILs is important not only from an economical point of view, but for environmental reasons also. Among the currently described methods for the recovery of ILs, melt crystallization seems to be the most efficient. However, this method is not fully studied and is very energy intensive due to the controlled cooling and heating over a wide temperature range. The advantages, disadvantages, and the maximum reported recovery efficiencies for the described methods are summarized in Table 5.3

Taking into account the efficiency, and the economic and environmental sustainability of each method, the optimal IL recovery technology has not yet been found. However, the combination of a few methods can allow the successful separation of ILs from waste streams. Meanwhile, the fast development of the methods for IL removal shows the increasing interest in the environmentally friendly disposal of IL waste streams. Most of the described degradation methods allow the production of nontoxic waste products and can degrade more than 95% of IL. In order to make the unique properties of ILs available on a wide scale, future studies should be directed into the following:

- Physical properties, such as melting and freezing, to propose more energy-saving solutions for such a promising recovery method like melt-crystallization.
- Stability of ILs at high temperatures and under pressure, to optimize the possible distillation of ILs.
- Phase behavior of ILs to optimize the conditions for environmental friendly separation processes using $scCO_2$.
- Specific properties of ILs to discover new separation pathways.
- Understanding the toxicity of ILs, as it influences the properties of the waste streams.
- Other complex research into the physical and chemical properties of ILs and possible ways of their recovery.

Table 5.3 Overview on key disadvantages and advantages of discussed separation methods

Method	Maximum recovery efficiencies reported (%)	Advantages	Disadvantages	Applicable to separate IL from (some examples)
Distillation	95.9	• Distilled compounds are 100% pure • No secondary waste products	• Insufficient pure IL • Very energy-intensive • Not always suitable for thermally unstable ILs	• Water and other low molecular weight fractions
Molecular distillation	Not reported	• Distilled compounds are 100% pure • Suitable for thermally unstable ILs	• Remaining ILs still contain other substances • Requires the use of other substances, VOCs • High temperature and vacuum make this method economically unfavorable.	• Free fatty acids and other high boiling point compounds
Phase separation using $scCO_2$	98.8	• $scCO_2$ is a "green solvent"; • Lack of cross-contamination between CO_2 and ILs	• Still not optimized method • Sometimes requires the introduction of amines	• Aqueous solutions and alcohols
Membrane separation	95	• Suitable for separation from nonvolatile compounds	• Slow rate	• Nonvolatile compounds • Water and some organic solutes • Dyes • Lactose • Alcohols • $FeCl_3$

(Continued)

Table 5.2 The state of art in absorption methods of IL by different organic solids *(cont.)*

Method	Maximum recovery efficiencies reported (%)	Advantages	Disadvantages	Applicable to separate IL from (some examples)
Melt crystallization	99.9	• High efficiency • No secondary waste products	• Lack of detailed study • Very energy intensive	• Organics • $AlCl_3$
Salting-out	98	• Inexpensive	• Creates secondary waste products	• Aqueous solutions
Electrodialysis	85.2	• Good for pretreatment	• Very energy intensive • Not currently efficient enough	• Water and other diluents
Liquid-liquid separation	Not reported	• Energy-saving	• ILs mainly used as the extractants	• Aqueous solutions • Extraction of organics from ILs
Force field separation	Not reported	• No secondary waste products	• Very task-specific • Currently poorly studied	• Water and salt solutions
Absorption	Not reported	• Highly efficient in removing of solvent • Suitable for separation from nonvolatile compounds	• Currently poorly studied • Mainly used to remove ILs	• Water streams

As more scientists start to study ILs the more unknown details appear. There will always be a possibility and need for the discovery of new recovery methods. Regardless of the chosen method of separation and/or removal, the proper risk assessment should be performed in order to avoid environmental pollution and its impact on our lives.

REFERENCES

Abu-Eishah, S.I., 2010. Ionic Liquids Recycling for Reuse, Ionic Liquids – Classes and Properties. In: Handy, S. (Ed.), InTech, Rijeka. Available from: http://www.intechopen.com/books/ionic-liquids-classes-and-properties/ionic-liquids-recycling-for-reuse.

Alvarez-Guerra, E., Irabien, A., Ventura, P.M., Coutinho, A.P., 2014a. Ionic liquid recovery alternatives in ionic liquid-based three-phase partitioning (ILTPP). Am. Inst. Chem. Eng. 60 (10), 3577–3586.

Alvarez-Guerra, E., Ventura, P.M., Coutinho, A.P., Irabien, A., 2014b. Ionic liquid-based three phase partitioning (ILTPP) systems: ionic liquid recovery and recycling. Fluid Phase Equilib. 371, 67–74.

Anthony, J.L., Maginn, E.J., Brennecke, J.F., 2001. Solution thermodynamics of imidazolium-based ionic liquids and water. J. Phys. Chem. B 105 (44), 10942–10949.

Arzephoni, A.Y., Naimi-Jamal, M.R., Sharifi, A., Abaee, M.S., Mirzaei, M., 2013. [Omim][BF4] ionic liquid, a green and recyclable medium for one-pot aminomethylation of electron-rich aromatic compounds. J. Chem. Res. 37 (4), 216–218.

Bai, L., Wang, X., Nie, Y., Dong, H., Zhang, X., Zhang, S., 2013. Study on the recovery of ionic liquids from dilute effluent by electrodialysis method and the fouling of cation-exchange membrane. Sci. China Chem. 56 (12), 1811–1816.

BASF, 2013. Frequently asked questions: how can ionic liquids be recycled? How can ionic liquids be disposed? Available from http://www.intermediates.basf.com/chemicals/ionic-liquids/faq.

Beaulieu, J.J., Tank, J.L., Kopacz, M., 2008. Sorption of imidazolium-based ionic liquids to aquatic sediments. Chemosphere 70 (7), 1320–1328.

Billard, I., Ouadi, A., Jobin, E., Champion, J., Gaillard, C., Georg, S., 2011. Understanding the extraction mechanism in ionic liquids: $UO_2^{2+}/HNO_3/TBP/C4\text{-}mimTf_2N$ as a case study. Solvent Extr. Ion Exch. 29, 577–601.

Birdwell, J.F., McFarlane, J., Hunt, R.D., Luo, H., DePaoli, D.W., Schuh, D.L., Dai, S., 2006. Separation of ionic liquid dispersions in centrifugal solvent extraction contactors. Sep. Sci. Technol. 41 (10), 2205–2223.

Blahušiak, M., Schlosser, Š., Cvengroš, J., Marták, J., 2011. New approach to regeneration of an ionic liquid containing solvent by molecular distillation. Chem. Pap. 65 (5), 603–607.

Blanchard, L.A., Brennecke, J.F., 2001. Recovery of organic products from ionic liquids using supercritical carbon dioxide. Ind. Eng. Chem. Res. 40 (1), 287–292.

Bridges, N.J., Gutowski, K.E., Rogers, R.D., 2007. Investigation of aqueous biphasic systems formed from solutions of chaotropic salts with kosmotropic salts (salt–salt ABS). Green Chem. 9 (2), 177–183.

Chen, Z.-J., Xi, H.-W., Lim, K.H., Lee, J.-M., 2013. Distillable ionic liquids: reversible amide O-alkylation. Angew. Chem. 125 (50), 13634–13638.

Choudhury, A.R., Winterton, N., Steiner, A., Cooper, A.I., Johnson, K.A., 2005. *In situ* crystallization of low-melting ionic liquids. J. Am. Chem. Soc. 5 (2), 16792–16793.

Choudhury, A.R., Winterton, N., Steiner, A., Cooper, A.I., Johnson, K.A., 2006. *In situ* crystallization of ionic liquids with melting points below −25 °C. Cryst. Eng. Comm. 8 (10), 742–745.

Cláudio, A.F.M., Ferreira, A.M., Freire, C.S.R., Silvestre, A.J.D., Freire, M.G., Coutinho, J.A.P., 2012. Optimization of the gallic acid extraction using ionic-liquid-based aqueous two-phase systems. Sep. Purif. Technol. 97, 142–149.
Cláudio, A.F.M., Marques, C.F.C., Boal-Palheiros, I., Freire, M.G., Coutinho, J.A.P., 2014. Development of back-extraction and recyclability routes for ionic-liquid-based aqueous two-phase systems. Green Chem. 16 (1), 259–268.
Czerwicka, M., Stolte, S., Müller, A., Siedlecka, E.M., Gołebiowski, M., Kumirska, J., Stepnowski, P., 2009. Identification of ionic liquid breakdown products in an advanced oxidation system. J. Hazard. Mater. 171 (1–3), 478–483.
Dal, E., Lancaster, N.L., 2005. Acetyl nitrate nitrations in [bmpy][N(Tf)$_2$] and [bmpy][OTf], and the recycling of ionic liquids. Org. Biomol. Chem. 3 (4), 682–686.
Deng, Y., Long, T., Zhang, D., Chen, J., Gan, S., 2009. Application for [Amim]Cl recovery. J. Chem. Eng. Data 883, 2470–2473.
Dibble, D.C., Li, C., Sun, L., George, A., Cheng, A., Çetinkol, Ö.P., Benke, P., Holmes, B.M., Singh, S., Simmons, B.A., 2011. A facile method for the recovery of ionic liquid and lignin from biomass pretreatment. Green Chem. 13 (11), 3255.
Docherty, K.M., Kulpa, C.F., 2005. Toxicity and antimicrobial activity of imidazolium and pyridinium ionic liquids. Green Chem. 7, 185.
Docherty, K.M., Dixon, J.K., Kulpa, C.F., 2007. Biodegradability of imidazolium and pyridinium ionic liquids by an activated sludge microbial community. Biodegradation 18 (4), 481–493.
Earle, M.J., Seddon, K.R., 2001. Preparation of imidazole carbenes and the use thereof for the synthesis of ionic liquids, Quill patents WO 01/77081.
Earle, M.J., Esperança, J.M.S.S., Gilea, M., Lopes, J.N.C., Rebelo, L.P.N., Magee, J.W., Sedon, K.R., Widegren, J., 2006. The distillation and volatility of ionic liquids. Nature 439 (7078), 831–834.
Esquivel-Viveros, A., Ponce-Vargas, F., Esponda-Aguilar, P., Prado-Barragán, L.A., Gutiérrez-Rojas, M., Lye, G.J., Huerta-Ochoa, S., 2009. Biodegradation of [bmim][PF6] using *Fusarium* sp. Rev. Mex. Ing. Quim. 8, 163–168.
Farooq, A., Reinert, L., Levêque, J.-M., Papaiconomou, N., Irfan, N., Duclaux, L., 2012. Adsorption of ionic liquids onto activated carbons: effect of pH and temperature. Microporous Mesoporous Mater. 158, 55–63.
Fernández, J.F., Waterkamp, D., Thöming, J., 2008. Recovery of ionic liquids from wastewater: aggregation control for intensified membrane filtration. Desalination 224 (1–3), 52–56.
Galiński, M., Lewandowski, A., Stępniak, I., 2006. Ionic liquids as electrolytes. Electrochim. Acta 51 (26), 5567–5580.
Gan, Q., Xue, M., Rooney, D., 2006. A study of fluid properties and microfiltration characteristics of room temperature ionic liquids [C_{10}-min][NTf$_2$] and N8881[NTf$_2$] and their polar solvent mixtures. Sep. Purif. Technol. 51 (2), 185–192.
Gathergood, N., Garcia, M.T., Scammells, P.J., 2004. Biodegradable ionic liquids: Part I. Concept, preliminary targets and evaluation. Green Chem. 6 (3), 166–175.
Gathergood, N., Scammells, P.J., Garcia, M.T., 2006. Biodegradable ionic liquids: Part III. The first readily biodegradable ionic liquids. Green Chem. 8 (2), 156–160.
Gorman-Lewis, D.J., Fein, J.B., 2004. Experimental study of the adsorption of an ionic liquid onto bacterial and mineral surfaces. Environ. Sci. Technol. 38 (8), 2491–2495. Retrieved from http://www.ncbi.nlm.nih.gov/pubmed/15116858.
Gutowski, K.E., Broker, G.A., Willauer, H.D., Huddleston, J.G., Swatloski, R.P., Holbrey, J.D., Rogers, R.D., 2003. Controlling the aqueous miscibility of ionic liquids: aqueous biphasic systems of water-miscible ionic liquids and water-structuring salts for recycle, metathesis, and separations. J. Am. Chem. Soc. 125, 6632–6633.
Haerens, K., Van Deuren, S., Matthijs, E., Van der Bruggen, B., 2010. Challenges for recycling ionic liquids by using pressure driven membrane processes. Green Chem. 12 (12), 2182–2188.

Han, S., Wong, H.-T., Livingston, A.G., 2005. Application of organic solvent nanofiltration to separation of ionic liquids and products from ionic liquid mediated reactions. Chem. Eng. Res. Des. 83 (3), 309–316.
Harjani, J.R., Singer, R.D., Garcia, M.T., Scammells, P.J., 2008. The design and synthesis of biodegradable pyridinium ionic liquids. Green Chem. 10 (4), 436–438.
Hayashi, S., Hamaguchi, H., 2004. Discovery of a magnetic ionic liquid [bmim]$FeCl_4$. Chem. Lett. 33 (12), 1590–1591.
Itakura, T., Hirata, K., Aoki, M., Sasai, R., Yoshida, H., Itoh, H., 2008. Decomposition and removal of ionic liquid in aqueous solution by hydrothermal and photocatalytic treatment. Environ. Chem. Lett. 7 (4), 343–345.
IUPAC, 2012. Compendium of Chemical Terminology. Gold Book/Version 2.3.2. IUPAC, p. 1622.
Jeapes, A.J., Thied, R.C., Seddon, K.R., Pitner, W.R. Rooney, D.W., 2001. Process for recycling ionic liquids. British Nuclear Fuels plc, WO 01/15175.
Kanel, J.S., Associates, LLC, 2003. Overview: industrial application of ionic liquids for liquid extraction. In: Chemical Industry Vision2020 Technology Partnership Workshop, New York, September 11, 2003.
King, A.W.T., Asikkala, J., Mutikainen, I., Järvi, P., Kilpeläinen, I., 2011. Distillable acid-base conjugate ionic liquids for cellulose dissolution and processing. Angew. Chem. Int. Ed. Eng. 50 (28), 6301–6305.
König, P.A., Wasserscheid, P.P., 2006. Ultra purification of ionic liquids by melt crystallisation. BIWIC 2006 Thirteenth International Workshop on Industrial Crystallization, Delft, September 2006, pp. 79–84
Kröckel, J., Kragl, U., 2003. Nanofiltration for the separation of nonvolatile products from solutions containing ionic liquids. Chem. Eng. Technol. 26 (11), 1166–1168.
Kroon, M.C., Buijs, W., Peters, J., Witkamp, G., 2006. Decomposition of ionic liquids in electrochemical processing. Green Chem. 8, 241–245.
Kuzmina, O., 2012. Research of dissolution ability of ionic liquids for polysaccharides such as cellulose. PhD Thesis. Institute for Organic Chemistry and Makromolecular Chemistry, Friedrich-Schiller-University of Jena.
Lee, S.H., Ha, S.H., You, C.Y., Koo, Y.M., 2007. Recovery of magnetic ionic liquid [bmim] $FeCl_4$ using electromagnet. Korean J. Chem. Eng. 24, 436–437.
Lemus, J., Palomar, J., Heras, F., Gilarranz, M.A., Rodriguez, J.J., 2012. Developing criteria for the recovery of ionic liquids from aqueous phase by adsorption with activated carbon. Sep. Purif. Technol. 97, 11–19.
Li, G., Shen, J., Zhu, Y., 1998. Study of pyridinium-type functional polymers. II. Antibacterial activity of soluble pyridinium-type polymers. J. Appl. Polym. Sci. 67, 1761–1768.
Li, X., Zhao, J., Li, Q., Chi, S., 2007. Ultrasonic chemical oxidative degradations of 1, 3-dialkylimidazolium ionic liquids and their mechanistic elucidations. Dalton Trans 19, 1875–1880.
Li, C., Han, J., Wang, Y., Yan, Y., Pan, J., Xu, X., Zhang, Z., 2010. Phase behavior for the aqueous two-phase systems containing the ionic liquid 1-butyl-3-methylimidazolium tetrafluoroborate and kosmotropic salts. J. Chem. Eng. Data 55 (3), 1087–1092.
Maase, M., 2005. BASF AG, Distillation of ionic liquids, WO 2005/068404.
Maase, M., Massonne, K., 2005. Method for the production of purified 1,3 substituted imidazolium salts. BASF, WO 2005/019183.
Maase, M., Budich, M., Grossmann, G., Szarvas, L., Method for extracting impurities using ionic liquids. WO 2005019137 A8.
Markiewicz, M., Jungnickel, C., Markowska, A., Szczepaniak, U., Paszkiewicz, M., Hupka, J., 2009. 1-Methyl-3-octylimidazolium chloride: sorption and primary biodegradation analysis in activated sewage sludge. Molecules 14 (11), 4396–4405.
Michud, A., Tanttu, M., Asaadi, S., Ma, Y., Netti, E., Kaariainen, P., Sixta, H., 2015. Ioncell-F: ionic liquid-based cellulosic textile fibers as an alternative to viscose and Lyocell. Textile Res. J., http://doi.org/10.1177/0040517515591774.

Mukesh, C., Mondal, D., Sharma, M., Prasad, K., 2013. Rapid dissolution of DNA in a novel bio-based ionic liquid with long-term structural and chemical stability: successful recycling of the ionic liquid for reuse in the process. Chem. Commun. 49 (61), 6849–6851.
Mulder, M., 1996. Mulder, M. (Ed.), Basic Principles of Membrane Technology, 1996, Kluwer Academic Publishers, Dordrecht, p. 564.
Mulder, M.H.V., van Voorthuizen, E.M., Peeters, J.M.M., 2005. Membrane characterization. In: Schäfer, A.I., Fane, A.G., Waite, T.D. (Eds.), Nanofiltration: Principles and Applications. Elsevier, Oxford, pp. 89–117, Chapter 5.
OECD, 1981. Ready Biodegradability: Modified Sturm Test in OECD Guidelines for the Testing of Chemicals, Test Guideline 301B, Organization for Economic Cooperation and Development (OECD), Paris.
OECD Chemical Group, 1993. Ready Biodegradability: Modified OECD Screening Test. Method 301 E. OECD Revised Guidelines for Tests for Ready Biodegradability, Paris.
Palomar, J., Lemus, J., Gilarranz, M., Rodriguez, A.J.J., 2009. Adsorption of ionic liquids from aqueous effluents by activated carbon. Carbon 47 (7), 1846–1856.
Park, C., 2007. A Dictionary of Environment and Conservation. Oxford University Press. Oxford.
Parviainen, A., King, A.W.T., Mutikainen, I., Hummel, M., Selg, C., Hauru, L.K.J., Kilpeläinen, I., 2013. Predicting cellulose solvating capabilities of acid-base conjugate ionic liquids. ChemSusChem 6 (11), 2161–2169, http://doi.org/10.1002/cssc.201300143.
Parviainen, A., Wahlström, R.M., Liimatainen, U., Liitiä, T., Rovio, S., Helminen, J.K.J., 2015. Sustainability of cellulose dissolution and regeneration in 1,5-diazabicyclo[4.3.0]non-5-enium acetate: a batch simulation of the IONCELL-F process. RSC Adv. 5, 69728–69737, http://doi.org/10.1039/C5RA12386K.
Pernak, J., Branicka, M., 2004. Synthesis and aqueous ozonation of some pyridinium salts with alkoxymethyl and alkylthiomethyl hydrophobic groups. Ind. Eng. Chem. Res. 43 (9), 1966–1974.
Pernak, J., Rogoza, J., Mirska, I., 2001. Synthesis and antimicrobial activities of new pyridinium and benzimidazolium chlorides. Eur. J. Med. Chem. 36, 313.
Ranke, J., Molter, K., Stock, F., Bottin-Weber, U., Poczobutt, J., Hoffman, J., Ondruschka, B., Filser, J., Jastorff, B., 2004. Biological effects of imidazolium ionic liquids with varying chain lengths in acute *Vibrio fischeri* and WST-1 cell viability assays. Ecotoxicol. Environ. Saf. 58, 396–404.
Ren, Q., Yang, Q., Yan, Y., Xing, H., Bao, Z., Su., B., Yang, Y., 2011. Ionic liquid-mediated liquid–liquid extraction. Applications of Ionic Liquids in Science and Technology. In: Handy, S. (Ed.), InTech, Rijeka. Available from: http://www.intechopen.com/books/applications-of- ionic-liquids-in-science-and-technology/ionic-liquid-mediated-liquid-liquid-extraction.
Schäfer, T., Rodrigues, C.M., Afonso, C.A.M., Crespo, J.G., 2001. Selective recovery of solutes from ionic liquids by pervaporation – a novel approach for purification and green processing. Chem. Commun. 17, 1622–1623.
Scott, M.J., Jones, M.N., 2000. The biodegradation of surfactants in the environment. Biochim. Biophys. Acta 1508 (1–2), 235.
Scurto, A.M., Aki, S.V.K., Brennecke, J.F., 2002. CO_2 as a separation switch for ionic liquid/organic mixtures. J. Am. Chem. Soc. 124, 10276–10277.
Scurto, A.M., Aki, S.N.V.K., Brennecke, J.F., 2003. Carbon dioxide induced separation of ionic liquids and water. Chem. Commun. 5, 572–573.
Seddon, K.R., Stark, A., Torres, M.-J., 2000. Influence of chloride, water, and organic solvents on the physical properties of ionic liquids. Pure Appl. Chem. 72 (12), 2275–2287.
Siedlecka, E.M., Stepnowski, P., 2009. The effect of alkyl chain length on the degradation of alkylimidazolium- and pyridinium-type ionic liquids in a Fenton-like system. Environ. Sci. Pollut. Res. 16, 453–458.

Siedlecka, E.M., Gołębiowski, M., Kaczyński, Z., Czupryniak, J., Ossowski, T., Stepnowski, P., 2009. Degradation of ionic liquids by Fenton reaction; the effect of anions as counter and background ions. Appl. Catal., B 91, 573–579.

Siedlecka, E.M., Czerwicka, M., Neumann, J., Stepnowski, P., 2011. Ionic liquids: methods of degradation and recovery. Ionic Liquids: Theory, Properties, New Approaches. In: Kokorin, A. (Ed.), InTech, Rijeka. Available from: http://www.intechopen.com/books/ionic-liquids-theory-properties-new-approaches/ionic-liquids-methods-of-degradation-and-recovery.

Sixta, H., Michud, A., Hauru, L., Asaadi, S., Ma, Y., King, A.W.T., Beadle, C., 2015. Ioncell-F: a high-strength regenerated cellulose fibre 30 (1), 043–057.

Slack, G.C., Snow, N.H., Kou, D., 2003. Mitra, S. (Ed.), Extraction ofVolatile Organic Compounds From Solids and Liquids, in Sample Preparation Techniques in Analytical Chemistry, 162, John Wiley & Sons, Inc, Hoboken, NJ, USA.

Solà Cervera, J.L., König, A., 2010. Recycling concept for aluminum electrodeposition from the ionic liquid system EMIM[Tf_2N]-$AlCl_3$. Chem. Eng. Technol. 33 (12), 1979–1988.

Stasiewicz, M., Mulkiewicz, E., Tomczak-Wandzel, R., Kumirska, J., Siedlecka, E.M., Gołebiowski, M., Gajdus, J., Czerwicka, M., Stepnowski, P., 2008. Assessing toxicity and biodegradation of novel, environmentally benign ionic liquids (1-alkoxymethyl-3-hydroxypyridinium chloride, saccharinate and acesulfamates) on cellular and molecular level. Ecotoxicol. Environ. Saf. 71 (1), 157–165.

Stepnowski, P., 2005. Preliminary assessment of the sorption of some alkyl imidazolium cations as used in ionic liquids to soils and sediments. Aust. J. Chem. 58 (3), 170–173.

Stepnowski, P., Zaleska, A., 2005. Comparison of different advanced oxidation processes for the degradation of room temperature ionic liquids. J. Photochem. Photobiol., A 170 (1), 45–50.

Stepnowski, P., Mrozik, W., Nichthauser, J., 2007. Adsorption of alkylimidazolium and alkylpyridinium ionic liquids onto natural soils. Environ. Sci. Technol., 41 (2), 511–516. Retrieved from http://www.ncbi.nlm.nih.gov/pubmed/17310715.

Stolte, S., Abdulkarim, S., Arning, J., Blomeyer-Nienstedt, A.-K., Bottin-Weber, U., Matzke, M., Ranke, J., Jastorff, B., Thöming, J., 2008. Primary biodegradation of ionic liquid cations, identification of degradation products of 1-methyl-3-octylimidazoliumchloride and electrochemical wastewater treatment of poorly biodegradable compounds. Green Chem. 10, 214–224.

Studzińska, S., Sprynskyy, M., Buszewski, B., 2008. Study of sorption kinetics of some ionic liquids on different soil types. Chemosphere 71 (11), 2121–2128.

Swatloski, R.P., Spear, S.K., Holbrey, J.D., Rogers, R.D., 2002. Dissolution of cellose with ionic liquids. J. Am. Chem. Soc. 124 (18), 4974–4975.

Titze-Frech, K., Ignatiev, N., Uerdingen, M., Schulz, P.S., Wasserscheid, P., 2013. Highly selective aromatic alkylation of phenol and anisole by using recyclable Brønsted acidic ionic liquid systems. Eur. J. Org. Chem. 30, 6961–6966.

Trinh, L.T.P., Lee, Y.J., Lee, J.-W., Bae, H.-J., Lee, H.-J., 2013. Recovery of an ionic liquid [BMIM]Cl from a hydrolysate of lignocellulosic biomass using electrodialysis. Sep. Purif. Technol. 120, 86–91.

Vijayaraghavan, K., Pham, T.P.T., Cho, C.-W., Won, S.W., Choi, S.B., Juan, M., Kim, S., Kim, Y.-R., Chung, B.-W., Yun, Y.-S., 2009. An assessment on the interaction of a hydrophilic ionic liquid with different sorbents. Ind. Eng. Chem. Res. 48 (15), 7283–7288.

Wang, X., Nie, Y., Zhang, X., Zhang, S., Li, J., 2012. Recovery of ionic liquids from dilute aqueous solutions by electrodialysis. Desalination 285, 205–212.

Wasserscheid, P., Welton, T., 2002. In: Wasserscheid, P., Welton, T. (Eds.), Ionic Ionic Liquids in Synthesis. Wiley-VCH Verlag GmbH & Co. KGaA, Weinheim, Copyright © 2002.

Wasserscheid, P., Kragl, U., Kröckel, J., 2003. Method for separating substances from solutions containing ionic liquids by means of a membrane. Solvent Innovation Gmbh. WO 2003/039719.

Wells, M.J.M., 2003. Mitra, S. (Ed.), Principles of Extraction and the Extraction of Semivolatile Organics from Liquids, in Sample Preparation Techniques in Analytical Chemistry-John Wiley & Sons, Inc, Hoboken, NJ, USA, vol. 162.

Welton, T., 1999. Room-temperature ionic liquids. Solvents for synthesis and catalysis. Chem. Rev. 99 (8), 2071–2083.

Won, S.W., Choi, S.B., Mao, J., Yun, Y.-S., 2013. Removal of 1-ethyl-3-methylimidazolium cations with bacterial biosorbents from aqueous media. J. Hazard. Mater. vol. 244–245, 130–134.

Wu, B., Liu, W., Zhang, Y., Wang, H., 2009. Do we understand the recyclability of ionic liquids? Chemistry 15 (8), 1804–1810.

Xiong, D., Wang, H., Li, Z., Wang, J., 2012. Recovery of ionic liquids with aqueous two-phase systems induced by carbon dioxide. ChemSusChem 5 (11), 2255–2261.

Zhang, Z.F., Wu, W.Z., Liu, Z.M., Han, B.X., Gao, H.X., Jiang, T., 2004. A study of tri-phasic behavior of ionic liquid–methanol–CO_2 systems at elevated pressures. Phys. Chem. Chem. Phys. 6, 2352–2357.

Zhou, H., Lv, P., Shen, Y., Wang, J., Fan, J., 2013a. Identification of degradation products of ionic liquids in an ultrasound assisted zero-valent iron activated carbon micro-electrolysis system and their degradation mechanism. Water Res. 47 (10), 3514–3522.

Zhou, H., Shen, Y., Lv, P., Wang, J., Fan, J., 2013b. Degradation of 1-butyl-3-methylimidazolium chloride ionic liquid by ultrasound and zero-valent iron/activated carbon. Sep. Purif. Technol. 104, 208–213.

CHAPTER 6

Economical Aspects of Ionic Liquid Application

Olga Kuzmina
Department of Chemistry, Imperial College London, UK

INTRODUCTION

Ionic liquids (ILs) have gained a lot of attention from researchers as a result of their unique properties. Since the second half of the last century, ILs have been promoted as "green chemicals." The reasons behind ILs being given such a term are:

- Their negligible vapor pressure. This property minimizes possible inhalation of IL vapors by persons handling them and the release of IL waste into the atmosphere. Thus, gas waste can be completely avoided using ILs.
- Their inflammability. This property removes the risk of accidental fires.
- They are claimed to be relatively nontoxic. This means that they are less toxic than conventional solvents; however, their toxicity is still under investigation.

In the last couple of decades, scientists have found numerous applications for these advantageous substances. However, as noted by Schubert (2011), time is needed for innovations to be developed into inventions. Despite technical advantages being obvious, they do not immediately result in technical application. Nowadays, ILs are being applied industrially at both pilot and commercial scales within chemical and electrochemical processes. The most famous industrial process employing ILs is biphasic acid scavenging utilizing ionic liquids (BASIL). ILs are the product of introducing a starting material, namely 1-methylimidazole, to acid scavenging. BASIL was established in 2004 for the synthesis of alkoxyphenylphosphines by BASF (a chemicals company) and ILs play an auxiliary part in this process. By introducing ILs into acid scavenging instead of using tertiary amines, BASF benefited in the following ways (Maase, 2008):

- Ease of handling as only chemicals are involved (i.e., no solids).
- Time efficiency (i.e., quick reaction time).
- Higher productivity.

Application, Purification, and Recovery of Ionic Liquids
http://dx.doi.org/10.1016/B978-0-444-63713-0.00006-7

- Lower investment cost (i.e., small reactors can be used instead of large ones).
- Higher process sustainability confirmed by ecoefficiency analysis.
- IL recycling capability.

One of the main advantages of the new process is the possibility of recovering and reusing worked-up ILs. In the BASIL process, 1-methylimidazole forms 1-methylimidazolium chloride by reacting with hydrochloric acid. 1-Methylimidazolium chloride can be switched off by deprotonation (as discussed in Chapter 5). The resulting 1-methylimidazole is a conventional neutral liquid that is distillable for reuse.

Other examples of the commercialization of ILs include:

- The chlorination process developed by BASF in which ILs are used as solvents (Maase, 2008).
- The Hi-Tech-Cleaning process developed by IoLiTec, as implemented by Wandres (a micro-cleaning company), in which ILs are used as antistatic cleaning agents (Bösmann & Schubert, 2004).
- The use of ILs as compatibilizers (secondary dispersants) for pigment pastes (Weyershausen and Lehmann, 2005).

These examples highlight the advantageous properties of ILs over less efficient and environmentally hazardous conventional substances. However, there are a number of factors that could prevent broad implementation of ILs. According to Joglekar et al. (2007), the main issues and barriers to commercialization of ILs are the lack of:

- physicochemical, thermodynamic, and toxicity data;
- characterization tools;
- process-engineering studies;
- lifetime and recyclability data; and
- scale-up studies.

In addition, there is industrial apathy due to uncertainty surrounding IL usage and the high prices of ILs (Joglekar et al., 2007).

These barriers have to be overcome before ILs can be commercialized. In addition to in-depth studies into IL properties and possible methods of scaling up production, their economic sustainability has to be predicted. Based on the first production runs carried out at BASF, the company predicted the prices for large-scale production of ILs would drop below €30 (approx. $39.85 in 2013) per kilogram[1] (Kazemi, 2013). IoLiTec (Schubert, 2011) also claimed that the production of ILs in larger batches

[1] BASF. Ionic liquid FAQs at http://www.intermediates.basf.com/chemicals/ionic-liquids/faq#12

or by novel techniques such as microreaction technology will lead to a significant drop in their price due to similarity in the number of working hours needed to produce 10, 100, or 1000 kg. Thus, despite prices varying according to the amount of raw materials used, other costs will only change slightly.

Current catalog prices leave much to be desired from this idealistic assumption. According to Sigma-Aldrich (a life science and high-tech company), the cheapest available ILs are 1-butyl-3-methylimidazolium chloride and its current price is $343 per kilogram and less known pyridinium chlorochromate with the price of approximately $277/kg.[2] Sigma-Aldrich's chemicals (as well as those of all other catalog suppliers) are often used in academia, R&D departments, etc. These customers mainly request small volumes. Hence, Sigma-Aldrich provides such customers with a variety of different substances in small volumes but high purity. This generates a service cost (personnel, packaging material, storage, etc.). On the other hand, large suppliers look to larger production volumes to benefit from economies of scale, and consequently are unable to produce such a large portfolio of products. However, the production of large quantities will always result in reduced prices. Therefore, there are two main reasons catalog prices are always higher than those of a commercial supplier of chemical products:

1. Wider range of products.
2. The extra cost involved in the supplier obtaining chemicals from large suppliers (logistics, packaging, etc.).[3]

There is always the option to synthesize cheap ILs using widely available inexpensive starting materials, which allows IL prices to fall into the range of specialty organic solvents such as dimethyl sulfoxide (DMSO) or dimethyl formamide (DMF) (Wasserscheid, 2003). However, it should be noted that most ILs, which consist of cheap anions and cations, have some noticeable disadvantages that limit their application.

The great advantage of ILs is their recyclability. Recovery and the possibility of reusing ILs has been studied intensively, and the availability of a highly efficient recovery technology for ILs is just a question of time. However, open-access studies on the economics of IL implementation, the main and most uncertain aspect, are limited in number. Economics questions on how the high cost of ILs will affect its applicability and will the recyclability of ILs increase the profitability of its implementation need to be answered.

[2] Sigma-Aldrich online catalog (2013) at http://www.sigmaaldrich.com

[3] E-mail consultation with Dr Christian Seitz, Business Development Manager, BASF (2013).

THE COST OF ILS

The potential of ILs has not yet been fully explored. The commercial availability of ILs depends on advances being made in research (Wagner & Hilgers, 2008). Some classes of ILs are readily available from a number of suppliers, such as Sigma-Aldrich, Merk, BASF, IoLiTec, and Solvionic. However, ILs are still costly and this prevents their large-scale application (Visak, 2012).

Figure 6.1 shows how the prices of commercially available ILs have changed over the last few years. This example is for imidazolium and pyridinium ILs which have different anions: chloride and hexafluorophosphate. It has been suggested that pyridinium-based ILs should be cheaper than imidazolium-based ILs (Kralisch et al., 2005a; Kröckel and Kragl, 2003; Visak, 2012) due to the additional energy costs of production of imidazole, while pyridine is readily available directly from charcoal. However, analysis of IL prices shows that it is not always the case. 100 g of 1-butyl-3-methylimidazolium chloride costs similarly to 5 g of hexafluorophosphate ILs while approximately 9 times cheaper than would cost 100 g of 1-butyl-4-methylpyridinium chloride. The reason for this incoherence could be lower demand for pyridinium ILs than imidazolium ones has led to

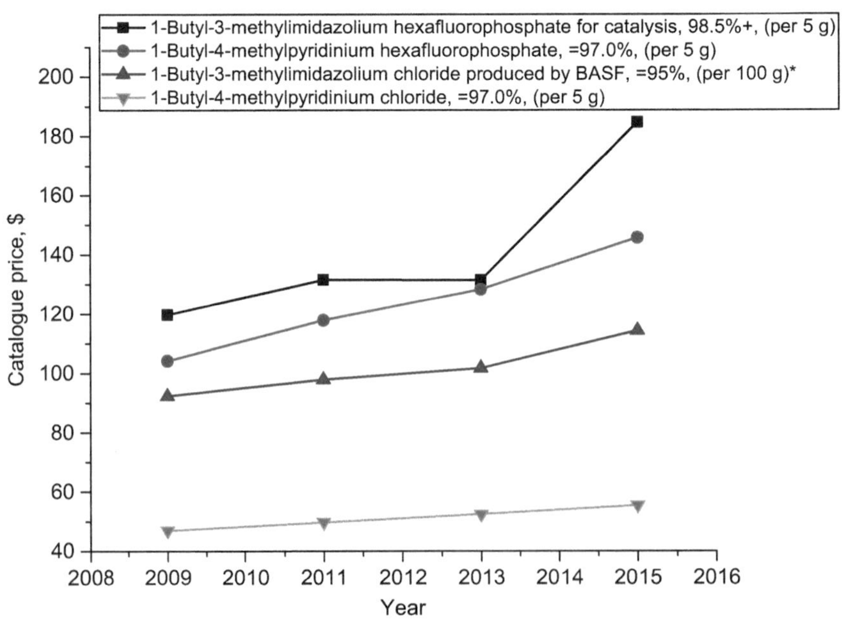

Figure 6.1 ***Price Development of Some ILs over the Last Few Years.*** *Data kindly provided by Sigma-Aldrich. *Withdrawn in 2014, replaced by Aldrich, purity ≥98%.*

decreased production and hence increased costs. Furthermore, demand always directs research, and the demand for imidazolium-based ILs will have meant greater research into efficient production methods, therefore allowing lower prices. From 2009 to 2015 prices increased by an average of 33%. When inflation over this period in UK is taken into account (19.18%),[4] it becomes clear that ILs are becoming increasingly costly.

The type of cation influences the cost of ILs dramatically, as ILs that have an ethyl alkyl chain in most cases are more expensive than ILs containing cations with butyl chains. Comparing the price of imidazolium chloride ILs with different alkyl chains, price will increase in the following order: butyl < proton < ethyl < hexyl < decyl < benzyl < octyl < allyl. The type of anion also affects the price of ILs. As shown in Figure 6.2 the chloride anion is relatively cheap compared with hexafluorophosphate and acetate anions.

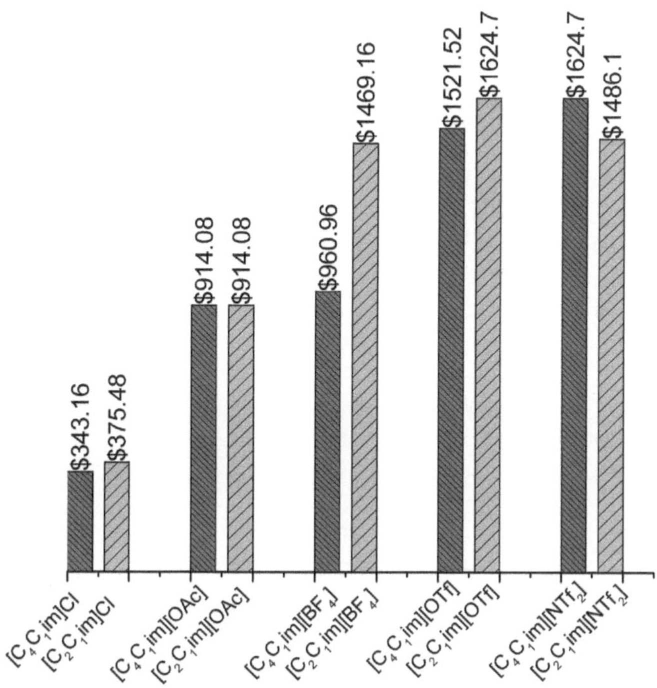

Figure 6.2 ***Price Comparison of Imidazolium ILs and Their Dependence on Anion and Cation Type.*** *Data from Sigma-Aldrich catalogue http://www.sigmaaldrich.com/technical-service-home/product-catalog.html.*

[4] http://www.thisismoney.co.uk/money/bills/article-1633409/Historic-inflation-calculator-value-money-changed-1900.html

There is some evidence that ammonium and phosphonium cations along with chloride and alkyl sulfate anions are the cheapest, whereas imidazolium cations and fluorinated anions are the most expensive (Visak, 2012). Tetrafluoroborate and hexafluorophosphate anions result in higher prices for the ILs produced, because the sources of these anions are not cheap (Wasserscheid, 2003). Thus, the difference in IL prices can be explained by the cost of starting materials and the complexity of synthesis. However, catalog prices for such small quantities should not be used as a guide to predict IL availability for large-scale production (Rakita, 2006). The question then arises as to what is included in the price of IL formation?

The prices of ILs are mostly determined by the prices of starting materials. ILs are not readily accessible from cheap industrial process streams in a simple isolation or conversion step. Moreover, their nonvolatility makes it difficult to isolate and purify them. However, a rough estimate of their price can be made by adding up the prices of all starting materials that are needed (Wasserscheid, 2003). Figure 6.3 gives rough estimates of the relative price of ILs.

The cost of starting materials depend on fixed costs (personnel, rent, insurance, accountancy, etc.) and variable manufacturing costs (production volume, materials, labor, utilities, delivery personnel, etc.). Some include manufacturing overheads (such as energy costs) and nonmanufacturing costs (selling and administrative expenses).

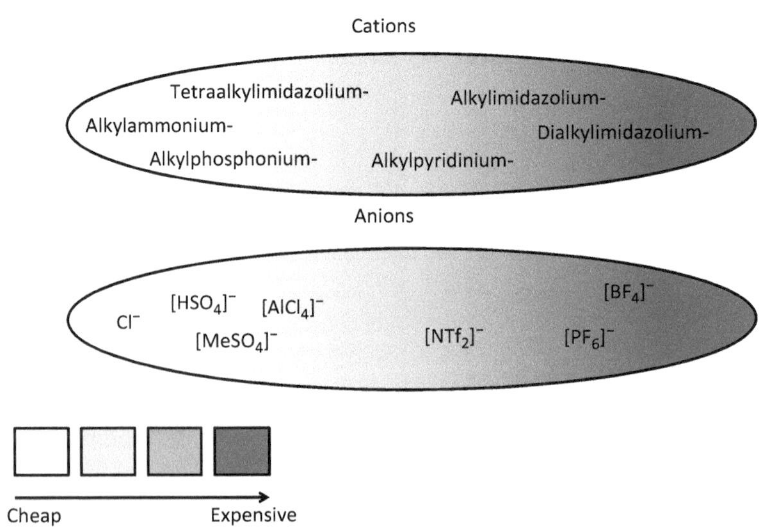

Figure 6.3 ***Rough Estimate of the Relative Price of ILs.***

Kralisch et al. (2005b) used life cycle assessment (LCA) to evaluate the energy, environmental, and economic balance necessary for the manufacture and work-up of 1-butyl-3-methylimidazolium tetrafluoroborate ILs. LCA is used to predict the optimal process from an environmental, economic, and ecological point of view. Kralisch et al. calculated an "economical factor" that allows rough estimation of the profitability of the process. Ideally, this parameter should include all the costs described above (i.e., the cost of starting compounds, energy, personnel, etc.). However, due to limited information at the R&D stage, the only comparison of bulk starting substances made by the authors was qualitative. Nevertheless, the large energy burden found in the manufacture of ILs meant the overall balance of their synthesis and work-up was four times worse than for the conventional solvent dichloromethane, and sustainable application of ILs in the metathesis of 1-octene (chosen by the authors as an example of IL implementation) was found to be only possible with guaranteed solvent recycling.

More optimistic results are given by Kazemi in his PhD thesis (Kazemi, 2013). Kazemi compared two conventional processes: the epoxidation reaction of cinnamyl alcohol with 3-phenylglycidol and the Friedel–Crafts acylation reaction of ferrocene to acetylferrocene with ILs/CO_2. The variable and fixed costs of both processes were compared. Prices predicted by BASF for ton scale–produced ILs (€30/\$39.85 kg^{-1}) were used in all estimates. The recyclability of ILs was taken into account with ILs being assumed to be replaced once a year. Even when the cost of purchasing a new reactor was included, the predicted savings for both processes exceeded several hundred euros in just the first year assuming production of 100 t. In addition, environmental evaluation showed the advantage of IL-based processes since waste emission was reduced compared with conventional processes. Energy costs also confirmed the attractiveness of switching to IL/CO_2 processes. Hence, despite Kazemi's calculations in favor of IL processes being somewhat idealistic, they are nevertheless extremely promising.

There are a number of methods that can be used to estimate the sustainability of any process: from the simplest one (used by Kazemi) to the most complex and broad (e.g., the wiki-based computer model of the biofuels research community[5]). The availability of required parameters and the stage at which the evaluation is performed (R&D, pilot plant, small scale) influence choice of the appropriate method. All these models consider environmental aspects from the very beginning, but not all of them take process

[5] "A wiki for the biofuels research community," Phys.org, October 29, 2010 at http://phys.org/news/2010-10-wiki-biofuels.html

profitability into account. This is key, however, since no process will be implemented unless it is economically beneficial.

As most IL processes are still at the R&D stage, Kralisch et al., (2007) suggest using simplified life cycle assessment (SLCA) for ecological and economic optimization (ECO). The authors defined the cost factor as a combination of:

- the supply costs of reactants, solvents, and auxiliaries (CS);
- costs resulting from synthesis (CR);
- work-up (CW);
- application (CA); and
- disposal (CD).

The cost factor was related to the molarity (or mass) of the product (*n*) in (Eq. (6.1)):

$$\mathrm{CF} = \frac{\sum_{i-1}^{\mathrm{XS}} \mathrm{CS}i + \sum_{i-1}^{\mathrm{XR}} \mathrm{CR}i + \sum_{i-1}^{\mathrm{XW}} \mathrm{CW}i + \sum_{i-1}^{\mathrm{XA}} \mathrm{CA}i + \sum_{i-1}^{\mathrm{XD}} \mathrm{CD}i}{n} \tag{6.1}$$

SLCA has been applied to the synthesis of 1-hexyl-3-methylimidazolium chloride (Kralisch et al., 2007). Parameters such as temperature, solvent, concentration, molar ratio, and reaction time have been optimized to strike a balance between the energy factors, environmental and human health factors, and cost factors involved in this process. Figure 6.4 shows cost factor dependence on time. Cost factor can be divided into a number of cost sources:

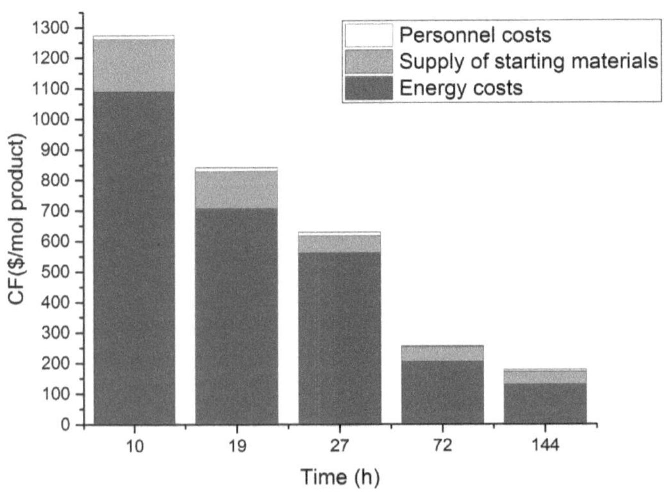

Figure 6.4 ***Time Dependence of CF, t = 10, 19, 27, 72, 144 h, nMIM:nC$_6$H$_{13}$Cl = 1:1, T = 70°C, solvent free (adapted from Kralisch et al., 2007).***

energy demand, supply of reactants (*N*-methylimidazole, *n*-hexyl chloride), solvent extraction, and personnel costs (preparation, monitoring, work-up).

The modulations given in Figure 6.4 help in making an assumption as to whether prolonged processing has a positive effect or not. Moreover, correlating these data with conversion yield helps in choosing the best possible parameters for the process.

The interaction between these factors has been looked at in much greater detail by Chen et al., (2013) in their evaluation of the attractiveness of screening of ILs for CO_2 capture. The sorption capacity (*C*), toxicity (*T*), and price (*P*) of ILs were taken into account when evaluating the capacity–toxicity green ratio ($R_g = C/T$), the capacity–price economic ratio ($R_e = C/P$), and the capacity–toxicity–price green economic ratio ($R_{ge} = C/TP$). Moreover, a special composite criterion $R_{i|C} > 0.5$ was created so that all criteria could be characterized together. Figure 6.5 depicts the relative value, R_g, R_e, R_{ge}, along with the absolute value *C*. Choosing the optimal IL involved taking two steps.

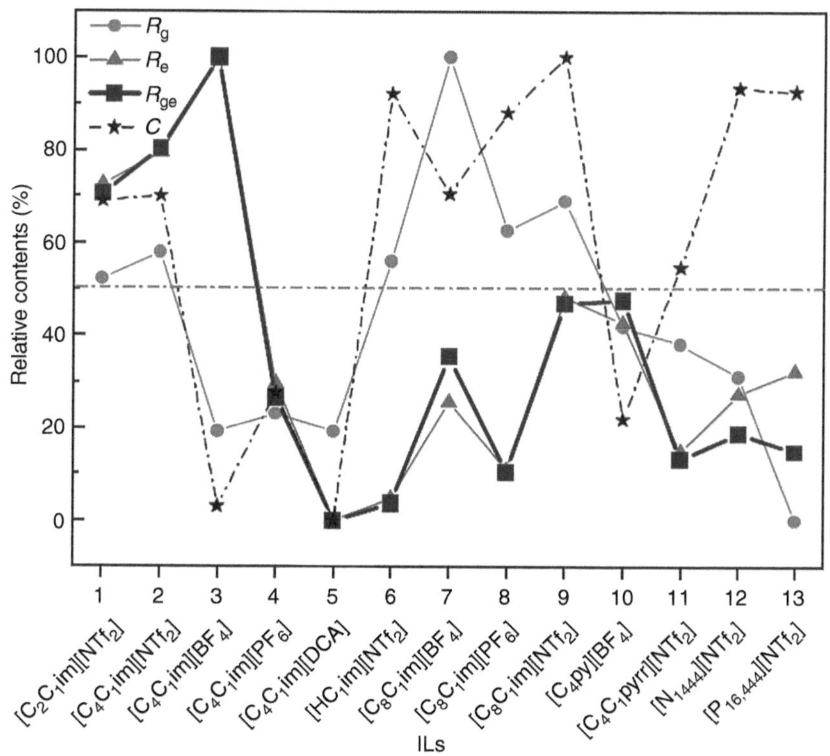

Figure 6.5 ***Capacity (C), Capacity–Toxicity Ratio (R_g), Capacity–Price Ratio (R_e), and Capacity–Toxicity–Price Ratio (R_{ge}) for CO_2 Capture by ILs (Chen et al., 2013).***

First, ILs that have a relatively high CO_2 sorption capacity were found, and then those with the highest R_g, R_e, and R_{ge} values were chosen. Numerous ILs with different anions, cations, and alkyl chain lengths were then compared regarding these criteria. Of all the ILs investigated, the authors found that an IL consisting of a 1-butylimidazolium cation together with the bis(trifluoromethylsulfonyl)imide anion $[NTf_2]^-$ was best for CO_2 capture. This was proven by its relatively high CO_2 capture capacity, relative low toxicity, and low price. However, the authors pointed out that the criteria used were not fully comprehensive. The economics involved in IL production should be taken into consideration even at the laboratory stage (Y. Chen et al., 2013).

Wiki-based software (mentioned earlier) allows researchers to perform technoeconomic analysis of lignocellulosic ethanol biorefinery with IL pretreatment (Klein-Marcuschamer et al., 2011) even at the R&D stage. The authors admit there are core issues preventing the commercialization of ILs in biofuel manufacturing. The main three issues are:

- the relative high cost of ILs;
- the requirement for significant IL recovery and recycling; and
- lack of knowledge in terms of process considerations.

Their research was based on previous studies of a wiki-based model for use in lignocellulosic biorefineries (Klein-Marcuschamer et al., 2010). Klein-Marcuschamer et al. (Klein-Marcuschamer et al., 2011) suggested that the main cost in the process was the amount of nonrecycled IL that was lost during pretreatment and burned in the boiler with other waste. After discussion with IL manufacturers and consideration of the scale of the operation, a fair IL cost was expected to vary from \$2.50 kg^{-1} to \$50 kg^{-1}. The authors claimed that the cost of ILs will depend in large part on the price of raw materials (Figure 6.6).

The key conclusion from this study was that reduction of the IL cost was a prerequisite to making the whole process viable economically. The authors also claimed that reducing IL loading was more important than increasing the IL recycling rate (Klein-Marcuschamer et al., 2011).

Assessments such as these are easier to carry out at small plants than in the laboratory. Moreover, there are a number of tools aimed at helping scientists with environmental, energy, and economic flows. Armed with the different theoretical economic coefficients and criteria discussed in this chapter, some tools come very close to delivering the necessary economic prediction data.

Technoeconomic evaluation of the conversion of biomass to sugars *via* IL hydrolysis and pretreatment of cellulosic biofuel production has been carried out by Sen et al. (2012) and Konda et al. (2014), who respectively

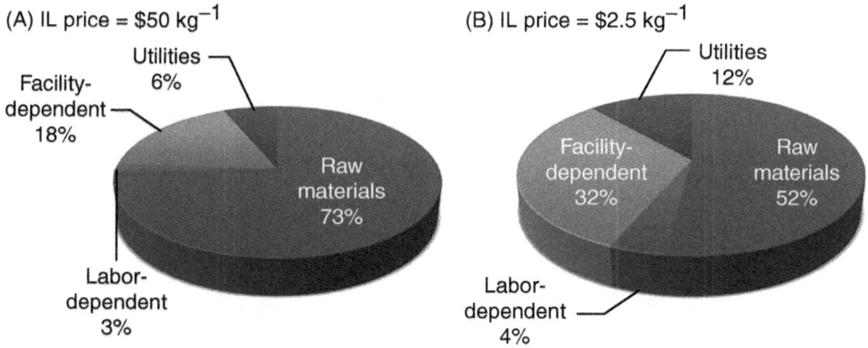

Figure 6.6 ***Distribution of the Annual Operating Cost of the Biorefinery Modeled at Two IL Purchase Prices: (A) \$50 kg^{-1} and (B) \$2.5 kg^{-1} for a Process with a Recycle Rate of 99.6 (Klein-Marcuschamer et al., 2011).***

agreed that raw materials and facility-dependent costs are significant contributors to process cost evaluation. Recently, technoeconomic assessment of the production cost of two ILs, triethylammonium hydrogen sulfate (IL1) and 1-methylimidazolium hydrogen sulfate (IL2), has been performed (L. Chen et al., 2014). Economic assessment in this study was carried out to ascertain the production cost of the two proposed syntheses at a plant with a capacity for 144,000 tons per year for a 10-year project life time. It was also carried out to evaluate both the capital investment and manufacturing costs required to produce these ILs in bulk. Estimated cost prices of IL1 and IL2 were \$1.24 kg^{-1} and \$2.96–5.88 kg^{-1}, respectively. These cost prices compare favorably with the selling price of conventional organic solvents such as acetone or ethyl acetate. These ILs are expected to be much less expensive than higher end organic solvents, such as acetonitrile (\$1.54 kg^{-1}), and are close to the price range of low-cost solvents such as toluene (\$1.03 kg^{-1}). The authors also suggested an equation to estimate the costs of other types of ILs made *via* acid–base neutralization and quaternization reactions (Eq. (6.2)) (L. Chen et al., 2014):

$$\text{IL price} = \frac{M1P1 + M2P2}{\text{M1} + \text{M2}} \times 1.25 \tag{6.2}$$

where M_1 and M_2 are the molecular weights of the two starting materials; and P_1 and P_2 are the prices of the two starting materials. This equation points out that the cost price of protic ILs will always be determined by the molecular weight of the more expensive component.

There are databases (e.g., Ecoinvent, SimaPro 752, and PROBAS) (Finnveden and Moberg, 2005) dedicated to energy issues for cost estimations. When a database cannot be used, as happens often during early-stage assessments, the energy demand for simple procedures can be determined using an energy-monitoring socket (e.g., Energy Monitor 3000, Voltcraft). The use of such tools can help fill information gaps and help choose the most appropriate process parameters.

HOW DOES RECYCLABILITY OF ILS AFFECT PROCESS PROFITABILITY?

A recent study pointed out that IL recycling will reduce product cost (Konda et al., 2014). The study compared two variants of IL-based processes: one based on complete removal of the IL prior to hydrolysis using a water wash (WW) step and the other based on a one-pot (OP) process that does not require IL removal prior to saccharification. The authors kindly provide representative models, developed in SuperPro Designer, free for download from their wiki[6] for noncommercial usage.

However, IL recyclability does not always mean the process is sustainable (Klein-Marcuschamer et al., 2011). IL recyclability is in most cases the factor justifying the use of such expensive solvents.

Some researchers assume that they can reuse ILs more than 10 times without any loss of properties (Kazemi, 2013). Others suggest a recycling rate of 90% (Kralisch et al., 2007). The basis for such assumptions is the claim made by BASF (one of the main chemical suppliers). They claim that they have successfully used an IL (in an extractive distillation process pilot plant) continuously for 3 months. They did admit that IL color changed to black, but this did not affect its performance.[7] Moreover, BASF is offering to design a recycling procedure for a chosen IL on request.[8]

Assuming that ILs are 10 times more expensive than common organic solvents (Table 6.1), it goes without saying that they should be recovered and recycled more than 10 times to make the process profitable.

[6] Klein-Marcuschamer, D., Konda, N.V.S.N.M., Oleskowicz-Popiel, P., Simmons, B.A., Blanch, H.W. (2014) A Wiki for the Technoeconomic Analysis (TEA) of Biorefineries, Joint BioEnergy Institute (JBEI) at http://econ.jbei.org/

[7] BASF. FAQs at http://www.intermediates.basf.com/chemicals/ionic-liquids/faq

[8] BASF intermediates at http://www.basionics.com/en/ionic-liquids/?getasset=file1&

Table 6.1 Prices of common organic solvents and some ILs

Solvent	Price/L ($)
Acetone	23.24
Toluene anhydrous	54.17
Benzene	57.71
N, *N*-dimethylformamide	106.95
Dimethyl sulfoxide	132.80
1-Butyl-3-methylimidazolium chloride	343.16
1-Butyl-3-methylimidazolium acetate	914.08

Available from the Sigma-Aldrich (2015) catalogue at http://www.sigmaaldrich.com/technical-service-home/product-catalog.html

A more realistic judgment assumes ILs can be recycled between 3–4 times (Atkins et al., 2003). The authors reported decreasing reactivity with each successive cycle, which indicates IL deactivation. However, in their study they confirmed IL recyclability to be three times higher than for the traditionally used catalyst known as "red-oil." Possible deactivation mechanisms have been suggested (deterioration in the presence of water, accumulation of residues from starting materials and products that have not reacted), but have not yet been studied (Atkins et al., 2003). An earlier research project suggested IL could be recycled between 10–15 times, without any loss, during the preparation of 1-phenyldodecane and cumene (Carter, 1999).

What is clear from these studies is that ILs should be continuously reloaded. A fresh or recycled IL should be added, since some amount can be wasted or deactivated during work-up. Nevertheless, researchers continue to pin their hopes on IL recyclability as a means of making them more widely available to industrial production sites (Freer and Curzons, 2003; Stojanovic et al., 2011). Other studies (Atkins et al., 2003) hark back to the statement that recyclability is less important than starting price (Klein-Marcuschamer et al., 2011).

CONCLUSION

There is always the possibility of creating task-specific ILs for every need. Hence, one approach to decreasing the price of ILs is to attach a needed functional group to a less expensive IL. An excellent example of this is the switching of relatively cheap, hydrophobic, ammonium-based ILs to hydrophilic ILs by adding hydroxyl groups to ammonium alkyl chains (Visak, 2012). Some scientists suggest focusing all research efforts into elaborating the potential of already existing ILs. Weyershausen and Lehmann

(Weyershausen and Lehmann, 2005) substantiate this point of view not only regarding the prices of ILs but also by limited listing of imidazolium-based ILs in EINECS (European Inventory of Existing Commercial Substances) or TSCA (Toxic Substances Control Act) which prevents their production and application.

However, all of these reasons will pale into insignificance as soon as certain criteria are fulfilled. For example, the price of a solvent is less important if the added value of the process is high enough or if its use leads to huge technological improvement (Wasserscheid, 2003). Therefore, research into new IL applications and into their properties is still the main focus of scientists worldwide.

ACKNOWLEDGMENTS

The author would like to acknowledge Dr Christian Seitz (BASF) for initial consultations and patient advice, and Sigma-Aldrich Sales Support team for their cooperation.

REFERENCES

Atkins, M.P., Bowlas, C., Ellis, B., Hubert, F., Rubatto, A., Wasserscheid, P., 2003. Ionic liquids as catalysts for ethylbenzene production. In: Rogers, R.D., Seddon, K.R., Volkov, S. (Eds.), Green Industrial Applications of Ionic Liquids. Kluwer Academic Publishers, Springer Netherlands, pp. 49–66.

Bösmann, A., Schubert, T., 2004. Identification of industrial applications for ionic liquids: high-performance additives for use in hi-tech cleaning solutions, IoLiTec. Poster presented at the Green Solvents for Synthesis. Bruchsal, Germany. October 3–6, 2004.

Carter, M.T., 1999. Ionic liquids as alternative solvents for industrial alkylation chemistry. Retrieved from http://cfpub.epa.gov/ncer_abstracts/index.cfm/fuseaction/display.abstractDetail/abstract/1265/report/F.

Chen, L., Sharifzadeh, M., MacDowell, N., Welton, T., Shah, N., Hallett, J.P., 2014. Inexpensive ionic liquids: $[HSO_4]^-$-based solvent production at bulk scale. Green Chem. 16 (6), 3098.

Chen, Y., Cao, Y., Sun, X., Yan, C., Mu, T., 2013. New criteria combined of efficiency, greenness, and economy for screening ionic liquids for CO_2 capture. Int. J. Greenhouse Gas Control 16, 13–20.

Finnveden, G., Moberg, Å., 2005. Environmental systems analysis tools – an overview. J. Cleaner Production 13 (12), 1165–1173.

Freer, R., Curzons, A., 2003. The pros and cons of using ionic liquids in the pharmaceutical industry. In: Rogers, R.D., Seddon, K.R., Volkov, S. (Eds.), Green Industrial Applications of Ionic Liquids. Kluwer Academic Publishers, Springer Netherlands, pp. 126–136.

Joglekar, H.G., Rahman, I., Kulkarni, B.D., 2007. The path ahead for ionic liquids. Chem. Eng. Technol. 30 (7), 819–828.

Kazemi, S., 2013. Sustainable ways of combining reactions and separations using ionic liquids and carbon dioxide. (PhD Thesis. Department of Process and Energy, Delft University of Technology).

Klein-Marcuschamer, D., Oleskowicz-Popiel, P., Simmons, B.A., Blanch, H.W., 2010. Technoeconomic analysis of biofuels: a wiki-based platform for lignocellulosic biorefineries. Biomass Bioenerg. 34 (12), 1914–1921.

Klein-Marcuschamer, D., Simmons, A.B., Blanch, W.H., 2011. Techno-economic analysis of a lignocellulosic ethanol biorefinery with ionic liquid pre-treatment. Biofuels Bioprod. Biorefin. 5 (5), 562–569.

Konda, M., Shi, J., Singh, S., Blanch, H.W., Simmons, B.A., Klein-Marcuschamer, D., 2014. Understanding cost drivers and economic potential of two variants of ionic liquid pretreatment for cellulosic biofuel production. Biotechnol. Biofuels 7 (1), 86.

Kralisch, D., Reinhardt, D., Kreisel, G., 2007. Implementing objectives of sustainability into ionic liquids research and development. Green Chem. 9 (12), 1308.

Kralisch, D., Stark, A., Körsten, S., Kreisel, G., Ondruschka, B., 2005a. Energetic, environmental and economic balances: spice up your ionic liquid research efficiency. Green Chem. 7 (5), 301.

Kralisch, D., Stark, A., Körsten, S., Kreisel, G., Ondruschka, B., 2005b. Energetic, environmental and economic balances: Spice up your ionic liquid research efficiency. Green Chem. 7 (5), 301–309.

Kröckel, J., Kragl, U., 2003. Nanofiltration for the separation of nonvolatile products from solutions containing ionic liquids. Chem. Eng. Technol. 26 (11), 1166–1168.

Maase, M., 2008. Industrial application of ionic liquids. In: Wasserscheid, P., Welton, T. (Eds.), Ionic Liquids in Synthesis, vol. 2, Wiley-VCH Verlags GmbH&Co. KGaA, Weinheim, pp. 663–687.

Rakita, P.E., 2006. Opportunities for commercializing ionic liquids. Chimica Oggi/Chemistry Today 24 (6), 10–12.

Schubert, T.J.S., 2011. Critical essay: availability of ionic liquids. Ionic Liquids Today (2), 4–6.

Sen, S.M., Binder, J.B., Raines, R.T., Maravelias, C.T., 2012. Conversion of biomass to sugars via ionic liquid hydrolysis: process synthesis and economic evaluation. Biofuel Bioprod Bior 6 (4), 444–452.

Stojanovic, A., Morgenbesser, C., Kogelnig, D., Krachler, R., Keppler, B.K., 2011. Quaternary ammonium and phosphonium ionic liquids in chemical and environmental engineering. In: Kokorin, A. (Ed.), Ionic Liquids: Theory, Properties, New Approaches. InTech, New York, pp. 657–680, Retrieved from http://cdn.intechopen.com/pdfs/13936/InTech-Quaternary_ammonium_and_phosphonium_ionic_liquids_in_chemical_and_environmental_engineering.pdf.

Visak, Z.P., 2012. Some aspects of ionic liquids as diverse and versatile sustainable solvents. J. Solution Chem. 41 (10), 1673–1695.

Wagner, M., Hilgers, C., 2008. Quality aspects and other questions related to commercial ionic liquid production. In: Wasserscheid, P.P., Welton, T. (Eds.), Ionic Liquids in Synthesis, vol. 1, Wiley-VCH Verlags GmbH&Co. KGaA, Weinheim, pp. 26–45.

Wasserscheid, P., 2003. Potential to apply ionic liquids in industry. In: Rogers, R.D., Seddon, K.R., Volkov, S. (Eds.), Green Industrial Applications of Ionic Liquids. Kluwer Academic Publishers, Springer Netherlands, pp. 29–47.

Weyershausen, B., Lehmann, K., 2005. Industrial application of ionic liquids as performance additives. Green Chem. 7 (1), 15–19.

SUBJECT INDEX

C

D

I

T